Sustainable Food Waste Management

Monika Thakur • V. K. Modi •
Renu Khedkar • Karuna Singh
Editors

Sustainable Food Waste Management

Concepts and Innovations

Springer

Editors
Monika Thakur
Amity Institute of Food Technology
Amity University
Noida, Uttar Pradesh, India

V. K. Modi
Amity Institute of Food Technology
Amity University
Noida, Uttar Pradesh, India

Renu Khedkar
Amity Institute of Food Technology
Amity University
Noida, Uttar Pradesh, India

Karuna Singh
Amity Institute of Food Technology
Amity University
Noida, Uttar Pradesh, India

ISBN 978-981-15-8966-9 ISBN 978-981-15-8967-6 (eBook)
https://doi.org/10.1007/978-981-15-8967-6

This Springer imprint is published by the registered company Springer Nature Singapore Pte Ltd.
The registered company address is: 152 Beach Road, #21-01/04 Gateway East, Singapore 189721, Singapore

Foreword

This book 'Sustainable Food Waste Management: Concepts and Innovations' provides a comprehensive overview of strategies for the mitigation and management of food waste. In fact, nearly one-third of all food produced in the world is discarded or wasted for various reasons and that equates to nearly 1.3 billion tons globally every year. Food losses arising at the retail, food services including both pre-consumer and postconsumer stages of the food chain have grown dramatically in recent years for a variety of reasons. Different strategies are required to curtail the type of food waste. Rising food prices or the prevalent food inflation in India can be effectively evaded by good food waste management practices. Also, food waste reduction would imply using less fuels for refrigeration and transportation purposes, thus preserving the environment. This book also reviews the concepts and innovations of food waste management strategies.

The book contains three parts: food industry waste: introduction and standards; utilization of waste from food processing industries and sustainable food waste management technologies. It is my hope and expectation that this book encompassing different areas of food waste management will provide an effective learning experience and referenced resource for young professionals, researchers, academicians, food industry and policymakers.

I take this opportunity to congratulate the editors and contributors for bringing out this book.

Entrepreneurship, Leadership and IT, Amity University
Noida, Uttar Pradesh, India

RBEF
New Delhi, India

Balvinder Shukla

Foreword

The book 'Sustainable Food Waste Management: Concepts and Innovations' provides a comprehensive overview of strategies used to reduce and manage food waste. It is estimated that on average 30–50% of the world's food is never consumed. Food losses arising at the retail, food services (pre-consumer) and at the consumer, difference stages in the food supply chain have grown dramatically in recent years, for a variety of reasons. Different strategies are required to tackle these types of food waste. Increasing food prices or the prevalent inflation in India can be effectively addressed by good waste management practices. Also, the waste reduction would imply using less fuels for the purpose of storage (refrigeration) and transportation, in turn preserving the environment. This book also reviews the concepts and innovations of food waste management strategies.

The book contains three parts, namely food industry waste: standards and management; utilization of waste from food processing industries and sustainable food waste management technologies. The topics have been treated in depth in 22 chapters from different areas of food waste management. This book will be a valuable reference source for young professionals, researchers, academicians, food industry and policymakers to appreciate and acknowledge various innovations for the utilization of food waste for value addition and sustainable food waste management techniques in detail.

I congratulate the editors and contributors for bringing such a useful book.

(Dr. KSMS Raghavarao)
Director CSIR-CFTRI
Mysore

CSIR-CFTRI
Mysore, Karnataka, India

K. S. M. S. Raghavarao

Foreword

The promotion of sustainability in food is an important aspect in order to end poverty which is one of the goals adopted by the UN General Assembly as the 2030 Agenda for Sustainable Development. The world population has been facing a severe threat as a result of the inadequate disposal of the waste generated from food which has been scientifically proven to have long-term irreversible implications on the environment. It also has a negative impact on the population both economically and socially. There is an urgent need to address the global concerns related to food security and its management. It has been observed that most countries still lack concrete management strategies and insights of diverse factors involved in the management as a result of which handling of significant amounts of food waste remains unidentified. Sustainable management of food in terms of reduction of wastage and its associated impacts is necessary for the existence of life on the planet.

This book on 'Sustainable Food Waste Management: Concepts and Innovations' is an effort aimed at addressing the issues related to food waste, associated losses and various strategies which can be employed for tackling this complex problem. The book also covers aspects of global importance such as radon pollution and food waste management practices which are rarely highlighted, but are very serious issues which need to be dealt with at the earliest. The book also will give a deep insight into research and innovations in the field of sustainable food management.

The book which comprises 22 chapters contributed by authors all over India is an endeavour to highlighting and finding solutions for food waste management, utilization of waste from food processing industries and sustainable food waste management technologies. It covers a broad selection of topics written by specialists in the field and will interest food scientists and technologists, food process engineers, researchers, faculty and students and various other stakeholders in the food industry who can be benefited from the application of suggested solutions in this volume.

I appreciate the efforts put forth by the authors for addressing an issue of global importance and enlighten the readers on the concept of 'Reduce, Reuse, Recycle', which will be the way forward in the effort to protect the planet in a sustainable manner.

Dr. W. Selvamurthy

W. Selvamurthy

Amity Science, Technology & Innovation
Noida, Uttar Pradesh, India

Amity Directorate of Science & Innovation
Noida, Uttar Pradesh, India

Amity University Chhattisgarh
Raipur, Chhattisgarh, India

Life Sciences, R&D (LS), DRDO
New Delhi, India
wselvamurthy@amity.edu

Preface

The food industry produces large volumes of wastes, both solids and liquids, resulting from the production, preparation and consumption of food. These wastes pose increasing disposal and can pose severe pollution problems and represent a loss of valuable biomass and nutrients. Many standard industrial waste treatment texts sufficiently address a few major technologies for conventional in plant environmental control strategies in the food industry. Food waste and land resource wastage add the carbon footprint which contribute to the greenhouse gas (GHG) emissions adding CO_2 in the atmosphere. Conventionally, this food waste, dumped in open area, causes severe health issues. Therefore, appropriate methods are required for the management of food waste. Environmental legislation has significantly contributed to the introduction of sustainable waste management practices worldwide. Considering the challenges in the area of food industry, efforts are to be made to optimize processing technologies to minimize the amount of waste. Food processing wastes have a potential for conversion into useful products of higher value as a by-product or even as raw material for other industries or for use as food or feed after biological treatment. There are many examples of utilizing waste materials from plant material processed by canneries, there are many other types of waste that can be utilized. In many canneries, the organic from the processing system is combined with the other types of non-usable wastes, such as hardware, glass, cans, nails, etc. Food industry should also have to concentrate on waste avoidance as well as utilization of process wastes. All the combined efforts of waste minimization during the production process, environmentally friendly preservation of the product and utilization of by-products would substantially reduce the amount of waste, as well as boost the environmental aspect of the food processing industry.

This book represents a comprehensive review of concepts and innovations on sustainable food waste management. It has three sections: food industry waste: introduction, standards and management; utilization of waste from food processing industries and sustainable food waste management technologies. The topics have been illustrated into 22 chapters from different areas of food science and technology, which balances perspectives and vision for new innovations in the developing world globally.

Part I deals with *Food Industry Waste: Introduction, Standards and Management* which has been covered in 4 chapters as introduction, environmental standards and regulations for waste management in food industries, characterization and treatment and advances in the treatment of food waste.

Part II dealing with *Utilization of Waste from Food Processing Industries* has 12 chapters from contributors discussing novel approaches for the valorization of fruit and vegetable industry waste, dairy industry by-products, cereals and pulses processing waste, oilseed industry waste, sugarcane industry by-products and meat industry waste utilization, production of bio-based packaging from food waste, etc.

Part III dealing with *Sustainable Food Waste Management Technologies* comprises 6 chapters which discuss microbial bioremediation, recovery of bioactive components, biofuels, etc.

There are many excellent books on food waste management. This book offers valuable guidance, both in approach and contents on the most important aspect, i.e. Food waste manageent and its sustaniablity. Readers, technical institution, food technologists, technocrats, existing industries and new entrepreneurs will find valuable material in this book. This book gives complete details on invaluable waste management concepts, utilization of by-products and the practical methods to implement them. It deals with the techniques and methods for food processing wastage. Comprehensive in scope, the book provides solutions that are directly applicable to the daily waste management problems specific to the food processing industry.

Noida, Uttar Pradesh, India Monika Thakur
Noida, Uttar Pradesh, India V. K. Modi
Noida, Uttar Pradesh, India Renu Khedkar
Noida, Uttar Pradesh, India Karuna Singh

Acknowledgements

'Sustainable Food Waste Management: Concepts and Innovations' is a comprehensive review of food industry waste, standards, management, utilization of waste from food processing industries and emerging technologies on sustainable food waste management in the food sector. We are extremely indebted to Respected Founder President, *Dr. Ashok K Chauhan*, for the blessings and constant encouragement. We have a great pleasure to acknowledge the whole-hearted support received from *Dr. Atul Chauhan*, Chancellor, Amity University, Uttar Pradesh and President RBEF, without their encouraging words, this endeavour is impossible. We are thankful to *Prof. Balvinder Shukla*, Vice Chancellor, Amity University, Uttar Pradesh for her constant motivation and support at all the stages of the progress.

We are thankful to *K S M S Raghavarao*, Director CFTRI for his kind blessings and guidance during the whole endeavour. We are also grateful to *Dr. A S Bawa* and *Prof. T N Lakhanpal* for their kind support and timely guidance in conceptualization of the work. Our sincere gratitude to *Dr. W Selvamurthy* and *Dr. Nutan Kaushik* for their valuable guidance and never-ending support. We also thank all the faculty members and staff members of Amity Institute of Food Technology, Amity University, Uttar Pradesh, Noida, for their full cooperation in this endeavour.

We are extremely indebted to all the authors who have contributed chapters and happily agreed to share their work on various aspects of food waste management. Without their painstaking efforts, it would not have been possible for us to bring this volume. Our sincere thanks to Springer Publishing for their full support and cooperation during the publication of volume. We are very thankful to the production team, Springer for all their efforts in publishing the book.

At last, we want to acknowledge the gratitude of our family members especially, *Ms. Anshu Modi, Mr. Harvansh Thakur, Mr. Deepak Khedkar* and *Mr. Neeraj Singh* for their astounding support and encouragements.

Contents

Editors and Contributors

About the Editors

Monika Thakur is faculty at the Amity Institute of Food Technology since 2009. She was awarded Ph.D. in 2006 by Himachal Pradesh University. She is Double Gold medalist (M.Sc & M.Phil) and received several prestigious awards as Post Graduate studies Himachal Pradesh (2004); Himalayan Research Scholar Association (2007); Wiley Research Academy (2017) and Young Scientist Award, IITT 2020. She has also worked on various research projects as SRF, and JRF at Himachal Pradesh University. Her area of expertise is Plant Sciences, Mycology, Functional foods & Nutraceuticals. Apart from her teaching assignments, she has been involved in various research and extension activities. She has guided more than 22 Dissertations and 2 students have completed Ph.D. under her guidance. She is a member of editorial board and reviewer in Journals of national and international repute. She has done two Industrial consultancy and currently one DST sanctioned project is running. She has published more than 40 research publications, 1 patent, 20 book chapters, 5 books, 01 compendium and presented more than 35 papers. She has also organized various Training programme, Faculty Development programmes, National conferences and national seminar. She has completed certificate course in Food Safety Management systems ISO: 22000:2004. She is also a certified Food Safety Trainer for the FoSTaC (Food Safety Training & Certification) programme conducted by FSSAI, GoI.

V. K. Modi is the Head of Institute at Amity Institute of Food Technology, Amity University, Noida. He joined Amity Institute of Food Technology after his superannuation as Chief Scientist and Professor AcSIR from CSIR-Central Food Technological Research Institute, Mysore. He graduated from Punjab Agricultural University, Ludhiana with a degree in Food Science and Technology. He joined the CSIR-Central Food Technological Research Institute, Mysore in 1987 and spent the next 30 years working in the area of Food Science and Technology. He was awarded a Post Doctorate UNU fellowship to carry out research work on flavour at the University of Nottingham, UK, University of Reading, UK, and Queen's University of Belfast, Northern Ireland. He has received several prestigious national awards such as Fellow of AFST-2013 by the Association of Food Scientists and

Technologists India (AIFT), Laljee Godhoo Smarak Nidhi Award (2011) by the AIFT, and Best Technology Award by the CSIR- CFTRI. Dr Modi has published more than 100 articles in national and international journals, holds 13 granted patents, and has developed 33 technologies, which have since been licensed to more than 130 entrepreneurs for successful commercialization.

Renu Khedkar is an M. Tech. and Ph.D. in Food Technology. She is faculty at Amity Institute of Food Technology since 2008. She was awarded Ph.D. for her research on "Development of process technologies and shelf life studies on traditional food adjuncts native to Central India". Her research areas are traditional foods, food engineering and food packaging. She has more than 12 years of experience in teaching and industry. She has published 7 research papers in reputed journals, 5 book chapters and 35 papers presented in conferences. She has guided 7 M.Tech. projects and 8 M.Sc. dissertations and is guiding two Ph.D. scholars at present. She has completed certificate course in Food Safety Management systems ISO: 22000:2004. She is also a certified Food Safety Trainer for the FoSTaC (Food Safety Training & Certification) programme conducted by FSSAI, GoI.

Karuna Singh graduated in Home Science Honors in Foods and Nutrition from Vanasthali Vidyapeeth , Rajasthan in 1995. Subsequently she passed her M.Sc. Foods and Nutrition with gold medal from Vanasthali Vidhyapeeth in 1997. She started her career at GDMG (PG) college as lecturer Foods and Nutrition 2000. She has done her PhD from CCS university in 2008. She joined Amity Institute of Food Technology in 2011. She is Certified Food Safety Management System Professional, Training conducted by British Standard Institute in 2013. She has a teaching experience of more than 16 years. She has supervised dissertations, major projects, Summer trainings, internships, seminar, term paper for M.Sc. Foods and Nutrition, M. Tech. (Food. Tech.) and B. Tech (Food. Tech.) students. She is a supervisor for 6 PhD scholars. She has more than 35 publications National and International journals and 20 paper presentations in conferences, 8 book chapters, 01 patents (filed). She is a member of the Editorial Board and reviewer of various reputed journals. She has been engaged in various industrial projects and an active member in organizing Faculty Development programmes and national seminar.

Contributors

Anushka Agarwal National Sugar Institute, Kanpur, India

Divya Agarwal Department of Environmental Science, Jesus and Mary College, University of Delhi, Delhi, India

Anshika Department of Food Technology, Shaheed Rajguru College of Applied Sciences for Women, University of Delhi, Delhi, India

Saumya Chaturvedi Department of Food Technology, Shaheed Rajguru College of Applied Sciences for Women, University of Delhi, Delhi, India

Parul Chugh Amity Institute of Biotechnology, Amity University, Noida, India

Praveen Dahiya Amity Institute of Biotechnology, Amity University, Noida, Uttar Pradesh, India

Raj Kumar Duary Department of Food Engineering and Technology, Tezpur University, Napaam, Assam, India

Devinder Kaur Centre of Food Technology, University of Allahabad, Prayagraj, India

Manish Kaushik Department of Chemistry, Noida Institute of Engineering and Technology, Greater Noida (Dr. APJ Technical University, Lucknow), Noida, Uttar Pradesh, India

Renu Khedkar Amity Institute of Food Technology, Amity University, Noida, Uttar Pradesh, India

Shweta Kulshreshtha Amity Institute of Biotechnology, Amity University, Rajasthan, Jaipur, Rajasthan, India

Rajendra Kumar Division of Genetics, IARI, PUSA, New Delhi, India

Simple Kumar Amity International Centre for Post-Harvest Technology and Cold Chain Management, Amity University, Noida, Uttar Pradesh, India

Girish N. Mathad Amity Institute of Horticulture Studies and Research, Amity University, Noida, Uttar Pradesh, India

Komal Mathur Amity Institute of Biotechnology, Amity University, Noida, Uttar Pradesh, India

Alok Mishra Department of Dairy Science and Food Technology, Institute of Agricultural Sciences, Banaras Hindu University, Varanasi, Uttar Pradesh, India

Narendra Mohan National Sugar Institute, Kanpur, India

Sumedha Mohan Amity Institute of Biotechnology, Amity University, Noida, Uttar Pradesh, India

Ram Mohan Mutharasu National Institute of Food Technology Entrepreneurship and Management, Sonipat, Haryana, India

Venkatachalapathy Natarajan Department of Food Engineering, Indian Institute of Food Processing Technology, Thanjavur, Tamil Nadu, India

Rongsenyangla Ozukum Amity Institute of Horticulture Studies and Research, Amity University, Noida, Uttar Pradesh, India

Ankit Paliwal Western Dairy, Nadi, Fiji

Falguni Patra Mansinhbhai Institute of Dairy and Food Technology (MIDFT), Mehsana, Gujarat, India

Amrita Poonia Department of Dairy Science and Food Technology, Institute of Agricultural Sciences, Banaras Hindu University, Varanasi, Uttar Pradesh, India

Ashokan Poovazhahi Amity Institute of Food Technology, Amity University, Noida, Uttar Pradesh, India

Manoj Kumar Pulivarthi Department of Grain Science and Industry, Kansas State University, Manhattan, KS, USA

Naleeni Ramawat Amity Institute of Organic Agriculture, Amity University, Noida, Uttar Pradesh, India

Chandrakala Ravichandran Department of Food Processing Technology, Karunya Institute of Technology and Sciences, Coimbatore, India

Isha Sai Department of Biosciences, Sri Sathya Sai Institute of Higher Learning, Anantapur, AP, India

Pinki Saini Centre of Food Technology, University of Allahabad, Prayagraj, India

Maysam Sarafrazy Horticultural Research in Urdu Khan Agricultural Research Station, Herat, Afganistan

Rukhsaar Sayeed Amity Institute of Food Technology, Amity University, Noida, Uttar Pradesh, India

Dipti Sharma Department of Food Technology, Shyama Prasad Mukherji College, University of Delhi, Delhi, India

Stuti Sharma Department of Food Technology, Shaheed Rajguru College of Applied Sciences for Women, University of Delhi, Delhi, India

Vatsala Sharma Amity Institute of Food Technology, Amity University, Noida, Uttar Pradesh, India

Vineet Shyam Food Safety and Standards Authority of India, New Delhi, India

Urba Shafiq Sidiqi Islamic University of Science and Technology, Awantipora, Pulwama Kashmir, Jammu & Kashmir, India

Ashmita Singh Amity Institute of Food Technology, Amity University, Uttar Pradesh, Noida, UP, India

Karuna Singh Amity Institute of Food Technology, Amity University, Noida, Uttar Pradesh, India

Deepshikha Thakur Amity Institute of Organic Agriculture, Amity University, Noida, Uttar Pradesh, India

Monika Thakur Amity Institute of Food Technology, Amity University, Noida, Uttar Pradesh, India

Suka Thangaraju Department of Food Engineering, Indian Institute of Food Processing Technology, Thanjavur, Tamil Nadu, India

Pratibha Tiwari Maharishi Markandeshwar University, Ambala, Haryana, India

Ashutosh Upadhyay National Institute of Food Technology Entrepreneurship and Management, Sonipat, Haryana, India

Vandana Department of Food Technology, Shaheed Rajguru College of Applied Sciences for Women, University of Delhi, India

Ayushi Varshney Amity Institute of Biotechnology, Amity University, Uttar Pradesh, Noida, Uttar Pradesh, India

Neelam Yadav Centre of Food Technology, University of Allahabad, Prayagraj, India

Renu Yadav Amity Institute of Organic Agriculture, Amity University, Noida, Uttar Pradesh, India

Anila Zahid Amity Institute of Food Technology, Amity University, Uttar Pradesh, Noida, Uttar Pradesh, India

Abbreviations

%	Percent
&	and
/	per
µg	Microgram
µl	Microlitre
µm	Micrometre
AC	Affinity chromatography
AI	Artificial intelligence
AP	Apple pomace
ASP	Apple skin powders
AWMS	Agricultural waste management system
BOD	Biochemical oxygen demand
BOD5	Biological oxygen demand 5 days
cm	Centimetre
CMA	Cornmeal agar
CNG	Compressed natural gas
COD	Chemical oxygen demand
conc.	Concentrated
DF	Dietary fibre
dil.	Dilute
DNA	Deoxyribonucleic acid
e.g.	For example
EAE	Enzyme-assisted extraction
EC	Extraction conditions
Ed.	Edition
ed.	Editor
eds.	Editors
EFSA	European Food and Safety Authority
EP	Extraction performance
EPA	Environmental Protection Agency
et al.	et. alia and others
etc.	et. cetera
FAO	Food and Agriculture Organization

Fig.	Figure
FLC	Food supply chain
FLW	Food loss and waste
FOG	Fats, oil and grease
FSSAI	Food Safety and Standards Authority of India
FVWs	Fruit and vegetable wastes
g	Gram
GAE	Gallic acid equivalent
GDP	Gross domestic product
GMP	Good manufacturing practices
HACCP	Hazard analysis and critical control points
HPP	High-pressure processing
hrs.	Hours
HVED	High voltage electrical discharges
i.e.	That is
kg	Kilogram
KTIS	Kaset Thai International Sugar Corporation
l	Litre
LDL	Low-density lipoprotein
m	Metre
MAE	Microwave-assisted extraction
MAP	Modified atmosphere packaging
MBT	Mechanical biological treatment
MF	Microfiltration
mg	Milligram
min	Minutes
ml	Millilitre
mln tons	Million tonnes
mm	Millimetre
MMT	Million metric tonnes
MoFPI	Ministry of Food Processing Industries
MSWM	Municipal solid waste management
MW	Microwave
NF	Nanofiltration
NH4+−N	Ammonium nitrogen
°C	Degree Celsius
PDA	Potato dextrose agar
PEF	Pulsed electric field
PLE	Pressurized liquid extraction
PUFA	Polyunsaturated fatty acid
RO	Reverse osmosis
SCF	Supercritical fluid extraction
SDA	Sabouraud dextrose agar
SEPA	Scottish Environment Protection Agency
SFE	Supercritical fluid extraction

sp.	Species (singular)
spp.	Species (plural)
sq km	Square kilometre
SSR	Sample solvent ratio
TCD	Tonnes crushed per day
TDS	Total dissolved solids
TKN	Total Kjeldahl nitrogen
TN	Total nitrogen
TP	Total phosphorus
TS	Total solids
TSS	Total suspended solids
UAE	Ultrasound-assisted extraction
UF	Ultrafiltration
ULBs	Urban local bodies
UNICA	União da Indústria de Cana-de-Açúcar
US	Ultrasonic
USDA	United States Department of Agriculture
var.	Variety
viz.	Videlicet; namely
vol.(s)	Volume(s)
VSS	Volatile suspended solids
w.r.t.	With respect to
WHO	World Health Organization
WPC	Whey protein concentrates
WRI	World Resources Institute

List of Figures

List of Tables

Part I

Food Industry Waste: Introduction, Standards and Management

Sustainable Food Waste Management: A Review

1

Karuna Singh

Abstract

Food wastage and loss is the major problem faced globally and affects developing and developed countries equally. Lost and wasted food represents a missed opportunity to feed the growing world population. Food is wasted at all the levels of food supply chain. Food products are heterogenic in nature and types of food waste and its composition also varied so it is difficult to apply a waste hierarchy to food products. Therefore, the waste hierarchy must be assessed for each type of food waste, rather than for "food waste" as a whole. This chapter will discuss in detail various preventive measures, impact of food waste and along with this various technology that can be used for the treatment of food waste. Best way to prevent food waste is using sustainable food waste management approaches to reduce and prevent food waste.

Keywords

Food loss and waste · Food supply chain · Prevention and treatment of food waste

1.1 Introduction

"Waste" is the term which denotes discarding a material after use at the end of their intended lifespan. Management of waste started from segregation, collection, transportation and is a collective activity utilising its reprocessing, recycling and disposal of waste. Waste management should be environmentally sound, socially satisfactory and must use techno-economically viable methods then only it would be considered

K. Singh (✉)
Amity Institute of Food Technology, Amity University, Noida, Uttar Pradesh, India
e-mail: ksingh11@amity.edu

M. Thakur et al. (eds.), *Sustainable Food Waste Management*,
https://doi.org/10.1007/978-981-15-8967-6_1

sustainable waste management. Two terms i.e. food losses and food waste, can be defined as "food losses" is one that take place at production, postharvest and processing stages in the food supply chain and "food waste" is one that occurs at the end of the food supply chain i.e. at retail and final consumption stage by consumer (Parfitt et al. 2010).

"Food" waste or loss is measured only for products that are directed to human consumption, excluding feed and parts of products which are not edible. Food losses or waste are the masses of food lost or wasted in the part of food chains and therefore food that gets out of human food chain which was originally meant to human consumption is considered as food loss or waste.

According to definition given by Food and Agriculture Organisation (2013), food loss is considered as the decrease in dry matter (mass) or nutritional value (quality) of food that was originally intended for human consumption. This is caused due to various inefficiencies throughout the food supply chain like poor infrastructure, technological insufficiencies, lack of knowledge and skills of participants at different levels of FSC. Whereas food waste refers to food appropriate for human consumption being discarded, whether or not after it is kept beyond its expiry date or left to spoil.

1.2 Types of Food Losses/Waste

Food-related waste (including edible and non-edible parts) represents an important proportion of total waste. Food losses can be grouped into two categories as Avoidable losses and Unavoidable losses.

Unavoidable food losses are those that cannot be in generally eaten by human beings like bones of fishes and the skin of pineapple. On the other hand, avoidable FLW occurs for the types of foods that are edible but neither used nor eaten. Food policies and researches should focus on avoidable FLW.

All the five system boundaries of food supply chain (FSC) as depicted in Fig. 1.1 include both vegetable and animal commodities which can be wasted or lost at each stage. Starting from production where losses occur due to mechanical damage during threshing, picking of fruits and vegetables during harvesting each stage of FSC depicted food loss and waste. At postharvest and storage stage losses occur due to

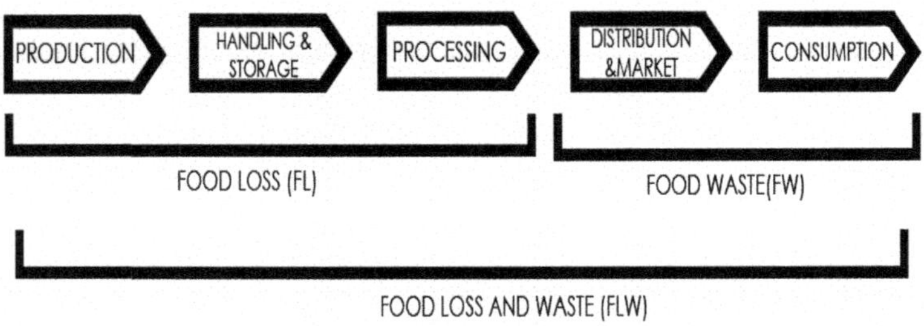

Fig. 1.1 Food supply chain (FSC) stages related to food loss and waste

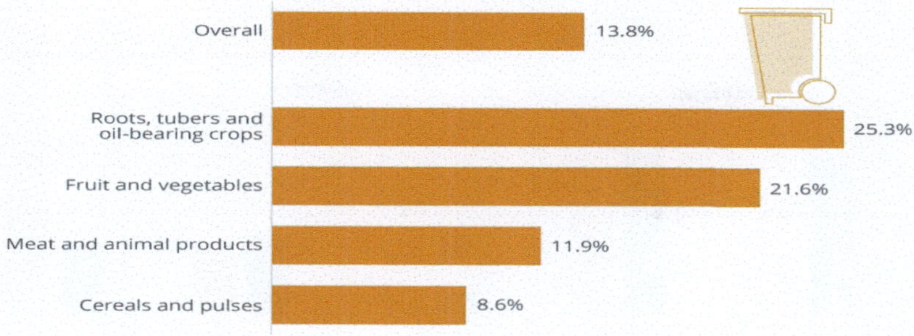

Fig. 1.2 Percentage of food loss globally. Source: FAO (2016)

Table 1.1 Food loss and waste among developed and developing countries

	Food loss	Food waste	Food loss and waste
Developing countries	30%	14%	44%
Developed countries	21%	35%	56%

spillage and mishandling of products. Along with this during storage and transportation between farm and distribution also cause some losses. During processing at industrial or domestic level and finally losses can take place in the market and consumption of food at hotel or household.

As per estimations globally about 1.3 billion tons of edible food produced for human consumption get wasted per year which is roughly around one-third of the edible parts produced throughout the FSC. In medium- and high-income countries food is wasted by thrown away even if it is still suitable for human consumption, whereas among low-income countries food is mainly lost during the early and middle stages of the food supply chain than at the consumer level (FAO 2011). Figure 1.2 showed the global food loss of various type of food group and commodities.

The UN, FAO and World Resources Institute (WRI) on global FLW highlight the significant differences in per capita FLW between economies (Gustavsson et al. 2011). As depicted in Table 1.1 56% of the FLW occurs in developed countries, while the other 44% occurs in developing countries. Figure 1.3 depicted the generated FLW that varies in each stage of FSC among developed and developing countries. There is relatively greater amount of food loss among developing countries, while developed countries have a higher portion of food waste.

There are various food waste/loss drivers in the food system such as industrialisation, urbanisation, globalisation, economic growth, socioeconomic factors, Dietary transations and diversification, product and packaging characteristics, retail market strategies, consumer skills, behaviour, habit motivation, poor education, lack of technology. Detail description of various causes of FLW at different level of food supply chains (FSC) has been presented in Table 1.2.

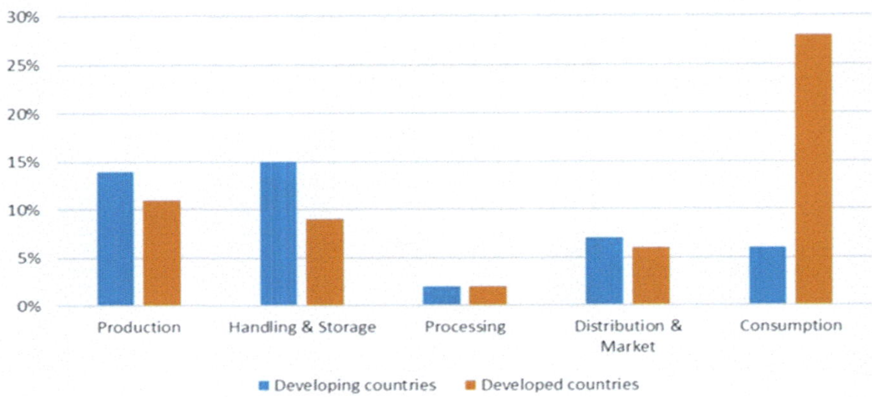

Fig. 1.3 Percent food loss and waste in food supply chain (Source: Lipinski et al. (2014))

1.3 Principles of Food Waste Management

1.3.1 4Rs: Refuse, Reduce, Reuse & Recycle

- *Refuse*: Refuse is the best method to manage waste of any type. Reduce suggests cutting the amount and the kind of products which we buy. The best way to reduce is to consume less and generate less waste. Avoid buying any food without its use.
- *Reduce*: Reduce the amount of garbage generated. Waste reduction is a good method of prevention of waste. Elimination of the source producing waste demands for large scale treatment and disposal facilities. Some protocols should be followed to reduce source that generates waste, for example, use scrapers and other equipment like high pressure spray to clean floors, donation of unwanted items, backyard composting, etc. Reducing the amount is the most significant of all the options to manage waste. Other method of reducing waste is to donate extra food to people who need them and also to use it for animal feeding. Along with this rendering fat for industrial use can also be used to reduce waste at source.
- *Reuse*: The term reuse refers to putting products and materials back into use before they become waste. It involves no physical change in the product and therefore keeps the products in use for longer. Reusing products reduce raw material consumption and energy use and associated costs. It also reduces the requirements for waste collection, treatment, and disposal.
- *Recycle*: Recycling means manufacturing of a new product by using items derived as raw material from the waste. Recycling occurs in three phases: first the waste is sorted and recyclables collected, secondly raw material is created out of sorted recyclable material and finally in the third phase a new product is produced using derived raw materials. Both recovery and recycling reprocess

Table 1.2 Possible causes of food loss and waste during different stages of food supply chain

Stage of FSC	Causes of FLW	Reference
Production stage	Infrastructural limitation	Gustavsson et al. (2011) and HLPE (2014)
	Over production	Kaipia et al. (2013) and Garrone et al. (2014)
	Harvesting timing and method (manual/mechanical)	Kumar and Kalita (2017), Grover and Singh (2013) and Kannan et al. (2013)
	Pesticides and fertilisers	Thompson (2008)
	Economic problems	FAO (2014)
	Quality standards and norms	Garrone et al. (2014) and Stuart (2009)
Postharvest handling and storage stage	– Degradation and spillage of product composition	Yusuf and He (2011), Bett and Nguyo (2007), Willersinn et al. (2015) and FAO (2014)
	– Loss during transportation from farm to distribution	
	– Storage infrastructure	
Processing and packaging stage	– Unavoidable losses	HLPE (2014), Beretta et al. (2013), Schieber et al. (2001) and Darlington and Rahimifard (2006)
	– Technical malfunctions	
	– Methods and changes in processing lines	
	– Contamination in processing lines	
	– Legislation restrictions	Gustavsson et al. (2011) and Papargyropoulou et al. (2014)
	– Packaging system	Stuart (2009)
	– Overproduction	Khedkar and Singh (2018) and Murthy et al. (2009)
Distribution and marketing stage	– Inappropriate transport conditions (temperature-controlled aircrafts and ships)	Martinez et al. (2014), Fox and Fimeche (2013), Kaipia et al. (2013), HLPE (2014), Papargyropoulou et al. (2014), Khedkar and Singh (2015) and Garrone et al. (2014)
	– Contamination of transportation	
	– Transportation and market facilities	
	– Road and distribution vehicles	
	– Packaging management	
	– Commercial conditions	
	– Consumer reference	
Consumption stage	– Composition unit and size of household	Parizeau et al. (2015), Pingali and Khwaja (2004), Neff et al. (2015) and Schanes et al. (2018)
	– Income group	
	– Demographics and culture	
	– Individual attitude	
	– Cooking practices and methods	

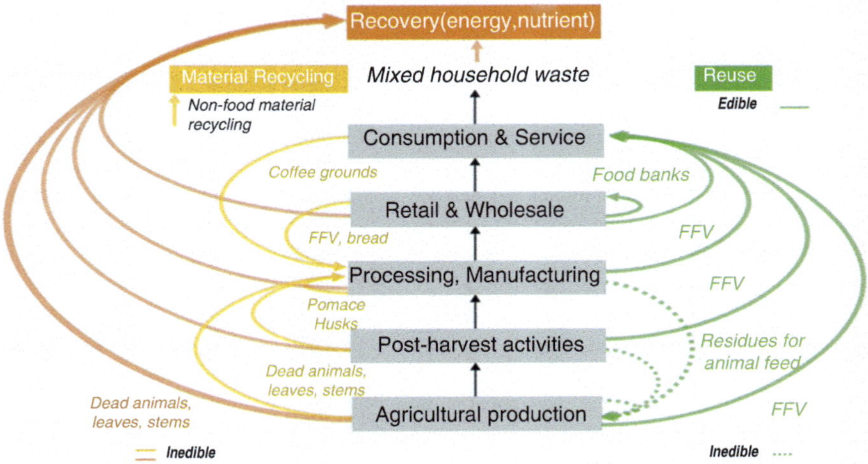

Fig. 1.4 Food waste and food loss management at different level of FSC (adapted from Teigiserova et al. 2020)

food waste materials into usable material. Recovery differs from recycling in which energy production also included. Recycling creates new opportunities of employment in areas of collection, treatment and reprocessing of recyclable materials particularly where new uses and applications for materials can be developed. The need is to increase the demand and markets for recycled products.

Food products are heterogenic in nature and types of food waste and its composition also varied so it is difficult to apply a waste hierarchy to food products. Therefore, the waste hierarchy must be assessed for each type of food waste, rather than for "food waste" as a whole.

Figure 1.4 describes the food waste produced at each level of food supply chain both edible or inedible. The diagram subsequently followed by arrows and making further indicator selections. The figure denoted various types of waste management alternatives which differ as per indicators for that food type. Food waste sub-grouped on the basis of treatment applied for both type of waste, i.e. plant-based waste and animal-based waste. Both types of waste should be assessed independently hence segregated and collected separately. On that basis more targeted management practices can be carried out on the different food waste streams. When separate collection is not possible, a thorough waste sorting is still recommended.

Various food waste management alternatives have been discussed in Fig. 1.5. Some of these alternatives have been grouped and applied, for example, various bio-compounds from food products can be extracted through various physical and chemical process and also there are various possibilities of utilising various food waste for industrial applications (Khedkar and Singh 2018). It is therefore not possible to apply all options explicitly for all the food waste categories. Each type of food waste needs to be assessed independently and studied so that suitable

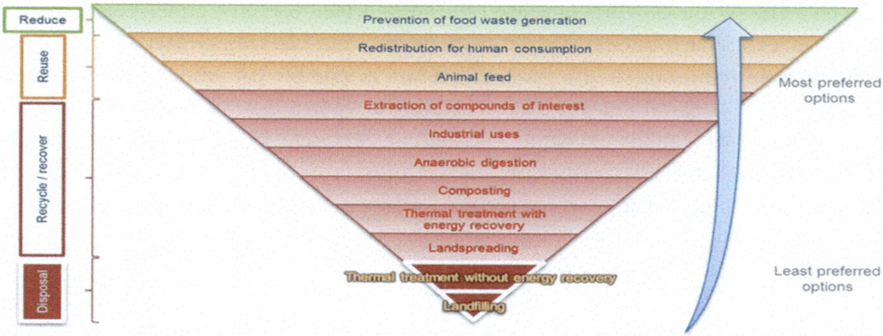

Fig. 1.5 Waste hierarchy for surplus food and food waste (adapted from Garcia-Garcia et al. 2015)

opportunities can be used to treat that waste and various compounds of interest of industry can be extracted out of it.

1.4 Prevention of Food Waste

Prevention of food waste generation is at the top of the food waste hierarchy and the least preferred options are landfilling and thermal treatment without energy recovery. Prevention of food waste can be applied to all types of edible food wastes likewise land spreading can be used with the majority of food waste types, but according to the food waste hierarchy (Fig. 1.5) this alternative is less beneficial than composting. Since food is such an incredibly valuable resource so this resource should not be wasted. Food waste utilisation for the generation of valuable commodities can be utilised as a way to overcome loss and waste of food at various levels of food system and must be considered before putting in a landfill or incinerator (Khedkar and Singh 2018). Various methods used for prevention of food waste are described in detail as below:

1.4.1 Redistribution for Human Consumption

Best way to prevent food loss is redistribution for human consumption. Food charities and food banks are good examples of this method. Food that is used for this purpose must be edible, eatable and improves resource utilisation.

1.4.2 Animal Feeding

The best alternative and sustainable solution of utilising food waste is animal feeding. The foods which are not fit for human consumption but can be given for animal (farmed animals like cow, goat, sheep, etc.) feeding after assessing it properly. The products which are used for animal feeding must either be eatable or uneatable for humans but eatable for animals and non-catering waste. Mixed waste containing animal products from manufacturers is suitable for animal feeding when the animal product is not the main ingredient. Meat (or plant-based products containing meat) cannot be sent for animal feeding. Eggs and egg products (or plant-based products containing them) must come from the agricultural or manufacturing stage when used for animal feeding after utilising specific treatments.

1.4.3 Quantification and Characterisation of Food Waste

Quantification helps in identification of the source and type of food waste. It also helps in implementing suitable targeted preventive measures. On the basis of characterisation of food waste standardised protocols can be formulated to prevent and utilise food waste. These protocols can be further used at different stages of food supply chain to develop food waste and loss inventories and management system to overcome losses.

1.4.4 Awareness and Communication Policies

A tailored communication instruments are required to deal with the changes occured in consumption pattern and social behaviour over a period of time. With the change in food choice, availability and type of food (RTE, RTC), the type of waste produced also changed and hence the prevention and management techniques need to be updated. Packaging waste increases, while food waste falls as people cook less at home (Tara Slade 2016). Awareness of stakeholders to overcome such waste needs to be communicated. Various types of information education communication material (IEC) can be used for awareness campaigns like printed materials, public meetings pamphlets, etc. Along with this other methods like seminars, surveys can be used for collecting information, constant monitoring and reporting.

1.4.5 Reporting and Monitoring

The management of food wastes is easier and more sustainable approach to prevent and it can be used if the amounts and quality of food waste produced are regularly monitored and accounted for. Good manufacturing practices and dissemination of best practices and its reporting prevent food waste. Businesses or industry that generate large quantities of food waste may be required to report the origin, volume and disposal methods of such waste. This helps in monitoring the amount and cost of their waste and thus encouraging its reduction.

1.4.6 Regulatory Initiatives

According to the Indian Constitution, the state government and the urban local bodies (ULBs) are responsible for solid waste management. MSWM is governed by the Municipal Solid Waste Management and Handling Rules, 2016. The rules designate ULBs responsible for the management of only solid waste and directed ULBS responsible for the management of municipal solid waste within their territorial area and also for the generation of infrastructure for implementation.

1.4.7 Research and Innovations

Innovating new ways of reducing food waste and also utilisation of food waste for production of more valuable commodities like extraction of pectin from fruit waste is hot topic for researchers. Along with this researches should be initiated to minimise food loss and waste at every stage of food handling which is the demand of this era.

1.5 Food Waste Treatment Technologies

1.5.1 Anaerobic Digestion (AD)

AD is the process in which organic matter gets decomposed using bacteria in an oxygen-free environment. Optimal conditions are required to conduct this process like alkaline pH, controlled temperature and minerals like nitrogen, phosphorous and potassium. Various benefits have been demonstrated by this method like reducing the health impacts which may occur otherwise due to poor waste management and it helps in recovery of energy. Anaerobic digestion recovers 60% more energy than direct combustion (Valorgas 2014). It also helps in producing a fertiliser that is rich in nutrients and hence replenishes soils.

1.5.2 Composting

Composting is relatively simple, predictable and naturally occurring process in which decomposition of organic matter using oxygen is done by the action of microorganisms and small invertebrates in controlled conditions. During composting heat is created by the biodegrading mass itself and its temperature may rise to 70 °C which further accelerate the biodegradation process and natural fermentation of the biomass. There are many composting techniques which can be utilised at small scale like backyard compositing and for larger volumes of waste aerated like in-vessel composting.

Various advantages of this technique include production of high organic matter compost that can be used to increase yield of plant and restore soil. It helps in stabilising and sanitising food waste. Various disadvantages of this techniques include that it does not recover energy and requires careful management of contaminants and odour.

1.5.3 Liquefaction

Liquefaction is the process where food waste gets converted into liquid effluent by using multiple methods like mechanical, biological or hydrothermal liquefaction and resultant effluent may be discharged into the drainage as in household or in municipal wastewater system. This method is simple to use but utilised where there is a wastewater treatment system and also requires energy input.

1.5.4 Rendering

Rendering is a process that converts waste animal tissue and by-products into usable material like high-quality fat and protein products. It is of two types: wet and dry rendering. Separation of fat from raw materials by boiling in water is done in wet rendering, whereas dehydrating the raw material to release the fat is done in dry method (Parry et al. 2015). Along with fat commodity rendering also produces highly valued protein meal. Rendering food waste provides a very good substitute for conventional animal feed which provides high protein supplement. This process requires close regulation and stringent legislation as it deals with food waste containing animal by-products. One big disadvantage of this process is that it requires energy input.

1.5.5 Thermal Treatment Can Be Used for Energy Recovery

These treatments can be applied to every type of food waste. Thermal treatments with energy recovery, which includes various techniques like gasification, pyrolysis and incineration, are the only alternative available to treat packaged food (non-separable from packaging) in non-biodegradable packaging, except the cases when the product is also edible, eatable and processed, and therefore can be redistributed for human consumption. It can also be used for mixed food waste also

- *Gasification*

Gasification is a process which uses a high temperature (>700 °C) with a controlled use of oxygen or steam to decompose organic materials like food wastes or combinations of organics and inorganics into a combustible gas called syngas. It is therefore a technology that involves thermochemical conversion, like incineration or pyrolysis. One big disadvantage of this method is lack of nutrient recovery. But according to the World Energy Council 2016, both gasification and pyrolysis are

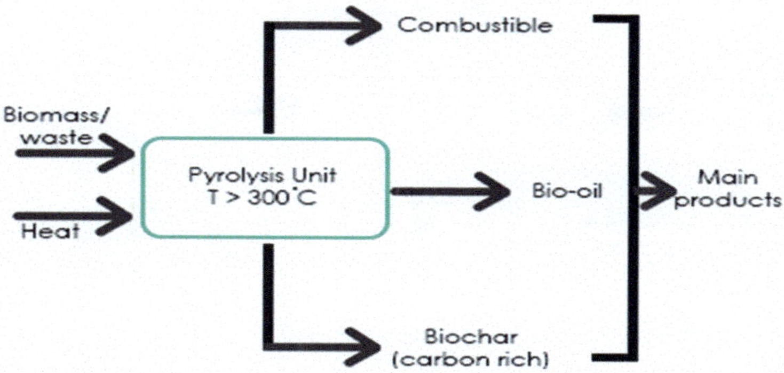

Fig. 1.6 Pyrolysis of food waste

more efficient and score better in environmental impacts than incineration with energy recovery.

- *Incineration*

Controlled exothermic combustion of mixed solid waste at extremely high temperatures is known as incineration. This method is applicable on municipal food waste and also generated from commercial and industrial sources ideally non-recyclable one. In this process food and other waste are not separated at the source hence save collection costs. This method is an effective approach that maximises recycling rates and recovers energy from non-recyclable residual waste. Depending on the treatment options for the bottom ash formed by the inorganic constituents of the waste, ferrous and nonferrous metals can be recovered and the remaining ash can be further enhanced to be for road construction and buildings.

- *Pyrolysis*

Pyrolysis is a thermo-chemical decomposition of organic material which gets converted into gases and biochar which are the main products of pyrolysis unit. Heating is done in the absence of oxygen. Compared to combustion in other methods, pyrolysis causes less emissions of air pollutants and biochar produced can be used to increase agriculture productivity (Fig. 1.6).

1.5.6 Mechanical Biological Treatment (MBT)

Mechanical biological treatment (MBT) is the combination of both biological and physical processes. This is a waste processing system in which sorting of food waste is further treated with various biological treatments like anaerobic digestion or compositing to produce biogas (Fig. 1.7).

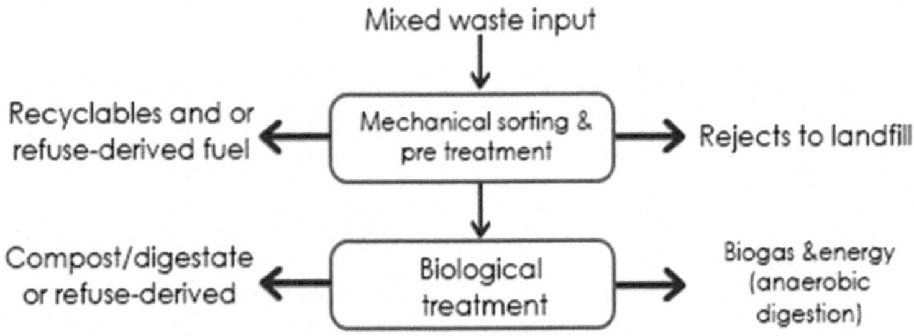

Fig. 1.7 Process of MBT

This technique can be used for mixed household, commercial as well as industrial waste. It helps in recycling of food waste which if not treated will be inefficiently combusted or landfilled. Other advantages of this method reduce the cost associated with the separation and segregation of food waste and allows energy recovery via the anaerobically digested organic fraction.

1.6 Impact of Food Waste

Along with squandering of resources needed for the production of food (including energy, carbon, water and nutrients), poorly managed food waste adversely affects our climate due to the emission of greenhouse gasses due to decomposition, water contamination, leaching of nutrients and place of breeding vector for diseases and other health hazard. Various impacts of food waste are discussed below:

1.6.1 GHG Emissions and Climate Change

Greenhouse gases are emitted at all stages of the food life cycle which contribute to global warming and climate change (Eco Watch 2017). Greenhouse gases CO_2, CH_4, N_2O are produced by the burning of fuels for energy production, emissions from manure and slurries, pasteurisation, refrigeration and transport of food. Landfill sites or dumpsites where waste food decomposes emitted these gases to the atmosphere. Composting or anaerobic digestion was considered as better option as compared to landfills and open dumps as they prevent methane emission to the environment. From long back fossil fuel has been used to produce biogas-based energy that can be used as an option to solid domestic fuel. This not only helps in improving indoor air quality but also helps to mitigate climate (Food and Agriculture Organisation of the United Nations 2015). Along with this if inorganic fertilizers can be substituted with bio fertilisers using different treatments like compost or digestate on food waste, will also help in managing carbon dioxide emissions. It is estimated that 580 kg CO_2 eq. can be saved per each tonne of food waste diverted from landfill to an anaerobic digester when the resulting biogas is used to replace natural gas (Ellen MacArthur Foundation 2013).

1.6.2 Water Footprint

Water is essential for life. In places where rainfall is not adequate or seasonal, water is extracted by the plant irrigation from groundwater aquifers and surface water bodies. For the growth of plant water is major essentiality hence the uncontrolled disposal of food waste has an impact on surface water as well as groundwater bodies. Along with this overuse and subsequent run-off of fertilisers and pesticides have an adverse impact on the water quality of both types of waterbodies: ground and surface. Dumpsites and landfills can also pollute the groundwater and surface water as leachate from them can reach that level of water. In various food processing industry untreated wastewater is released to water bodies which pollutes the surface water. Hence preventing food waste can reduce the pollution of water.

1.6.3 Nutrient Loss of Agriculture Land

Changes in agriculture practices from decades have resulted in nutrients depletion of the soil. Agriculture land usage is not increasing as the population increasing day by day and hence the demand of more food to feed that population is creating more loss of nutrient and organic matter. This leads to more usage of synthetic fertilisers to increase yield of the land. Food, agriculture waste and even human excreta production increase in cities due to urbanisation did not returned back to farm which in past recycled and naturally move to farmland (UN 2014). Recycling urban waste through the use of methods like digestate and compost can be used to overcome this gap created by urbanisation and unsustainable agriculture practices. Various benefits of replenish agriculture land are that it maintains reserves of phosphorus, potassium, nitrogen and organic carbon back to soil and will reduce the use of inorganic fertilisers. Nutrient recycling also prevents run-off nutrients to surface water bodies and hence prevents water pollution and aquatic life and the livelihood of people who depend on it.

1.6.4 Hygiene and Sanitation

Globally, about 50% of food waste is sent to landfills, while 13 to 33% of waste is still being openly dumped in lower- and middle-income countries (The World Bank 2012). These landfills and dumpsites can lead to various health hazards in the populations living and working near these places. Organic waste in dumpsites becomes the centre of various communicable diseases and place of breeding for mosquitoes and flies which further increase the health risk transfer of food borne diseases (ISWA 2015). Treatment of food waste using suitable technology helps in preventing spread of rodents and various disease spreads. One good example favouring the prevention spread of disease and foul odour from food waste is anaerobic digestion. It also helps in promoting sanitation and hygiene.

1.6.5 Ecological Impacts

Various types of ecological changes take place to increase food production so as growing global population can be fed. These changes include cutting of forests to be used for other purposes like agriculture or houses building. There is a loss of various species of biodiversity which in turn lead to various transition in eating habits and development of various life style related diseases. Various indicators were used at global scale to find the impacts of this damage from food production like quality of soil, loss of biodiversity, etc. Various evidences have suggested that segregation of food waste is one of the approach to make it measurable and this allows effective targeted policy making and preventive measures. To avoid dumping of food waste in landfills and open dumping proper collection and management system is a necessity. Various methods like anaerobic digestion and composting are better techniques that can be utilised to convert food waste to more productive fertilisers for farmland and in turn can prevent environmental pollution like water and air.

1.6.6 Economic Impacts

Nearly one-third of the food that is produced each year goes uneaten, costing the global economy over $940 billion. Uneaten food is responsible for emitting about 8% of planet-warming greenhouse gases into the atmosphere. Economic cost of waste management includes the maintenance of landfills, transport cost, treatment plant operation cost and also separation and segregation cost. The total annual economic, environmental and social costs of food waste to the global economy are in the order of USD 2.6 trillion food waste which is not separately collected and disposed of in landfill. Along with this wasting food also rises financial crisis. Rising food prices and food shortage further increase cost of diseases and healthcare.

1.7 Conclusion

Food waste throughout food supply chain is complex and it imparted significant impact on various dynamics and factors. It imparts a great impact on economics, agriculture and food security of the country. Along with this the waste utilisation and management will determine the environmental conservation and human health. For resolving the problem of food waste there is a need to develop sustainable strategies which require cooperation between all stakeholders. The key to successful food waste management is to develop and utilise appropriate eco-friendly, sustainable technologies that can prevent and reduce the wastage. If wastage is generated, then use of appropriate technology that can reduce its impact on environment and generate energy out of it should be preferred. Consumer awareness and education is also very important and a great challenge. The challenge must be accepted by each stakeholder throughout food supply chain and use of sustainable eco-friendly reduction approach is the key to manage this vast problem faced globally by many countries.

References

Beretta C, Stoessel F, Baier U, Hellweg S (2013) Quantifying food losses and the potential for reduction in Switzerland. Waste Manag 33:764–773

Bett C, Nguyo R (2007) Post-harvest storage practices and techniques used by farmers in semi-arid eastern and central Kenya. In: Proceedings of the 8th African Crop Science Society Conference, El-Minia, Egypt, vol 27(31), pp 1023–1227

Darlington R, Rahimifard SA (2006) Responsive demand management framework for the minimization of waste in convenience food manufacture. Int J Comput Integr Manuf 19:751–761

Eco Watch (2017) World's soils have lost 133bn tonnes of carbon since the dawn of agriculture. Retrieved from https://www.ecowatch.com/soil-carbon-loss2478725457.html. Accessed 7 Mar 2020

Ellen MacArthur Foundation (2013) Towards the circular economy. Retrieved from https://www.ellenmacarthurfoundation.org/assets/downloads/publications/TCE_Report-2013.pdf. Accessed 23 Apr 2020

FAO (2011) Global food losses and food waste – extent, causes and prevention. Rome available at http://www.fao.org/3/a-i2697e.pdf. Retrieved from https://www.statista.com/chart/19672/global-shares-of-different-agricultural-products-thrown-away/. Accessed 12 Mar 2020

FAO (2013) Impacts on natural resources. FAO, Rome, Italy: 2013. Food wastage footprint. Summary Report

FAO (2014) Food wastage footprint: full-cost accounting, Final Report. FAO, Rome, Italy

FAO (2016) The State of Food and Agriculture 2016 – Climate Change, Agriculture and Food Security. Rome, FAO

Food and Agriculture Organisation of the United Nations (2015) Food wastage footprint & climate change. Retrieved from http://www.fao.org/3/a-bb144e.pdf. Accessed 10 Feb 2020

Fox T, Fimeche C (2013) Global food waste not, want not. Institution of Mechanical Engineers, London

Garcia-Garcia G, Woolley E, Rahimifard S (2015) A framework for a more efficient approach to food waste management. Int J Food Eng 1:65–72

Garrone P, Melacini M, Perego A (2014) Opening the black box of food waste reduction. Food Policy 46:129–139

Grover DK, Singh JM (2013) Post-harvest losses in wheat crop in Punjab: past and present. Agric Econ Res Rev:26

Gustavsson J, Cederberg C, Sonesson U, van Otterdijk R, Meybeck A (2011) Global food losses and food waste. FAO, Rome, Italy

HLPE (2014) Food losses and waste in the context of sustainable food systems. A Report by the High Level Panel of Experts on Food Security and Nutrition of the Committee on World Food Security. Committee on World Food Security, Rome, Italy. Retrieved from https://www.oecdilibrary.org/docserver/5js4w29cf0fen.pdf?expires=1525784803&id=id&accname=guest&checksum=097503A68F4EA992CADBEBC498B54F03. Accessed 15 Mar 2020

ISWA (2015) The tragic case of dumpsites. Retrieved from: https://www.iswa.org/fileadmin/galleries/Task_Forces/THE_TRAGIC_CASE_OF_DUMPSITES.pdf. Accessed 12 Feb 2020

Kaipia R, Dukovska-Popovska I, Loikkanen L (2013) Creating sustainable fresh food supply chains through waste reduction. Int J Phys Distrib Logist Manag 43:262–276

Kannan E, Kumar P, Vishnu K, Abraham H (2013) Assessment of pre and post-harvest losses of rice and red gram in Karnataka. Crops 44:61

Khedkar R, Singh K (2015) New approaches for food industry waste utilization. In: Neetu S (ed) Biologix, pp 51–65

Khedkar R, Singh K (2018) Food industry waste: a panacea or pollution hazard? In: Jindal T (ed) Paradigms in pollution prevention. Springer briefs in environmental science. Springer, Cham

Kumar D, Kalita P (2017) Reducing postharvest losses during storage of grain crops to strengthen food security in developing countries. Foods 6:8

Lipinski B, Hanson C, Lomax J, Kitinoja L, Waite R, Searchinger T (2014) Reducing food loss and waste. Retrieved from http://pdf.wri.org/reducing_food_loss_and_waste.pdf. Accessed 9 Mar 2020

Martinez Z, Menacho P, Pachón-Ariza F (2014) Food loss in a hungry world, a problem? Agron Colomb 32:283–293

Murthy DS, Gajanana T, Sudha M, Dakshinamoorthy V (2009) Marketing and post-harvest losses in fruits: its implications on availability and economy. Indian J Agric Econ 64:902-2016-67302

Neff RA, Spiker ML, Truant PL (2015) Wasted food: U.S. consumers reported awareness, attitudes and behaviors. PLoS ONE 10:e0127881

Papargyropoulou E, Lozano R, Steinberger J, Wright N, Ujang ZB (2014) The food waste hierarchy as a framework for the management of food surplus and food waste. J Clean Prod 76:106–115

Parfitt J, Barthel M, Macnaughton S (2010) Food waste within food supply chains: quantification and potential for change to 2050. Philos Trans R Soc B: Biol Sci 365:3065–3081

Parizeau K, von Massow M, Martin R (2015) Household-level dynamics of food waste production and related beliefs, attitudes, and behaviours in Guelph, Ontario. Waste Manag 35:207–217

Parry A, Bleazard P, Okawa K (2015) "Preventing Food Waste: Case Studies of Japan and the United Kingdom", OECD Food, Agriculture and Fisheries Papers, No. 76. OECD Publishing, Paris

Pingali P, Khwaja Y (2004) Globalisation of Indian diets and the transformation of food supply systems. In: Proceedings of the Inaugural Keynote Address to the 17th Annual Conference of the Indian Society of Agricultural Marketing; Hyderabad, India, pp 5–7

Schanes K, Dobernig K, Gözet B (2018) Food waste matters—a systematic review of household food waste practices and their policy implications. J Clean Prod 182:978–991

Schieber A, Stintzing FC, Carle R (2001) By-products of plant food processing as a source of functional compounds—recent developments. Trends Food Sci Technol 12:401–413

Stuart T (2009) Waste: uncovering the global waste scandal. Penguin, London

Tara Slade (2016) Could you live in a home without a kitchen? Retrieved from http://popupcity.net/could-you-live-in-a-home-without-a-kitchen/. Accessed 4 Mar 2020

Teigiserova DA, Hamelin L, Thomsen M (2020) Towards transparent valorization of food surplus, waste and loss: clarifying definitions, food waste hierarchy, and role in the circular economy. Sci Total Environ 706:136033

The World Bank (2012) What a waste – a global review of solid waste management. Retrieved from https://siteresources.worldbank.org/INTURBANDEVELOPMENT/Resources/336387-1334852610766/What_a_Waste2012_Final.pdf. Accessed 15 Feb 2020

Thompson K (2008) Fruit and vegetables: harvesting, handling and storage. Wiley, Hoboken

United Nations (2014) World urbanization prospects. Retrieved from https://esa.un.org/unpd/wup/publications/files/wup2014-highlights.pdf. Accessed 12 Feb 2020

Valorgas (2014) Valorisation of food waste to biogas: 33 http://www.valorgas.soton.ac.uk/Pub_docs/VALORGAS_241334_Final_Publishable_Summary_140110.pdf. Accessed 25 Feb 2020

Willersinn C, Mack G, Mouron P, Keiser A, Siegrist (2015) Quantity and quality of food losses along the Swiss potato supply chain: stepwise investigation and the influence of quality standards on losses. Waste Manag 46:120–132

World Energy Council (2016) World energy resources - waste to energy. Retrieved from https://www.worldenergy.org/wpcontent/uploads/2017/03/WEResources_Waste_to_Energy_2016.pdf. Accessed 1 Mar 2020

Yusuf BL, He Y (2011) Design, development and techniques for controlling grains post-harvest losses with metal silo for small and medium scale farmers. Afr J Biotechnol 10:14552–14561

Environmental Standards & Regulations for Waste Management in Food Industries

2

A. Poovazhahi and Monika Thakur

Abstract

The world is continuously changing and in recent times it was paced up with rapid environmental changes. The growing population has intensified the demands for agricultural production. Food waste is composed of biological matter, dissolved organic and inorganic solids which constitute to an environmental load if untreated. These wastes can be further processed and utilized as a by-product. Environmental legislation has significantly contributed to the introduction of sustainable waste management practices in food industries. The foremost aim of waste legislation governing body is to prevent or to minimize the waste generation in food industries.

Keywords

Waste management · Food waste · Environmental standards · Waste utilization

2.1 Introduction

Food waste management is a process of converting the material exempted during the food processing technique into a useful form, in order to reduce the losses accompanied during the process. The food waste composition from the industry is extremely varied and it depends upon the nature of the product and the processing technique used. Management of industrial waste in a sustainable manner is a promising and challenging task. It includes 4R (Reduce, Reuse, Recover, Recycle) concept to optimize utilization of solid and liquid waste in an industry. Food industry produces large volumes of wastes including both solid and liquid resulting from the

A. Poovazhahi (✉) · M. Thakur
Amity Institute of Food Technology, Amity University, Noida, Uttar Pradesh, India

M. Thakur et al. (eds.), *Sustainable Food Waste Management*,
https://doi.org/10.1007/978-981-15-8967-6_2

production, preparation, consumption (Russ and Meyer-Pittroff 2010). These waste disposals pose severe environmental pollution and also represent a loss of valuable biomass and nutrients. The Combined process in waste minimization during the production process, environmentally friendly preservation of the product and utilization of the byproducts would substantially reduce the amount of waste as an boost the environmental aspects of food processing industry.

2.2 Waste Management in Food Industries

Foods are the prime source for the survival of homo sapiens and it is found as sine qua non for the human health and safety. The food and beverage industry plays an inevitable role in the world economy. The food produce after being successfully processed into a valuable product based on the consumer needs and demands. In parallel, every processing comes with end product and waste product. The wastes thus produced from the food industries need to be treated accordingly so that it does not affect the environment and human race. Food industry waste and effluents are rich in biodegradable components with high biological oxygen demand (BOD) and chemical oxygen demand (COD) content (Kjeldsen et al. 2002). The uncontrolled decomposition of waste is hazardous to the environment due to presence or formation of methane and toxic materials. A range and scope of strategies for the management of waste have been adopted at national and international levels. In European community (EC), they have introduced a community for waste management in 1989 and it further amended in 1996 with a set of legal principles which includes Prevention Principle to minimize waste production, Polluter penalty Principle, Precautionary Principle, Proximity Principle. These principles create a backbone of European community Waste Policy. The three main inbound strategies for the waste management highlighted by (Sanders and Crosby 2004) include

(1) Hierarchy of waste management (Fig. 2.1)—Focus on 3R Principle (Reduce, Reuse, Recycle).

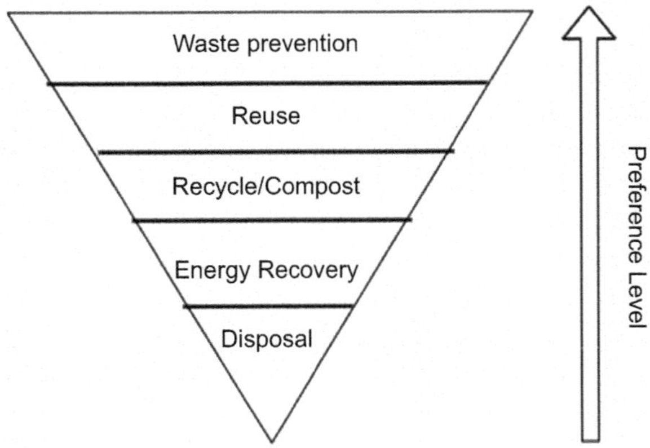

Fig. 2.1 Waste hierarchy

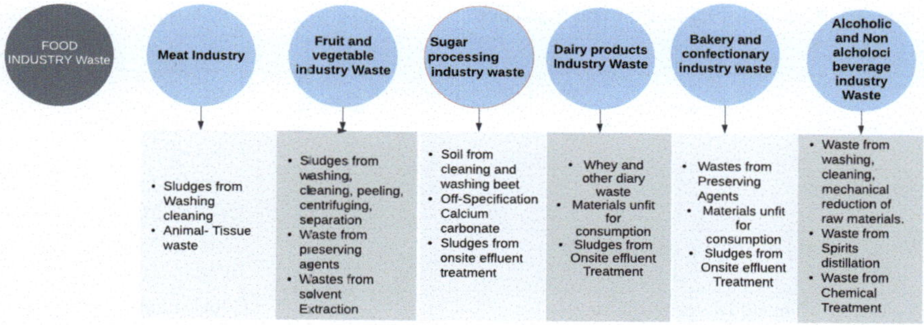

Fig. 2.2 Food industry waste

(2) Responsibility of the producer to takeback the end of life products.
(3) Control of shipment waste inside and outside the region.

Wastes from food industries include solid wastes, air pollutants, and waste waters. These separate categories of wastes are regulated by separate and distinct bodies of laws and regulations, i.e., solid wastes are regulated by the Comprehensive Environmental Response Compensation and Liability Act (CERCLA), Superfund Amendments and Reauthorization (SARA), and as well as certain federal laws and regulations; air pollutants are regulated by the Clean Air Act and wastewater discharges are regulated by the Clean Water Act (European Commission 2004). However, the three major categories of wastes are interlinked, both as they impact the environment and as they are generated and managed by individual industrial facilities. The total spectrum of food industrial wastes can be successfully managed as a system of interrelated activities and substances.

The legal definition of waste mentioned in EU commission (European Commission 2012) drafted from the waste Framework Directive 75/442/EEC as "Waste shall mean any substance or object which the holder discards or intends or is required to discard." The lists of wastes extracted from the EU commission directive related to food processing industries are listed in Fig. 2.2.

Food processing industries hold a significant position economically and generate large volumes of biodegradable wastes. However, hazardous waste is also generated rarely depending upon the situations such as contamination by pesticides, herbicides, and pathogens (AWARENET 2004). The unbalanced localization may induce unsuitable accumulation or putrefaction of organic wastes.

Table 2.1 Genesis of legislation and regulation

Category	Legislation
Solid and liquid waste	
Waste oil disposal	Council Directive 75/439/EEC
General waste	Council Directive 75/442/EEC
Waste harmful to aquatic environment	Council Directive 76/464/EEC
Protection of groundwater pollution	Council Directive 80/68/EEC
Urban waste water treatment	Council Directive 91/271/EEC
Hazardous waste	Council Directive 91/689/EEC
Shipments waste	Council Directive 259/93
Hazardous waste from incineration process	Council Directive 94/67/EEC
Landfill waste	Council Directive 1999/31/EC
Incineration waste	Council Directive 2000/60/EC
Waste statistics	Regulation 2150/2002/EC
Landfill waste acceptance criteria	Council decision 2003/33/EC
Value added products from food waste	
Health and safety of animal by-products for humans	Regulation 1774/2002/EC
Promotion concerning the use of biofuels or other renewable fuels for transport	Directive 2003/30/EC
Compound feeding products marketing	Council Directive 79/373/EEC
Veterinary rules for disposal and processing of animal waste	Council Directive 90/667/EEC
Official inspection in the field of animal nutrition	Council Directive 95/53/EC
Measures on animal waste protecting against Transmissible Spongiform Encephalopathies (TSE)	Council Directive 1999/534/EC
Animal protein protection measures	Council Directive 2000/7656/EC
Prohibition of animal by-products in animal feed	Decision 2001/25/EC
Prevention, control, eradication of TSE	Regulation 999/2001/EC
Ban on intra species for fish recycling, burying and burial of animal by-products	Regulation 811/2003/EC
Processing Std for type 3 material & manure in composting plants	Regulation 809/2003/EC
Processing Std for type 3 material & manure used in biogas plants	Regulation 810/2003/EC

(continued)

Table 2.1 (continued)

Category	Legislation
Novel foods	
Authorizing the marketing of coagulated potato proteins and hydrolysates as novel food ingredients	
Agro-food sectors	
Nutritional labelling for food stuffs	Council Directive 90/496/EEC
Health conditions for production and marketing of fishery products	Council Directive 91/493/EC
Sweeteners used in food stuffs	Council Directive 94/435/EC
Colors used in food stuffs	Council Directive 94/435/EC
Food additives used in food industry	Council Directive 95/2/EC
Wine market organization	Council Directive 1493/99/EC
Fruits and vegetable market organization	Council Directive 2200/96/EC
Manufacture of gelatin industry to maintain specific health conditions	Commission Decision 1999/724/EC
Fishery and aqua culture products market organization	Council Directive 104/2000/EC
General principles and requirements of food law	Council Directive 179/2002/EC
Health conditions for production and marketing of meat and poultry meat	Commission Decision 2001/471/EC
Substances that can be added in foods for nutritional purposes	Commission directive 2001/15/EC

2.3 Origination of Legislation and Regulations

The principles for community strategy management was adopted from Directive 91/156/EEC which was originally framed by Waste Framework Directive (WFD)-75/442/EEC in 1975 (European Commission 2004, 2012). Table 2.1 highlights the regulation respective to the food industries and it comprises

(1) Council Directive 96/61/EC—Integrated Pollution Prevention to prevent pollution at a source,
(2) Council Directive 1993/31/EC—Landfills to minimize the landfill of biodegradable municipal waste,
(3) Regulation 1774/2002—Health and safety rules on treating animal by-products,
(4) Council Directive 2000/76/EC—Incineration limitations to prevent negative effects on environment and human health.

2.4 Evolution of ISO 14000

These days consumers are attracted towards the eco-friendly products. An eco-friendly product is one when its life cycle from "cradle to grave" respects the needs of the environment (Russ and Meyer-Pittroff 2010). The socio-economic and environmental pressure creates an urge to establish the need for adopting a management system which is environment friendly and the food industries came up with their own Environmental Management System (EMS).

The organizations contemplating the implementation of ISO 14000 should evaluate the impact that an EMS will have on its internal structures and its ability to provide the comprehensiveness required to meet the external expectations. The implementation of ISO 14000 envisages the establishment of a proactive environmental management. The instigation of set of new worldwide standards flagging the environmental concerns, known as ISO 14000, is a serendipity which resolves the most complex and environmental issues in implementing the standards (Russ and Meyer-Pittroff 2010). The ISO comprised of 130 member countries and its main objective is to harmonize the existing technical standards. In the past years, ISO has implemented 9000 standards which established the global standard of quality management.

2.5 History of Environmental Awareness

Environmental management is the emerging concept towards the sustainable development, which is indicated by the voluntary adoption of codes that promote sustainability (Pap et al. 2004). Figure 2.3 defines the various environmental standards which led to development of ISO 14000.

The two important outcomes of the conference held in 1992 (UN conference on Environment and Development, Earth Summit) were the compilation of Agenda 21 and ISO 14000. The former compilation is a set of guidelines framed to promote the sustainability and was successfully adopted by 172 countries in the conference (European Commission 2004, 2012). The ISO 14000 is a concatenation of standards in which ISO: 14001 highlights Environmental management and pollution prevention. The first two published standards are ISO: 14001 and 14004 are the pillars of ISO: 14000 since they specify the requirements against the food industries environmental management systems to be judged. The documents related to auditing principles and procedures; Life cycle assessment principles are published since 1996. Numerous proposed standards are evolved in the merging of environmental standards such as BS 7750 and later the standards has been revised for future updates. Several organizations worked on the proposed standards in order to make it user friendly and to make it acceptable to all nations who have endorsed ISO: 14000. The ISO: 14001 is a series of environmental standards and it was published in the year 1996.

Fig. 2.3 Milestone to the development of ISO

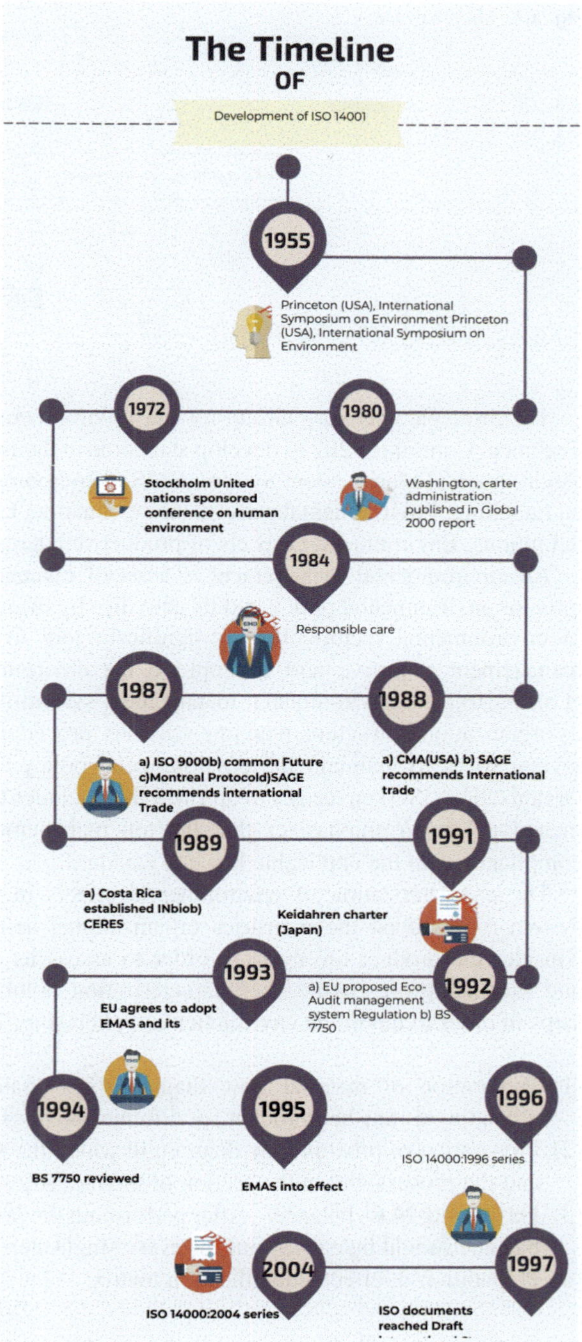

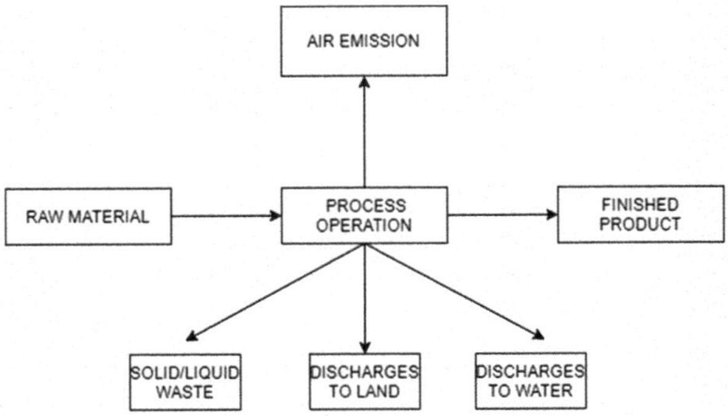

Fig. 2.4 Mass balance

The Strategic Advisory Group for the Environment (SAGE) has proposed new Technical Committee 207 to develop standards in the following areas which include Environmental Management System (EMS), Environmental Auditing, Environmental Labelling, Environmental performance evaluation, Life cycle analysis, Terms and definitions, Environmental aspects of product standards (Kjeldsen et al. 2002).

An environmental management is a set of management activities intended to monitor environmental policy and its activities by measuring the results and control of environmental factors. It is a significant tool to achieve the environmental management objective and to improve the environmental performance (Xiong et al. 2010). The environmental management system is a system which focuses on its organizational structure, planning schemes, procedures, process, implementation, review so as to maintain the environmental policy. The International Standards Organization (ISO) envisages of an EMS that includes the establishment of environmental policy to emphasizes that the top management of the company should compliance with the applicable law and standard.

The characterization of environmental effects in environmental management system is based on the industries organizational activities. It is also advised to consider the product life cycle in order to assess its environmental effects (Russ and Meyer-Pittroff 2010). The characterization involves the following necessary steps in order to devise an environmental policy (Fig. 2.4)

(1) Preparation of material flow diagram (raw materials, energy, water) from unloading to finished product for different sections of the industries.
(2) Preparation of process flow diagram in which the steps involved in the process and the process within the section of the industries should be sketched.
(3) Performing Mass balance—After performing the process flow diagram, the mass balance should be carried out for every single step in process flow diagram.
(4) Preparation of effects identification matrix.

2.6 Environmental Standards and Regulations—ISO 14000

ISO encapsulates a unique approach by demanding each industry to create its own objectives and targets and to commit itself to produce a reliable process for the environmental protection of the industry. It also provides a framework of organizational resources through reliable management processes to have environmental impacts. This entitles the regulatory compliance as a result of the new management strategy in environmental protection standards (Sanders and Crosby 2004; European Commission 2012). The environment management system can be documented in industries operational manual and it should clearly state the industries environmental goals. The effective environmental strategy involves the definition of proactive management process which supports sustainable approach in all process of the industry. The articles encompassed in ISO 14000 series are listed in Table 2.2.

The environmental management system will describe all the environmental objectives and targets in the industry and ISO 14001 entitles about how each objective will be achieved and it includes a specific plan that describes each action to be performed to meet the proposed objective. It is an inclusive of comprehensive and informative index and it is termed as specification document which provides specification for an environmental management system (Figs. 2.5 and 2.6).

Table 2.2 ISO 14001 Articles (European Commission 2004; Arvanitoyannis et al. 2006)

ISO 14000	Environment Management Guidelines
ISO 14001	EMS-Specification with guidance for use
ISO 14004	EMS-Guidelines on principle, system, and supporting techniques
ISO 14010	Environmental Auditing Guidelines—Principles of Auditing
ISO 14011/1	Environmental Auditing Guidelines—Audit Procedures—Audit of EMS
ISO 14012	Environmental Auditing Guidelines—qualification criteria for Auditors
ISO 14013	Environmental audit programs management
ISO 14014	Initial reviews
ISO 14015	Environmental Site Assessments
ISO 14020	General principles on Environmental Labelling
ISO 14021	Self-declaration on Environmental claims—Terms and conditions
ISO 14022	Environmental Labelling—Symbols
ISO 14023	Environmental Labelling—Testing and verification Methodologies
ISO 14024	Environmental Labelling—guide for certification
ISO 14031	Environmental Evaluation Performance
ISO 14040	Environmental Assessment—Life cycle Assessment—Principles and Guidelines
ISO 14041	Environmental Assessment—Life cycle Assessment—Goal and Inventory analysis
ISO 14042	Environmental Assessment—Life cycle impact Assessment
ISO 14043	Environmental Assessment—Life cycle Assessment-Interpretation
ISO 14050	Principles of ISO—Terms and definitions
ISO 14060	Guide 64 for the inclusion of Environmental actions

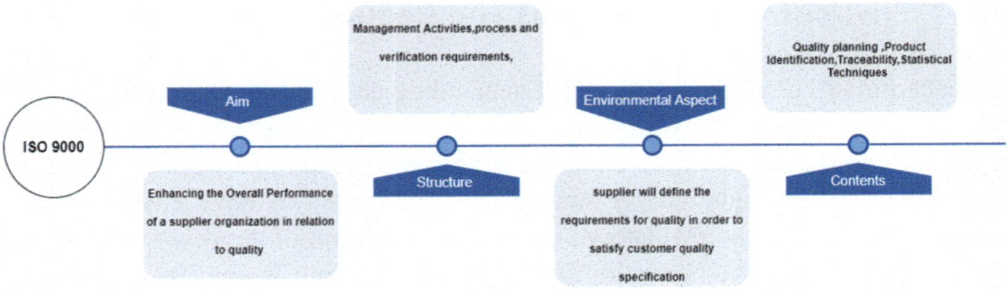

Fig. 2.5 Contents of ISO: 9000

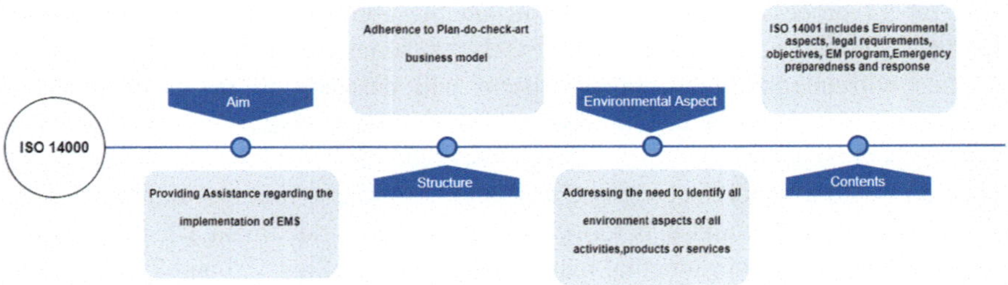

Fig. 2.6 Contents of ISO: 14000

In 1996, the EC directive 96/61/EC of 24 September 1996 concerning the pollution prevention and control was issued (European Commission 2004). The aim is to be utilized for any organization's environmental performance. This chapter will guide the food technologist a way to comply with those regulations. The contents of ISO: 14001 are

(1) Introduction and background,
(2) Policy making and planning,
(3) Implementation and operation,
(4) Checking and corrective action,
(5) Strategic planning.

The ISO directive has 22 articles in the mentioned sections and 4 annexures with sample environment policies.

2.7 Organization: Food Industry

The entire agro-food industry can be treated as an organization and it consists of primary production, post-harvest, production processing, packaging (Pap et al. 2004; Risse 2003). The general flow diagram which represents the organization interaction with the environment is shown in Fig. 2.7.

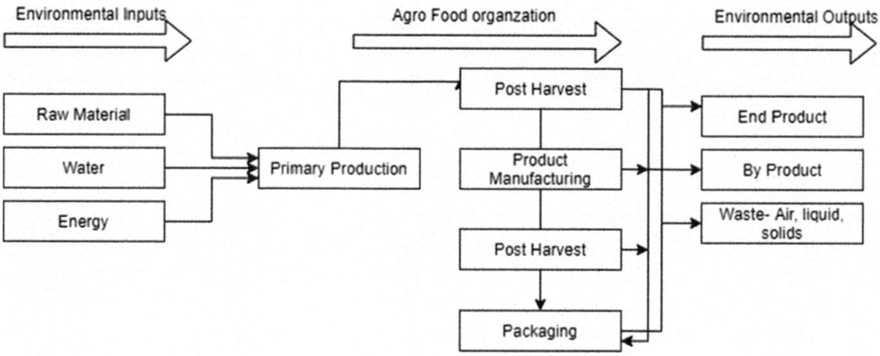

Fig. 2.7 Structure of typical food industry

In inputs section of the industry consists of raw material, water, energy. The basic criteria should be evaluated in order to ensure the effective environment management in the agro industry.

(1) The incoming water and its source should be monitored and periodical test should be taken to ensure the water quality conforming to specifications, because water plays a vital role in manufacturing as well as to maintain hygienic conditions in the industry.
(2) The incoming raw material should be free from pesticides residue, in order to ensure nil contamination in the primary stage of processing.
(3) The energy source can be driven from eco-friendly source, renewable source, conventional source.
(4) The life cycle analysis of the inputs should also be considered in order to maintain a healthy environment management system.
(5) The effective management is the one which utilizes the energy and water from the output and recycled to use as an input for the industry.
(6) The part of the waste can be used as biomass to produce biofuel which in turn can be used for the energy requirements of the organization.
(7) In the output section, the finished end product should be free from any line contamination starting from raw material to packaging process. The by-products obtained during the processing steps should also ensure nil contamination.
(8) The waste effluents should be treated and should be free from any toxic contaminants and it can be discharged to air, surface water, and soil.
(9) The waste generated can be in gaseous form, liquid phase, and solid.

In predicting the Environmental Performance (EP), the physical properties of food and its constituents play a significant role (Pap et al. 2004). The thermodynamic properties like partition coefficients, chemical potential are also considered in prediction process.

The knowledge on kinetic data K, Km for enzyme catalyzed reactions, Biological Oxygen demand (BOD), chemical oxygen Demand (COD), Suspended solids (SS) values to determine the presence of toxic and hazardous substance in the waste (Sanders and Crosby 2004) which will help in prediction of a given component between two or more phases in the process. For example, presence of universal solvent hexane in kernel oil.

The technical software Super Pro by the Intelligence Company includes the elaborative database of the above properties which is the outcome of DOPPOF (Database of Physical Properties of Foods).

The food industries are facing a difficult situation in handling wastes and by-product management which are the core phenomenon for environmental protection and to promote the sustainability (Xu et al. 2011). The two traditional methods of waste utilization includes, the following:

- Wastes can be used as animal feed (e.g., spent grains, distiller's wash).
- Wastes can be used as fertilizer (oil filtration sludge, carbonation sludge).

These methods of the existing agricultural solutions of waste utilization form the regulatory balancing act between the legal conditions and the ecological and economical solutions (AWARENET 2004). Waste disposal in food industries is challenging because

(a) Biological stability and pathogenic growth potential—Most of the food industries wastes contain microbes, if regulations concerning infectious diseases are not treated properly, it will lead to an environmental hazard and food waste hazard. The protein breakdown in waste is always characterized by the generation of strong odor which in turn causes air pollution.
(b) Water content availability—Water content, especially in meat and vegetable, lies between 70 and 90% and it promotes microbial growth.
(c) Oxidation—Waste with high fat content is susceptible to lipid oxidation, which in turn releases foul smelling fatty acids.
(d) Enzymatic changes—Waste with active enzymes will initiate the process of spoilage and intensify the odor associated with the reactions.

The paramount objective in food waste management is to avert the waste production in the industry. The waste produced during the process must be recycled in order to minimize the waste. Mostly the waste from food industries contain organic matter with less nutritional value and inedible components which cannot be used for human consumption. The production efficiency should be maintained in order to prevent the waste accumulation in the industry. However, the mass waste reduction does not represent true waste prevention but it increases the chances in product quality (Pal et al. 2014). The typical food waste which is of raw material origin should be recycled. There is certain type of production aids like immobilized yeast, adsorbents which are added in the processing technique and are subsequently removed with lesser functionality. The wastes from food industries are of biological matter, so the most effective method of disposal is the utilization as animal and plant feed (Litchfield 1987; Pal et al. 2014). The studies focused in increasing the efficiency of the utilization and fewer problems still persist in waste disposal and it includes certain factors like agriculture industry crisis, concentration of the food industries, food industry restructuring.

In the early 2000, there is a constant strive to minimize the prices for agricultural products and so the industries began to limits the finding sources for animal feed. Later in 2001, the consequences are exemplified in the outbreak of Bovine Spongiform Encephalopathy (BSE) and scandals of dioxins in animal feed that contained used oils in Belgium. These incidents created a strict regulation especially made it difficult to use waste from food industry with fat and protein (Litchfield 1987; Ockerman and Hansen 2000)

In recent years the usage of processed and semi-processed products increased for about two-thirds of the global food trade which in turn increases the amount of waste produced. Noticeably, the convenience foods increased from 18% in 19080 to 34% in 1998. Convenience foods are foods which are ready to cook and these wastes ae widely distributed throughout the food chain. Fewer collection sites and increasing waste production are the two prime factors which creates more pressure for food industries in its waste disposal (Christensen et al. 1996; Poli et al. 2011). The increase in incidents of irregular disposal of waste from food industries in 1990 has apparently urged to establish the governing body European Food Safety Authority (EFSA) in 2002 to ensure a higher level of consumer health protection in food safety. Sustainability in the cultivation of raw materials is another demand on food industries which required consideration, planning. Food waste is composed of highly complex molecules, in which a portion of the material extracted and the left over can still be used as raw material that can be utilized for another product.

2.8 The Product Specific Food Waste Assessment in Food Industries

Based on the evaluation from the divergent bough of the food industry, the actual amount of waste generated in each type of operation can be found. To further differentiate the types of operations within the individual branches of the industry, production process was studied and documented in (Russ and Meyer-Pittroff 2010). The specific waste index is defined as the mass of the accumulated waste to the mass of the finished product.

$$\text{Specific Waste Index} = \frac{\text{mass of the accumulated waste}}{\text{mass of the finished product}}$$

Table 2.3 represents an overview of different food industries, waste production, and specific waste index.

(1) Grain products—The most predominant grains in human consumption is wheat, rice, corn, millet, barley, rye, and oats. They are cultivated approximately 80% of the world's land supply around 50% of the world's population. Wheat and rye are ground to flour and grits to flour mills, oats are made into rolled oats in oat

Table 2.3 Food industry waste (Source: Russ and Meyer-Pittroff 2010)

Industry type	Waste generated	Specific waste index
Grain products	Bran	0.11–0.18
	Middlings	0.06–0.11
	Broken grains, seeds, shell, husks	<0.01
	Fine dust, chaff, stalks	<0.01
	Ergot	<0.01
	Oats refuse containing bran and husks	0.39
	Brown rice waste	0.11
	Rice bran	0.11–0.18
	Rice flour	<0.01
	Seedlings from malt	0.038
	Malt dust	<0.01
	Grain separator waste	0.01–0.04
Extrusion product-noodles	Dough waste	0.0012–0.0014
	Eggshells	0.02–0.08
Potato processing unit	Potato peels	0.3–0.5
Coffee roasters	Silver skins	0.02–0.04
Sugar processing unit	Molasses	0.191
	Beet pulp	0.517
	Beet leaves and stems	0.136
	Carbonation sludge	0.427
Dairy unit	Whey	4.0–11.3
	Cheese residue	0.01–0.04
	Waste from milk production	0.04
Meat industry	Slaughterhouse waste	0.1–0.87
Egg industry	Eggshells	0.03–0.12
Beer industry	Malt dust	<0.001
	Spent grains	0.192
	Break material	0.024
	Yeast	0.024
	Kieselguhr sludge	0.006
Wine cellars	Pomace	0.136–0.145
	Clarification sediment	0.015–0.050
	Yeast sediment	0.03–0.045

mill, and rice is hulled in rice mills. Corn is primarily processed to corn starch and corn oil, whereas barley is processed into malt (Schieber et al. 2001).

(2) Extrusion products—The main waste from this industry is unused dough. The most common extrusion products are pasta and noodles which usually consists of wheat flour, water, eggs, and spices. Eggs and spices are optional ingredient, if eggs are added to the dough, then eggshells raise the specific waste index.

(3) Coffee processing unit—The external skin and the silver skin are the waste generated in coffee roasting process (Murthy and Naidu 2010).

(4) Sugar processing unit—The sugar is generally produced by extracting sugar beets. There are several kinds of specific product waste like molasses, beet pulp can be utilized (Gu et al. 2011; Poli et al. 2011).
(5) Dairy processing unit—The primary waste from all the milk product processing industry is sludge from the centrifugation process. The waste generated per ton of centrifugation of milk is 275 g of sediments. This sludge must be pasteurized, since it is rich in microbial content. The rinsed milk from the tanks, lines, etc. and processing error milk are treated as waste and this type of waste is created during the production of fluid milk products like yoghurt, buttermilk at the rate 0.04 per mass of the finished product. During cheese production, whey is generated as a primary waste and it is result of protein coagulation in milk, followed by cheese residue. The generated amount of whey and cheese residue is greatly depending on the type of cheese produced and the nature of the raw milk source (Litchfield 1987; Pap et al. 2004).
(6) Meat processing unit—The prime source of waste from meat processing unit is slaughterhouse. The slaughterhouse waste includes bones, tendons, skin, and the contents of gastrointestinal tract, blood, and internal organs. The waste generated also varies for each type of animal slaughtered and processed (Ockerman and Hansen 2000; Westlake 1995).
(7) Egg processing unit—The waste generated in the processing units is eggshells, regardless of which egg products like frozen, spray dried, egg oil, egg derived lecithin are produced. The eggshell waste is highly fluctuating because of its size variation and thickness.
(8) Beer production—The waste generated in beer production process is mainly produced by filtration and separation which produces separation residue, mostly kieselguhr and organic material. The wort is the liquid extract separated from the spent grains which also stay behind as waste. The hot trub and cool trub also known as protein-tannin complex separated from the wort are also treated as waste. The yeast left behind the fermentation process is also treated as waste.
(9) Wine making process—The main category waste generated from wine processing unit is pomace, clarification sediment, and yeast sediment (Arvanitoyannis et al. 2006; Poli et al. 2011). The amount depends upon what processing methods are used, as well as the condition of the grapes at the time of harvest. Yeast sediment consists of yeast cells and tartar.

2.9 Contents of Food Waste

Before understanding how the waste can be treated it is important to understand the composition of food industry waste (Zeng et al. 2010). In general food waste from the industry is categorized as

(1) Waste rich in protein—It includes blood, seedlings from malt, protein-tannin complex—hot &cool trub, yeast.

(2) Waste with high cellulose content—It includes oat husks, pulp from sugar beets, spent grains, broken grains, seeds, peels, husks.
(3) Waste with high amount of carbohydrate—Dough waste in noodles, bran, middlings, waste from oats, oat bran, brown rice waste, rice bran, rice flour, silver skins, molasses, whey, potato peels.
(4) Waste rich in minerals—Eggshells, carbonation sludge, kieselguhr sludge.
(5) Waste with high fat content—Slaughterhouse waste.

2.10 Treatment for Food Industry Waste

The food industry waste can be used for animal feed, and apart from these the three most common ways used to treat food industry waste includes incineration, anaerobic fermentation, composting.

Incineration is the best choice for waste disposal only if the water content of the waste is less than 50% by mass (Gu et al. 2011; Schieber et al. 2001). Irrespective of its advantage, the air emission is a challenging factor which hinders its usage. For example, sulfur from waste can be oxidized to SO_2 and nitrogen is also susceptible to this reaction.

Anaerobic fermentation is a promising technique where the food waste with higher water content above 50% by mass can be disposed by anerobic fermentation process into methane and biogas (CO_2). Food industry waste with higher amount of cellulose and hemicellulose can predominantly affect the anerobic fermentation process because it takes longer time to breakdown into a biogas (CO_2) (Litchfield 1987; Xu et al. 2011).

Composting is a process where the food industry waste can be composed for a certain period of time into a soil conditioner which is highly rich in nutrients (Litchfield 1987; Westlake 1995). It speeds up the process of decomposition of organic matter by promoting the environment, ideal for microorganisms.

In traditional method of using food industry waste for agricultural means depends upon the contents of waste. Omnivores animal can digest protein and fat effectively than others (Westlake 1995). Hence these category waste can be used as feed for these animals but the waste should be sterilized in order to prevent microbial contamination (Litchfield 1987; Ockerman and Hansen 2000; Xu et al. 2011). These wastes are susceptible to spoilage very quickly. Ruminants can digest wastes that are rich in cellulose and hemicellulose because of its enzymatic activity of its digestive system which can easily breakdown cellulose and hemicellulose. Waste with high amount of mineral content like pomace and hop waste which are rich in phenolic components cannot be used as animal feed rather it can be used as fertilizer.

2.11 Sustainable Ways to Utilize Food Industry Waste

The best sustainable approach is to prevent the waste and to maximize the utilization of food waste from the industries. This method of approach involves the exploitation of fully traceable, food grade systems (Xiong et al. 2010). In the first step, the waste from the food industries should be stabilized against microbial deteriorations and the autolysis reaction to prevent the unpleasant odor and to prevent the loss in food grade status (Russ and Meyer-Pittroff 2010). Modernized enzyme technology is applied in order to disorient the biochemical properties of the food waste and it provides a range of components from high to low value, of all which would contribute to whole waste exploitation from food industries. In accordance with this to promote the safety of the performing operations Hazard Analysis Critical Control Point, Novel food legislation, should be developed. Though in certain cases, some co-products from the industry may be unsuitable for waste exploitation due to their complexity, uncontrolled spoilage, and lack of traceability (Waldron 2004). In this scenario, traditional methods for food waste like biogas production, composting can be used. The new front edge technologies converted the biomass into biofuels, thereby considerably reducing the landfill, and introducing the most cost-effective technique to the era of food waste utilization (Fig. 2.8).

The soluble and insoluble fibers from apple, tomato, carrot waste and from citrus peels can be extracted and used in food industries as a binding agent because of its absorptive nature and its ability to form gels.

Fibrous material from the spent grains can be used in building of fiber where it can be used as a filler and structural material.

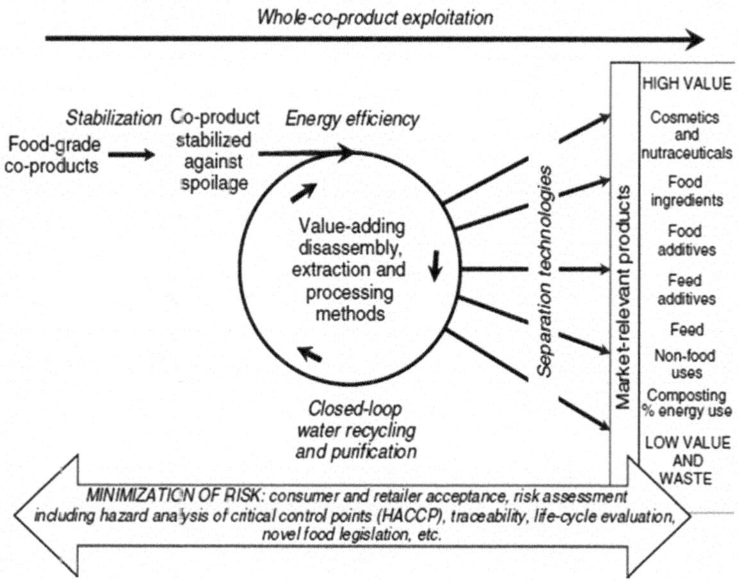

Fig. 2.8 Pathway of Co-Product Exploitation

Brewery waste spent grains can find application in kilning process of brick making where it helps to thermally insulate the bricks to improve its strength.

Slaughterhouse waste, fat which is partially removed, can be used in chemical and cosmetic industries.

Pomace after fermentation process can be used as a fertilizer and the grapeseed oil can be extracted which is having high nutritional value and finds numerous applications. Grappa can be produced from pomace.

Whey from milk industry can be utilized in beverage industry. Methods in which whey is fermented to produce alcohol and lactose can also be extracted from whey and it can be used in confectionary industries.

2.12 Future Trends

The world is facing loads of changes every day irrespective of its environmental changes. Certainly, in this rising population which will directly demand the fuel which in turn result in high cost and in other hand threat and pressure to the environment to reduce carbon emission are bound to have implicit effect on food chain (Kjeldsen et al. 2002). Over the next ten years, the costs of food processing and transport will increase, as well as demand to use land for non-food crop production. This in conjunction with waste hierarchy will substantially add pressure to ensure that majority food processing and food by-products are fully utilized. To tackle this complex sphere from economic and environmental perspectives, the demand for user friendly life cycle assessment is likely to increase.

References

Arvanitoyannis IS, Ladas D, Mavromatis A (2006) Wine waste treatment methodology. Int J Food Sci Technol 41:1117–1151

AWARENET (2004) Handbook for the prevention and minimization of waste and valorization of by-products in the European agro-food industries. Summary of the final report of the AWARENET Thematic network for 2002–2004. URL: http://www.zap.pw.plock.pl/english/EXESUM.htm

Christensen TH, Kjeldsen PA, Lindhardt B (1996) Gas generating processes in landfills. In: Christensen TH, Cossu R, Stegman R (eds) Landfilling of waste: biogas. E and FN Spon, London

European Commission (2004) Integrated pollution prevention and control. Draft reference document on the best available techniques for waste incineration. European Commission, Brussels

European Commission (2012) Waste prevention – handbook: guidelines on waste prevention programmes. URL: http://ec.europa.eu/environment/waste/prevention/pdf/Waste%20prevention%20guidelines.pdf

Gu G, Li Y, Ren W (2011) Reclamation technologies of food residue. Environ Sanitation Eng 19:1–6

Kjeldsen P, Barlaz MA, Rooker AP, Baun A, Ledin A, Christensen TH (2002) Present and long-term composition of MSW- landfill leachate: a review. Crit Rev Environ Sci Technol 32:297–336

Litchfield JH (1987) Microbiological and enzymatic treatments for agricultural and food processing wastes. Food Biotechnol 1(1):29

Murthy PS, Naidu MM (2010) Recovery of phenolic antioxidants and functional compounds from Coffee industry by-products. Food Bioprocess Technol. https://doi.org/10.1007/s11947-010-0363-z

Ockerman HW, Hansen CL (2000) Animal by-product processing and utilization. CRC Press, pp 544

Pal M, Kumar M, Gupta GR, Dwivedi KN (2014) Utilization of unused bio- waste for agricultural production - a review. Plant Arch 14:597–604

Pap N, Pongrácz E, Myllykoski L, Keiski R (2004) Waste minimization and utilization in the food industry: processing of arctic berries, and extraction of valuable compounds from juice-processing by-products. In: Pongrácz E (ed) Proceedings of the Waste Minimization and Resources Use Optimization Conference. June 10th 2004, University of Oulu, Finland. Oulu University Press, Oulu, pp 159–168

Poli A, Anzelmo G, Fiorentino G, Nicolaus B, Tommonaro G, Di Donato P (2011) Polysaccharides from wastes of vegetable industrial processing: new opportunities for their eco-friendly re-use. Biotechnol Biopolym:33–51

Risse M (2003) "Food waste composting", Public Service Associate, Biological and agricultural Engineering. The University of Georgia College of agricultural and Environmental Sciences

Russ W, Meyer-Pittroff R (2010) Utilizing waste products from the food production and processing industries. Crit Rev Food Sci Nutr 44:57–62

Sanders GB, Crosby KS (2004) Waste legislation and its impact on the food industry. In: Waldron KW, Faulds CB, Smith AC (eds) Total Food 2004 – Exploiting Co-Products, Minimising Waste. Proceedings volume. Institute of Food Research, pp 16–28. http://www.totalfood2004.com

Schieber A, Stintzing FC, Carla R (2001) By-products of plant food processing as a source of functional compounds-recent developments. Trends Food Sci Technol 12:401–413

Waldron KW (2004) Plant residues. In: Waldron KW, Faulds CB, Smith AC (eds) Total Food 2004 – Exploiting Co-Products, Minimising Waste. Proceedings volume. Institute of Food Research, pp 117–131. http://www.totalfood2004.com

Westlake K (1995) Landfill pollution and control. Albion Publishing, Chichester

Xiong T, Huo W, Dou L (2010) Research on necessity of restaurant garbage resourceful treatment in domestic cities. Environ Sci Manage 35:148–152

Xu J-L, Zhang G-X, Xu M-Y (2011) Research progress on reuse of food waste with microbial technology. Microbiology China 38:928–933

Zeng C, Li X, Chen P (2010) Discussion and research on management and disposal of food waste. Environ Sci Manage 35:31–35

Characterization and Treatment of Waste from Food Processing Industries

3

Manish Kaushik and Dipti Sharma

Abstract

Food waste is complex which consists of both liquid and solid waste. Solid waste from food processing units includes both organic waste and packaging waste. The food processing sector includes milk and milk products, fruits, vegetables and juices, grain and flour processing, poultry and meat, alcoholic beverages, fisheries, and many consumer-predicated products like chocolates, coffee, tea, confectionaries, soya-based products, mineral water, soft-drink, etc. Food waste from each type of food industry has specific characteristic and hence need treatments, e.g., aerobic and anaerobic lagoons, vermicomposting accordingly. Along with this the use of food waste could significantly reduce food waste levels and create new opportunities and benefits for all those involved in the food production system. Reducing food waste through the recovery of its valuable components is therefore an important way to increase sustainability. The need for a sustainable approach can no longer be overlooked by food production companies and other stackholders of food production system. Sustainable food waste management system is the only way to reduce the developed waste.

Keywords

Food waste · Food industry · Characterization · Treatment

M. Kaushik
Department of Chemistry, Noida Institute of Engineering and Technology, Greater Noida, Uttar Pradesh, India

D. Sharma (✉)
Department of Food Technology, Shyama Prasad Mukherji College, University of Delhi, Delhi, India

M. Thakur et al. (eds.), *Sustainable Food Waste Management*,
https://doi.org/10.1007/978-981-15-8967-6_3

3.1 Introduction

The term food industry is very familiar to us. Food industries cover several activities like processing, conversion, preparation, preservation, and packaging of foodstuff. The raw materials are extracted from agriculture, farming, and fishing. The food processing sector includes milk and milk products, fruits, vegetables and juices, grain and flour processing, poultry and meat, alcoholic beverages, fisheries, and many consumer-predicated products like chocolates, coffee, tea, confectionaries, soya-based products, mineral water, soft-drink, etc. (Garcia et al. 2019). Food processing units may cause soil, dihydrogen monoxide, air, and noise pollution. These wastes have an abundance of organic matter and thus often yarely degraded biologically (Nevárez-Moorillón et al. 2020). These wastes withal contain nutrients like nitrogen, phosphorus, potassium which are good for biological growth. These wastes cause sizably voluminous-scale contamination of land, water, and air. All of these are subject to an incrementation in environmental concern. Dihydrogen monoxide pollution is a major issue because aliment processing units consume an abundance of dihydrogen monoxide in processing operations like washing, evaporation, extraction, filtration (Table 3.1). The processed waters have high concentrations of suspended solids and soluble organic compounds such as carbohydrates, proteins, and lipids, which present arduous disposal quandaries. The primary issues of concern in wastewater are biochemical oxygen demand (BOD), total suspended solids (TSS), exorbitant nutrient, pathogen, and residual

Table 3.1 Typical values for the volume of water required to produce common foodstuff ("Use of water in food and agriculture - Lenntech," n.d.)

Foodstuff	Quantity	Water consumption, liters
Chocolate	1 kg	17,196
Beef	1 kg	15,415
Sheep meat	1 kg	10,412
Pork	1 kg	5988
Butter	1 kg	5553
Chicken meat	1 kg	4325
Fresh poultry	1 kg	6000
Cheese	1 kg	3178
Olives	1 kg	3025
Rice	1 kg	2497
Bread	1 kg	1608
Pizza	1 unit	1239
Milk	1 × 250 ml glass	255
Egg	1	196
Wine	1 × 250 ml glass	109
Beer	1 × 250 ml glass	74
Tea	1 × 250 ml cup	27

Table 3.2 Amount and strength of wastewater from industries

Industry	BOD (mg/L)	COD (mg/L)	TSS (mg/L)	pH	Protein (mg/L)	Fat (mg/L)
Brewery	850	17,000	90	4–6	–	–
Dairy	1000–4000	–	1000–2000	5	6–82	30–100
Fish	500–2500	–	100–1800	–	300–1800	100–800
Meat	1000–6500	–	100–1500	–	350–950	15–600
Poultry	500–800	600–1050	450–800	6.5–9.0	300–650	100–400

BOD, biological oxygen demand; COD, chemical oxygen demand; TSS, total suspended solids

chlorine and pesticide levels (Aniyikaiye et al. 2019). Table 3.2 summarizes the sundry pollutants of wastewater.

Solid waste from food processing units includes both organic waste and packaging waste. Organic waste includes fruit and vegetable peel, seeds, skin, and bones. Solid waste needs congruous management of organic waste by biodegradation because it may cause earnest health issues (Woodard and Curran 2006). The victuals packaging is to forfend aliment after processing till utilization by consumers. Food packaging wastes, such as plastic, glass, and metal, are needed to switch to more recycling (Beiras 2018). After the utilization, packaging needs to be disposed in a responsible manner (Dreher et al. 2004). Food and packaging/containers account for virtually 45% of the materials land-filled in the USA (US EPA 2017).

In general, food processing units do not contribute major sources of emissions into the air (Sun et al. 2017). The main sources are particulate matter and combustion gases like carbon dioxide, sulfur dioxide, and nitrogen oxides from combustion processes; volatile organic compound (VOC) and chemical emissions like ammonia, hydrogen chloride, sulfides from cooking of food, leakage from refrigeration system, or from fermentation processes.

Noise is undesired sound, and sound is a sensation engendered in the auditory perceiver due to variations of air pressure. There are many sources of generation of the air pressure changes; the following is a list of the main ones: air turbulence by moving surface like a fan, air turbulence by commixing of gases such as flue and exhaust, the material being dropped onto a plate, impacts between components in a machine, out of balance forces from a motor, electromagnetic excitation, turbulent fluid flow within a pipe (Walsh and Key 1994). The sound is a result of fluctuation of air pressure, so measures can be done in designing a caliber of the machine to minimize noise at source, noise enclosures, minimizing reverberant sound by embarking, etc. Ear defenders reduce noise at source. Sound enclosures lower re-verberation exposure of machine and lower noise to neighbors (Lindawati et al. 2018).

3.2 Definitions of "Food Waste" and "Food Loss"

During the food processing, food is disoriented or wasted. The definitions of "food waste" and "food loss" are matter of dispute among scientists. According to Food and Agriculture organization of United Nations the "food waste" is defined as "decrease in the quantity and quality of aliment resulting from decisions and actions by retailers, pabulum accommodation providers and consumers"("Food Loss and Food Waste," n.d.). According to United Nations Environment program the "food waste" is "food that completes the aliment supply chain up to a final product, of good quality and fit for consumption, but still doesn't get consumed because it is discarded"("Definition of food loss and waste," n.d.). According to European Coalescence Commission Council Directives "food waste" is "the victuals, which the holder discards or intends to discard" ("REGULATION (EU) 2016/ 679 OF THE EUROPEAN PARLIAMENT AND OF THE COUNCIL - of 27 April 2016 - on the protection of natural persons with regard to the processing of personal data and on the free movement of such data, and repealing Directive 95/ 46/ EC (General Data Protection Regulation)," n.d.). Food waste is the food which is fit for consumption but it is intentionally discarded by the owner whether or not after it is left to spoil or expire. Food waste typically takes place at retail and consumer stages in victuals supply chain.

"Food loss" refers as the qualitative and quantitative abbreviation in the amount and value of food (Nash et al. 2017). The qualitative loss corresponds to the loss of caloric and nutritive value, loss of quality, and loss of edibility. Quantitative loss refers to the decrementation in edible aliment mass throughout the component of the supply chain that concretely leads to edible aliment for human consumption. According to Food and Agriculture organization of United Nations the "food loss" is defined as "decrease in the quantity or quality of aliment resulting from decisions and actions by pabulum suppliers in the chain, omitting retailers, pabulum accommodation providers and consumers." According to United Nations Environment program the "food loss" is "the aliment which gets spilled, spoilt or otherwise lost, or incurs abbreviation of quality and value during its process in the victuals supply chain before it reaches its final product stage." Food loss typically takes place at engenderment, post-harvest, processing, and distribution stages in the victuals supply chain (Salihoglu et al. 2018).

3.3 Characteristics of Food Processing Waste

3.3.1 Fruit and Vegetable Processing Waste

Waste produced in India, the Philippines, China, and the USA is about 55 million tons from organized sector of fruit and vegetable production, packaging, distribution, and consumption business (Wadhwa and Bakshi 2013). Fruit and vegetable waste processing is essentially identical to food. Some methods create large quantities of waste, such as vegetable washing water, which only contain soil and

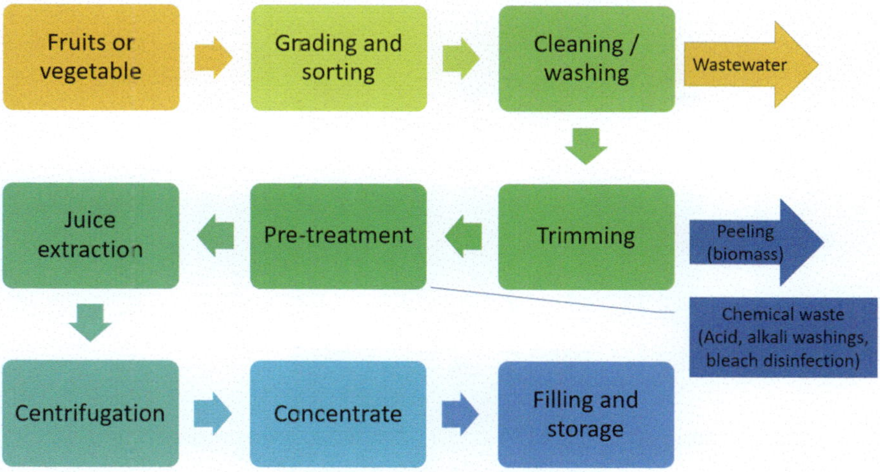

Fig. 3.1 Fruit and vegetable processing unit and waste discharge

organic matter. Cannery wastewater is practically similar to household kitchen waste (Valta et al. 2017). This waste comes from fruit and vegetables being cooked, culled, juiced, and whitened. It is rich in suspended solids and organic matter is dissolved, such as starch and fruit sugar. Some waste is dumped into areas or waterways and therefore creates environmental risks. Alternatively, the use of waste as compost, feed, renewable energy sources, or value-added goods may be a solution (Fig. 3.1).

3.3.2 Dairy Processing Waste

The dairy sector is one of the major wastewater producing industries. Wastewater is produced for dairy and butter processing, cheese processing, skim milk pulverization, ghee production, etc. The main by-products of these processes are skim milk, butter milk, and whey. Edible casein is made from skim milk. This is commonly used for milk and foodstuff. Discharging of whey as waste causes serious contamination issues owing to its high BOD (35–40 g/L). This strong BOD is primarily attributed to lactose present at amounts between 4.5 and 5% (Mansoorian et al. 2016). Within a milk production facility, the most noticeable cause of waste is the whey that comes from different manufacturing processes for cheese. Owing to its high capacity and quantity of contamination, whey is a special feature in milk wastewater production. With every 1000 kg of cheese made around 150–200 kg of whey is made (Mistry 2005).

In various phases of dairy operation, clean water is used, for example, for milk processing, cleaning, packing, and cleaning of the milk duct, can, and tanker. Dairy wastewater contains high amounts of organic liquid ingredients such as whey products, lactose, fat, and minerals. The milk waste is largely neutral or slightly alkaline and, because of milk sugar fermentation, is likely to become very acidic

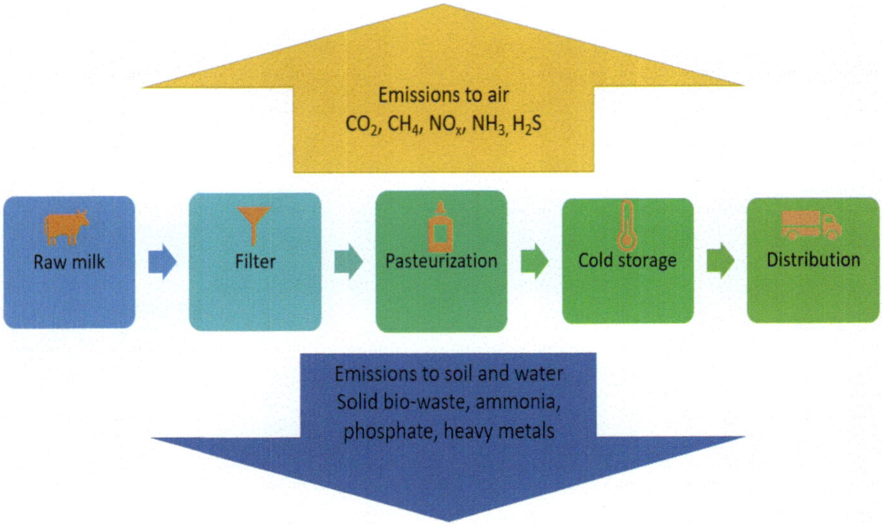

Fig. 3.2 Milk processing unit and waste discharge

very quickly (Hansen and Cheong 2019). Casein precipitation can occur with a lower pH (Fig. 3.2).

3.3.3 Meat, Fish, and Poultry Waste

Much of the pollution emerges from slaughterhouses in the meat industry. The important by-products of the animals are skin and fur. It is used in leather, athletics, cosmetics, edible gelatin, and glue sector. Bones are also used in the manufacture of gelatin. Brain, nervous system, and spinal cord are the sources of cholesterol as raw material for vitamin D3. Liver extract from pigs and cattle is a form of vitamin B12. Insulin is derived from the pancreas and used in the diabetes mellitus treatment (Jayathilakan et al. 2012). Mist emmanates from paunch contents and fluid are source of disgusting smell from slaughterhouses. Odors on routine recycling and composting of these products with bacteria and fungi is efficient way of handling these waste. The biodegradability of the manufacture of raw materials influences odor output directly (Okoro et al. 2017). The raw materials are stored at ambient temperature for longer time to boost the development of harmful odors and reduce the value of their resulting protein food and the fat component before being separated and stabilized through the rendering process. Fast processing eliminates unhealthy odors and optimizes consistency of the commodity.

Fish waste is a source of protein, mineral, fat. In addition to chemicals such as chitosan, fish protein hydrolysate and fish oil are important items obtained from fish waste (da Rocha et al. 2018). As by-products of the poultry sector, feathers are produced in vast amounts. Unless they are not treated, they can become a source of

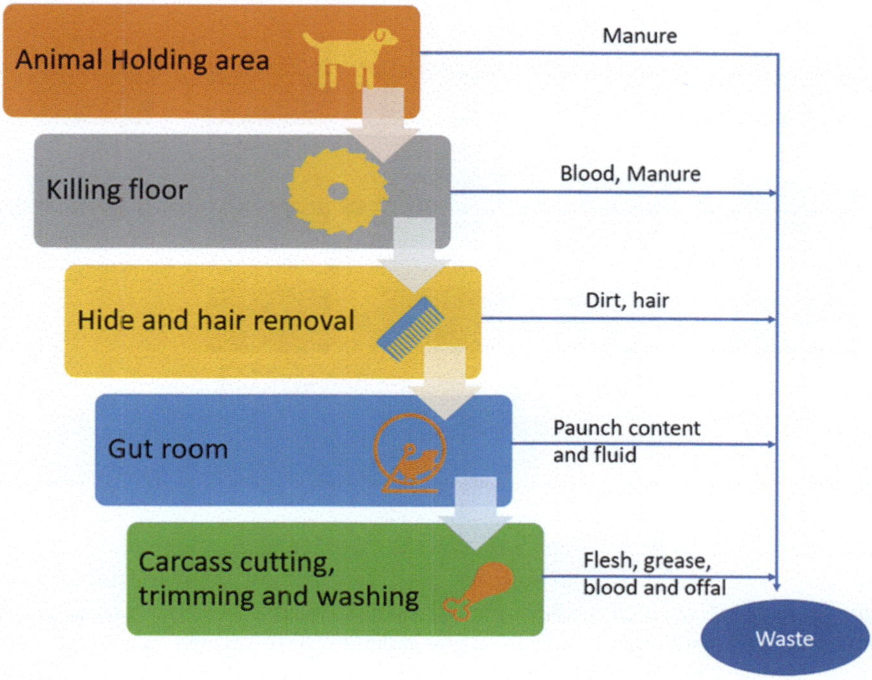

Fig. 3.3 Meat processing unit and waste discharge

emission threat. They consider a range of applications in feed, oxidation, heat separation, cooling systems, composites for biodegradation and fabrics. Poultry feathers are also used in biofuels development after hydrolyzation (Seidavi et al. 2019).

Meat, poultry, and fish waste are somewhat close. Such contaminants include a large volume of organic matter dissolved and stored in specific proteins and oils, a significant degree of organic nitrogen and fat and a considerable number of pathogens. Slaughterhouse and meat packing wastewaters smells strong and unpleasant, comprising feces and urine. Blood washings from carcasses, floors, utensils, and undigested food from paunches are source of waste water (Carpentier 2009) (Fig. 3.3).

3.3.4 Brewery and Distillery Waste

Brewery wastewater is very special. The brewing process generates alcohol, sugars, and proteins, all of which end up in their wastes. During the brewing cycle, malts, rice and maize starch are fed into the breeding tank and malted, as supplemental raw material. The malt liquid, then dosed with hops, is filtered out. After heat-processing in a boiling caldron, the filtrate is cooled down to 7–10 °C and dosed with yeast and

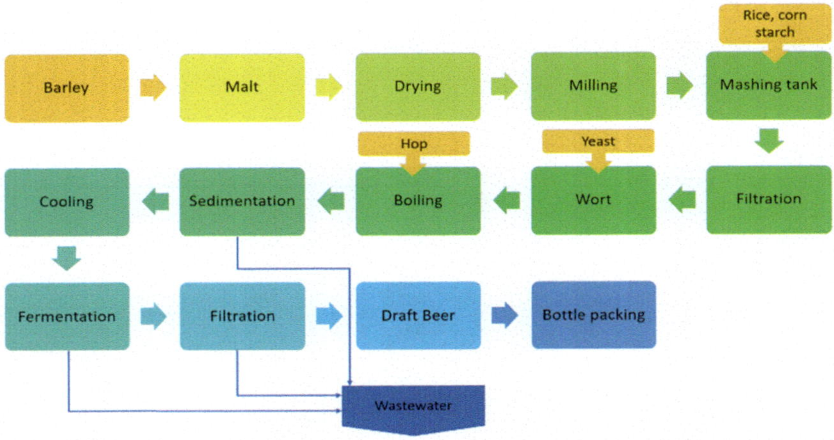

Fig. 3.4 Beer brewery processes and wastewater discharge

fermented for 7–10 days before it becomes the final product (Gunes et al. 2019). Brewery wastewater, rich in nutrients, will significantly conflict with natural environments if released without the appropriate treatment. The wastewater of breweries and distilleries is rich in solids including nitrogen and fermented stuff and their materials. The fermentation waste, particularly concentrated spent yeast in dissolved or colloidal form of the suspended solid content, is responsible entirely for BOD (2000–15,000 mg/L) total nitrogen (800–900 mg/L) and phosphate (20–140 mg/L) (Gunes et al. 2020) (Fig. 3.4).

3.4 Processes for Waste Treatment

Concept of 4-R involving reduce, reuse, recycle, and recovery should be goal to ascend utilization of solid and liquid waste to minimize environmental issues. Few prevalent food-waste treatment methods are given below.

3.4.1 Landfilling

Landfilling is most used method of waste disposal. Landfills obtain around 80% of the solid waste that has been discarded. A modern sanitary site is a site planned for collection, construction, and activity in a way that minimizes environmental impacts. Land filling refers to dumping of materials on land, which results in an ascension of ground level. Land filling is the least sumptuous method for untreated or partially treated waste. This method is utilized when inexpensive land is available nearby and local ascendant entities sanction it. Efficacious pretreatment of the waste is subsidiary. Waste high in organic matter content should be screened. The soil can do an excellent job of abstracting nutrients and pathogens from waste (Lu et al. 2012).

Oxidation takes place during the process and decomposition of the waste eventually leads to the production of methane (greenhouse gas) and groundwater pollution, due to the presence of organic compounds and heavy metals (Arvanitoyannis et al. 2008).

Development of new landfilling site is difficult not because of land lack, but rather because of public opposition in a suitable position where soil is appropriate and hydrogeological conditions are suitable. Citizens object to landfills due to nuisance issues (e.g., dust, noise, transportation, odor), esthetic issues, and ecological problems (e.g., groundwater contamination from landfill leachate, landfill gas movement to adjacent property, agricultural uses) (Aziz et al. 2007). A sanitary waste disposal is performed in order to reduce nuisances from mice, insects, moose, and birds induced by dirt, dust, fumes, and flames (Kjeldsen et al. 2002). The following is the typical mode of operation:

- Dissemination of waste to the confined area in thin layers.
- Minimum feasible amount of compacted waste.
- Cover the waste with soil every day.

With a heavy compactor vehicle, waste compaction is accomplished through repeated trips over a waste layer. The soil cover is required less frequently than at the end of each day of operation.

Bioreactors in waste management have many benefits over traditional waste disposal methods. Up to now, four types of bioreactor deposits have been developed: aerobic, anaerobic, hybrid, and facultative. Deposits of anaerobic bioreactors showed lower degradation than other, and another major finding was the high concentration of ammonium nitrogen. Typically, aerobic landfill shows more nitrogen extraction from ammonium. Methane yield rate from food waste processed by aerobic pretreatment was investigated to be in significant quantities (Wu et al. 2017). For the first time, a rapid beginning to digest anaerobic semi-pilot-scale food waste hydrolysate for the manufacture of biogas was undertaken. In the later stage of anaerobic digestion, more than 85% of the chemical oxygen demand was deteriorated to over 90%. Anaerobic digestion was initiated shortly, and anaerobic digestion of hydrolysate food waste has been finding viability and attractiveness for industrial food waste treatment and biogas processing (Huang et al. 2017).

Aerobic and anaerobic methods are combined in a series in the hybrid bioreactor system. The mechanism for nitrification and denitrification may involve full nitrate removal. Externally nitrified leachate is recirculated in the facultative bioreactor and the content of ammonium nitrogen in the effluent is observed to be greater (Kumar et al. 2011).

More attempts must be made for maximizing processing of leachate, gas output, and recalcitrant degradation (Mandal et al. 2017). Gas storage devices are designed to store the gases released in a bioreactor. Gasses are processed for useful uses, such as the supply of heating and power generation. Depositing and monitoring decomposition conditions are important to speed up the decomposing cycle (Herrmann et al. 2018).

3.4.2 Sedimentation, Precipitation, Coagulation, and Flocculation

These are the physical waste removal processes. Sedimentation or settling of solids can occur in amassing ponds or lagoons, or in basins designed for this. Sedimentation tanks are designed to operate on a perpetual-flow substructure. They are conventionally rectangular or circular in shape and are equipped with mechanical sludge-accumulating contrivances. With the exception of tanks designed for perpetual sludge abstraction (final sedimentation tanks in activated sludge plants), the bottom of sedimentation tanks is essentially flat (up to 15°) and has sludge hoppers with relatively steep sides. Settled sludge on the tank floor is moved by mechanical scrapers into the hoppers for subsequent withdrawal (Scholz 2016).

Precipitation in a stringently chemical sense is the conversion of a substance from the dissolved state to the non-dissolved state upon the integration of other reagents (coagulants) that lead to the formation of precipitates. The precipitation process is described by utilization of the equilibrium or the cessation-point of the reaction for designated boundary conditions. If precipitation is defined as the transformation of two or more dissolved components to a non-dissolved substance, called precipitate, then dissolution processes and precipitation processes are homogeneous reactions but of antithesis directions (Halling-Sorensen and Jorgensen 1993).

Chemical coagulants are added to prevail over the repulsive forces of the particles. The three types of coagulants are inorganic electrolytes, namely alum, lime, ferric chloride, and ferrous sulfate, organic polymers, and synthetic polyelectrolytes (Mo et al. 2007).

Flocculant settling implies to a dilute suspension of particles that intensify particles in mass and settle at a quicker rate by which fine particulates are caused to clump together into a floc. The amount of flocculation that arises hinges on the chance for contact, which varies with overflow rate, the depth of the basin, the velocity gradients in the system, the concentration of particles, and the range of particle sizes (Thomas et al. 1999).

3.4.3 Dissolved Air Flotation

Dissolved air flotation (DAF) has proven to be very efficacious for the abstraction of biological solids from wastewater streams. Dissolved air flotation is profoundly utilizable and efficient to abstract fats, oils, and grease (FOG) and can abstract paramount quantities of other pollutants. A DAF unit can abstract up to 90% of FOG and 50% of biochemical oxygen demand (BOD).

3.4.4 Wastewater Stabilization Ponds

Wastewater stabilizing ponds (WSP) are astronomically immense, man-made dihydrogen monoxide bodies in which wastewater treated by naturally occurring processes with the influence of solar light, wind, microorganisms, and algae (Mara

2009). Oxygen is provided by the algal population. These ponds have a relatively astronomically immense surface to volume ratio and thus sunlight reaches algae that engender oxygen by photosynthesis (Von Sperling 2007). Stabilization ponds require a relatively sizably voluminous land area and may be impractical for that reason. The effluent still contains nutrients (e.g., N and P) and is consequently appropriate for the reuse in agriculture, but not for direct recharge in surface waters (Yaser and Safie 2020).

3.4.5 Aerated Lagoons

Aerated lagoons are deep waste stabilization ponds in mechanical aerators that facilitate biological oxidation, rather than depending solely on algae-generated photosynthetic oxygen. Aerated lagoons are one of the easiest types of on-site application, with aeration by surface aerator denotes or dispersed bubble aeration. Treatment is occurred by chemical and biological oxidation. The results of increasing retention periods must be measured in detail in aerated lagoon system. In the most practical applications the retention time is shorter in winter and longer in summer (Mehmood et al. 2009).

3.4.6 Anaerobic Lagoons

Anaerobic lagoons or lagoon of manure are most commonly seen in dairy and pig farms for the disposal of animal waste. These are man-made earthen dam more similar to septic tanks usually deeper than 8 feet and built to retain and handle wastewater from 20 to 150 days (Pal 2017). Under the lack of oxygen, the anaerobic bacteria kill pollutants. Further treatment is required for the wastewater leaving an anaerobic lagoon.

3.4.7 Biofuel Conversion Methods

Food processing wastes contain an abundance of organic components that could be converted into energy and then recuperated in the form of heat or electricity (Lee et al. 2019). The major methods of biofuel processing are anaerobic fermentation and thermochemical treatments (e.g., oxidation, gasification, and pyrolysis) (Murugan et al. 2013); wastes containing less than 30% of moisture are felicitous for thermochemical processing, which transforms energy-opulent biomass into intermediate liquid or gaseous materials. For instance, incineration is a thermal process that happens by the oxidation of the waste combustible material for heat production. Incineration is a feasible choice for food waste of reasonably less than 50% moisture content and a hazardous waste choice. There are, however, several increasing questions regarding their pollution, detrimental effects on the atmosphere, and high costs (Chandrasekaran 2012). Anaerobic digestion is a commonly employed waste disposal technique with greater than 50% moisture content and organic content. In the absence of oxygen a number of microorganisms are used during this cycle to preserve food waste.

3.4.8 Vermicomposting

Vermicomposting is the very efficient process by which the earthworms transform agricultural waste into a humus-like substance. It produce a biomass capable of enhancing the soil structural properties, raising its water retaining ability, raising its nutrients, sustaining living soil species and eventually supplying organic materials back to the soil (Shilev et al. 2007). The subsequent requirements for maximizing biological behavior are temperature, pH, ratio of carbon/nitrogen, oxygen, and humidity (Roupas et al. 2007).

3.4.9 Membrane Processes

Dissolved and colloidal constituents from contaminants are isolated through membrane treatment processes. In membrane treatment, water is passed through a membrane using hydrological pressure as driving force. Reverse osmosis, diafiltration, electrodialysis, and ultrafiltration are some of the methods used. These processes may produce profitable products from certain wastes, such as whey protein.

Diafiltration is a diluting method eliminate, the ions or salts based on molecular size by application of micro-molecular-permeable filters (Shao and Zydney 2004). Ultrafiltration membrane systems are usually equipped for colloids with a size of 5–100 nm or higher and massive molecules. Due to the small pore size in ultrafiltration membranes, the waste streams to be treated with ultrafiltration require expensive processing. Ultrafiltration can produce effluent of good quality. When a better level of effluent also needs to be obtained, the output from an ultrafiltration process may be used as the feedstock for a reverse osmosis device.

Particles with size of 0.1 nm or higher may be removed easily through reverse osmosis. The word osmosis defines the normal passage of water from one solution to a more condensed solution, through a semipermeable membrane. The flow path may be reversed by applying pressure to the more condensed solution, such that it flows through the membrane, thus the term reverse osmosis. For this method, pressure requirements are usually between 2000 and 7000 kPa ("Reverse Osmosis - ScienceDirect," n.d.).

In Electrodialysis a membrane separation is done by electrical potential as a guiding factor for transferring salt ions through solutions. The membrane is restrictive as only anions or cations are permitted to get in, but not both. This is not because of the disparity in size but because of the fee. This can be used to isolate molecules with identical sizes of specific loads (Panagopoulos et al. 2019).

3.4.10 Chemical Methods

Chemical method is often employed in combination with other methods. It does have the downside of comparatively high cost when used alone, with mostly negligible efficacy in extracting organic matter. Metals such as calcium or iron may be used to support operations in other systems. The introduction of polyelectrolytes improves the efficacy of certain device processes like dissolved air flotation (DAF).

3.4.11 Cyclone Collector

The major sources of odors from food processing units emanate from manure, peeling of fruits and vegetable, storage of raw material. Modern plants enclose the entire rendering line, allowing odors to be trapped and treated, in order to minimize odor emissions. Microbial decomposition of the environmental contaminants consumed by the scrubber layer allows the process of microbial scrubber. In an existing boiler, burning rendering gasses present a simple and effective way to reduce noxious odors cantily. Steam from cookers, driers, and evaporators is collected via a cyclone to remove solids and a heat switch to dewater the damp air (ter Linden 1949). The condensate is released for treatment of wastewater. The remainder of the odor is burnt.

3.4.12 Disinfection

Disinfection is the elimination of pathogenic microorganisms or their inactivation. Disinfection operations aimed at killing pathogens. Sterilization, the complete destruction of all living matter, is not usually the object of disinfection. A healthy disinfectant must be toxic to microorganisms at concentrations far below the human and higher toxicity levels. It should also have a rapid kill rate and be persistent enough to prevent organisms from regenerating in the system of distribution. Chemical agents, commonly chloride or its derivatives, may be used or water may be exposed to light or radiation from ultraviolet (Rhyner et al. 2017). Ozone is being used perpetually as a disinfectant (Table 3.3).

3.4.13 Recovery and Valorization

Two key approaches are required to adopt a circular economy: abbreviate the amount of waste, and find the most efficient way to handle the waste that is recycled. Waste valorization is defined as the process of converting waste into more utilizable products (Arancon et al. 2013). It is a valuable solution to waste content management and ergo to improve the productivity of biorefineries by utilizing waste as feedstock to produce a broad variety of items (Venkata Mohan et al. 2016).

Table 3.3 Comparison of ozone, chlorine, and ultraviolet exposure characteristics of wastewater effluent disinfection

Characteristic	Ozone	Chlorine	Ultra violet
Effectiveness	Active against bacteria, known to be a stronger viricide than chlorine. Ozone breaks down the cell walls of the bacteria	Very effective against bacterial cells, much less against bacterial spores or the parasite oocysts. Chlorine kills pathogens such as bacteria and viruses by breaking chemical bonds in their molecules	Active for microbial cells; UV damage reproduction systems for microbe by destroying nucleic acid and disrupting their DNA
Toxicity of residue	Rapidly dissipates, little or no residual formation	Chlorine residue can develop potentially toxic halogenated organic substances unless dechlorination has been used	No toxic residue
Penetration	High	High	Moderate
Cost	Moderately high, but becoming competitive with chlorination/ dechlorination	Low if dechlorination is not required	Moderately high

3.5 Food Recovery: Sustainability of Food System

The use of food waste could significantly reduce food waste levels and create new opportunities and benefits for all those involved in the food production system. Reducing food waste through the recovery of its valuable components is therefore an important way to increase sustainability. Food processing organizations and food production systems in particular can no longer ignore the need to act in a sustainable manner. “Sustainability is the development that meets the needs of today, but does not compromise the ability of future generations to meet their individual requirements,” the Brundtland Commission stated. The description is, however, very broad and challenging to grasp and adapt for organizations. Corporate Social Responsibility (CSR) implies that companies have a social accountability that reaches beyond their legal and economic responsibilities. A food processing system includes all the aspects of the production, processing, distribution, food preparation & consumption, and their socio-economic and environmental outcomes (environment, people, inputs, processes, infrastructures, institutions, etc.). The United State Environment Protection Agency’s (EPA) Food Recovery Hierarchy provids guidance to start up a program that will offer the most environment friendly, society-related and food manufacturing benefits to food processors and beverage producers, it is an excellent resource to be followed [Fig. 3.5]. In particular, landfill is the least-favored waste disposal choice created by food and beverage producers

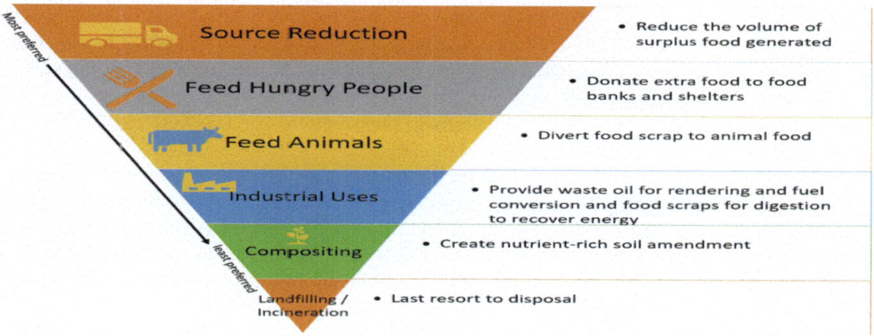

Fig. 3.5 EPA's Food Recovery Hierarchy (US EPA 2015)

worldwide. Sustainable, effective, and productive waste management solutions include: animal feed processing, composting for nutrient rich fertilizer generation, anaerobic digestion for biogas production, waste collection for use in other sectors, feeding surplus food to needy people.

References

Aniyikaiye TE, Oluseyi T, Odiyo JO, Edokpayi JN (2019) Physico-chemical analysis of wastewater discharge from selected paint industries in Lagos Nigeria. Int J Environ Res Public Health 16:1235. https://doi.org/10.3390/ijerph16071235

Arancon RAD, Lin CSK, Chan KM, Kwan TH, Luque R (2013) Advances on waste valorization: new horizons for a more sustainable society. Energy Sci Eng 1:53–71. https://doi.org/10.1002/ese3.9

Arvanitoyannis IS, Kassaveti A, Ladas D (2008) Food waste treatment methodologies. In: Arvanitoyannis IS (ed) Waste management for the food industries, food science and technology, vol 1. Academic Press, Amsterdam, pp 345–410. https://doi.org/10.1016/B978-012373654-3.50009-2

Aziz HA, Alias S, Adlan Mohd N, Faridah Asaari AH, Zahari Mohd S (2007) Colour removal from landfill leachate by coagulation and flocculation processes. Bioresour Technol 98:218–220. https://doi.org/10.1016/j.biortech.2005.11.013

Beiras R (2018) Plastics and other solid wastes. In: Beiras R (ed) Marine pollution. Elsevier, pp 69–88. https://doi.org/10.1016/B978-0-12-813736-9.00006-4

Carpentier B (2009) 14 - biofilms in red meat processing. In: Fratamico PM, Annous BA, Gunther NW (eds) Biofilms in the food and beverage industries, Woodhead Publishing series in food science, technology and nutrition. Woodhead Publishing, pp 375–395. https://doi.org/10.1533/9781845697167.4.375

Chandrasekaran M (2012) Environmental concerns and sustainable development. In: Valorization of food processing by-products, vol 1. CRC Press, pp 768–785

da Rocha M, Alemán A, Romani VP, López-Caballero ME, Gómez-Guillén MC, Montero P, Prentice C (2018) Effects of agar films incorporated with fish protein hydrolysate or clove essential oil on flounder (Paralichthys orbignyanus) fillets shelf-life. Food Hydrocoll 81:351–363. https://doi.org/10.1016/j.foodhyd.2018.03.017

Definition of food loss and waste [WWW Document] (n.d.) ThinkEatSave. URL http://www.unenvironment.org/thinkeatsave/about/definition-food-loss-and-waste (accessed 6.1.20)

Dreher P, Faulstich M, Weber-Blaschke G, Berninger B, Keilhammer U (2004) Recycling of plastic waste, rubber waste and end-of-life cars in Germany. In: Twardowska I (ed) Waste management series solid waste: assessment, monitoring and remediation. Elsevier, pp 815–863. https://doi.org/10.1016/S0713-2743(04)80035-7

Food Loss and Food Waste [WWW Document] (n.d.) Food Agric Organ U N URL http://www.fao.org/food-loss-and-food-waste/en/ (accessed 6.1.20)

Garcia NH, Mattioli A, Gil A, Frison N, Battista F, Bolzonella D (2019) Evaluation of the methane potential of different agricultural and food processing substrates for improved biogas production in rural areas. Renew Sustain Energy Rev 112:1–10. https://doi.org/10.1016/j.rser.2019.05.040

Gunes B, Stokes J, Davis P, Connolly C, Lawler J (2019) Pre-treatments to enhance biogas yield and quality from anaerobic digestion of whiskey distillery and brewery wastes: a review. Renew Sustain Energy Rev 113:109281. https://doi.org/10.1016/j.rser.2019.109281

Gunes B, Carrié M, Benyounis K, Stokes J, Davis P, Connolly C, Lawler J (2020) Optimisation and modelling of anaerobic digestion of whiskey distillery/brewery wastes after combined chemical and mechanical pre-treatment. Processes 8:492. https://doi.org/10.3390/pr8040492

Halling-Sorensen B, Jorgensen SE (eds) (1993) Precipitation. In: Studies in environmental science studies in environmental science, vol 1. Elsevier, pp 355–392. https://doi.org/10.1016/S0166-1116(08)70533-8

Hansen CL, Cheong DY (2019) Agricultural waste management in food processing. In: Kutz M (ed) Handbook of farm dairy and food machinery engineering, vol 1, 3rd edn. Academic Press, pp 673–716. https://doi.org/10.1016/B978-0-12-814803-7.00026-9

Herrmann C, Ramm P, Murphy JD (2018) The relationship between bioreactor design and feedstock for optimal biogas production. In: Liao Q, Chan J, Herrmann C, Xia A (eds) Bioreactors for microbial biomass and energy conversion green energy and technology, vol 1. Springer, pp 163–197. https://doi.org/10.1007/978-981-10-7677-0_5

Huang C, Zhao C, Guo HJ, Wang C, Luo MT, Xiong L, Li HL, Chen XF, Chen XD (2017) Fast startup of semi-pilot-scale anaerobic digestion of food waste acid hydrolysate for biogas production. J Agric Food Chem 65:11237–11242. https://doi.org/10.1021/acs.jafc.7b04005

Jayathilakan K, Sultana K, Radhakrishna K, Bawa AS (2012) Utilization of byproducts and waste materials from meat, poultry and fish processing industries: a review. J Food Sci Technol 49:278–293. https://doi.org/10.1007/s13197-011-0290-7

Kjeldsen P, Barlaz MA, Rooker AP, Baun A, Ledin A, Christensen TH (2002) Present and long-term composition of MSW landfill leachate: a review. Crit Rev Environ Sci Technol 32:297–336. https://doi.org/10.1080/10643380290813462

Kumar S, Chiemchaisri C, Mudhoo A (2011) Bioreactor landfill technology in municipal solid waste treatment: an overview. Crit Rev Biotechnol 31:77–97. https://doi.org/10.3109/07388551.2010.492206

Lee SY, Sankaran R, Chew KW, Tan CH, Krishnamoorthy R, Chu DT, Show PL (2019) Waste to bioenergy: a review on the recent conversion technologies. BMC Energy 1:4. https://doi.org/10.1186/s42500-019-0004-7

Lindawati L, Fitriadi N, Afdhal A (2018) Analysis of noise level generated by stone cutter machine a case study in marble production unit, South Aceh. J Inotera 3:53–58. https://doi.org/10.31572/inotera.Vol3.Iss1.2018.ID48

Lu Q, He ZL, Stoffella PJ (2012) Land application of biosolids in the USA: a review [WWW Document]. Appl Environ Soil Sci. https://doi.org/10.1155/2012/201462

Mandal P, Dubey BK, Gupta AK (2017) Review on landfill leachate treatment by electrochemical oxidation: drawbacks, challenges and future scope. Waste Manag 69:250–273. https://doi.org/10.1016/j.wasman.2017.08.034

Mansoorian HJ, Mahvi AH, Jafari AJ, Khanjani N (2016) Evaluation of dairy industry wastewater treatment and simultaneous bioelectricity generation in a catalyst-less and mediator-less membrane microbial fuel cell. J Saudi Chem Soc 20:88–100. https://doi.org/10.1016/j.jscs.2014.08.002

Mara D (2009) Waste stabilization ponds: past, present and future. Desalin Water Treat 4:85–88. https://doi.org/10.5004/dwt.2009.359

Mehmood MK, Adetutu E, Nedwell DB, Ball AS (2009) In situ microbial treatment of landfill leachate using aerated lagoons. Bioresour Technol 100:2741–2744. https://doi.org/10.1016/j.biortech.2008.11.031

Mistry VV (2005) Salt whey product and method of making. EP1565075A1

Mo J, Hwang JE, Jegal J, Kim J (2007) Pretreatment of a dyeing wastewater using chemical coagulants. Dyes Pigments 72:240–245. https://doi.org/10.1016/j.dyepig.2005.08.022

Murugan K, Chandrasekaran SV, Karthikeyan P, Al-Sohaibani S (2013) Current state of the art of food processing by products, valorization of food processing by-products. CRC Press, Boca Raton

Nash J, Peña O, Galford GL, Gurwick N, Pirolli G, White JM, Wollenberg EK (2017) Reducing food loss in agricultural development projects through value chain efficiency (Working Paper). CGIAR Research Program on Climate Change, Agriculture and Food Security

Nevárez-Moorillón GV, Prado-Barragán A, Martínez-Hernández JL, Aguilar CN (2020) Food microbiology and biotechnology: safe and sustainable food production. CRC Press

Okoro OV, Sun Z, Birch J (2017) Meat processing waste as a potential feedstock for biochemicals and biofuels – a review of possible conversion technologies. J Clean Prod 142:1583–1608. https://doi.org/10.1016/j.jclepro.2016.11.141

Pal P (2017) Biological treatment technology. In: Pal P (ed) Industrial water treatment process technology, vol 1. Butterworth-Heinemann, pp 65–144. https://doi.org/10.1016/B978-0-12-810391-3.00003-5

Panagopoulos A, Haralambous KJ, Loizidou M (2019) Desalination brine disposal methods and treatment technologies - a review. Sci Total Environ 693:133545. https://doi.org/10.1016/j.scitotenv.2019.07.351

Regulation (Eu) 2016/ 679 Of The European Parliament And Of The Council - of 27 April 2016 on the protection of natural persons with regard to the processing of personal data and on the free movement of such data, and repealing Directive 95/ 46/ EC (General Data Protection Regulation), n.d. 88

Reverse Osmosis ScienceDirect [WWW Document] (n.d.). URL https://www.sciencedirect.com/science/article/pii/B9780128137222000200 (accessed 6.6.20)

Rhyner CR, Schwartz LJ, Wenger RB, Kohrell MG, Schwartz LJ, Wenger RB, Kohrel MG (2017) Solid waste source reduction and recycling [WWW Document]. Waste Manag Resour Recovery. https://doi.org/10.1201/9780203734278-4

Roupas P, De Silva K, Smithers G, Ferguson A (2007) Waste management and co-product recovery in red and white meat processing. In: Waldron K (ed) Handbook of waste management and co-product recovery in food processing. Woodhead Publishing Series in Food Science Technology and Nutrition. Woodhead Publishing, pp 305–331. https://doi.org/10.1533/9781845692520.4.305

Salihoglu G, Salihoglu NK, Ucaroglu S, Banar M (2018) Food loss and waste management in Turkey. Bioresour Technol Bioconversion of Food Wastes 248:88–99. https://doi.org/10.1016/j.biortech.2017.06.083

Scholz M (2016) Primary treatment. In: Scholz M (ed) Wetlands for water pollution control, 2nd edn. Elsevier, pp 27–35. https://doi.org/10.1016/B978-0-444-63607-2.00006-X

Seidavi A, Zaker-Esteghamati H, Scanes CG (2019) Poultry byproducts. In: Byproducts from agriculture and fisheries. Wiley, pp 123–146. https://doi.org/10.1002/9781119383956.ch6

Shao J, Zydney AL (2004) Optimization of ultrafiltration/diafiltration processes for partially bound impurities. Biotechnol Bioeng 87:286–292. https://doi.org/10.1002/bit.20113

Shilev S, Naydenov M, Vancheva V, Aladjadjiyan A (2007) Composting of food and agricultural wastes. In: Utilization of by-products and treatment of waste in the food industry. Springer, pp 283–301

Sun F, Dai Y, Yu X (2017) Air pollution, food production and food security: a review from the perspective of food system. J Integr Agric 16:2945–2962. https://doi.org/10.1016/S2095-3119(17)61814-8

ter Linden AJ (1949) Investigations into cyclone dust collectors. Proc Inst Mech Eng 160:233–251. https://doi.org/10.1243/PIME_PROC_1949_160_025_02

Thomas DN, Judd SJ, Fawcett N (1999) Flocculation modelling: a review. Water Res 33:1579–1592. https://doi.org/10.1016/S0043-1354(98)00392-3

US EPA O (2015) Food recovery hierarchy [WWW Document]. US EPA. URL https://www.epa.gov/sustainable-management-food/food-recovery-hierarchy (accessed 6.2.20)

US EPA O (2017) Containers and packaging: product-specific data [WWW Document]. US EPA. URL https://www.epa.gov/facts-and-figures-about-materials-waste-and-recycling/containers-and-packaging-product-specific-data (accessed 6.1.20)

Use of water in food and agriculture - Lenntech [WWW Document] (n.d.). URL https://www.lenntech.com/water-food-agriculture (accessed 6.1.20)

Valta K, Damala P, Panaretou V, Orli E, Moustakas, Loizidou M (2017) Review and assessment of waste and wastewater treatment from fruits and vegetables processing industries in Greece. Waste Biomass Valoriz 8:1629–1648. https://doi.org/10.1007/s12649-016-9672-4

Venkata Mohan S, Nikhil GN, Chiranjeevi P, Nagendranatha Reddy C, Rohit MV, Kumar AN, Sarkar O (2016) Waste biorefinery models towards sustainable circular bioeconomy: critical review and future perspectives. Bioresour Technol Waste Biorefinery Advocating Circular Economy 215:2–12. https://doi.org/10.1016/j.biortech.2016.03.130

Von Sperling M (2007) Wastewater characteristics, treatment and disposal. IWA publishing, London

Wadhwa M, Bakshi MPS (2013) Utilization of fruit and vegetable wastes as livestock feed and as substrates for generation of other value-added products

Walsh P, Key M (1994) Noise and air pollution in the food industry: sources, control and cost implications. In: Dalzell JM (ed) Food industry and the environment: practical issues and cost implications. Springer, pp 106–136. https://doi.org/10.1007/978-1-4615-2097-9_5

Woodard & Curran Inc (2006) Wastes from industries (case studies). in Woodard & Curran Inc (ed) Industrial waste treatment handbook, 2nd edn. Butterworth-Heinemann, Burlington 409–496. doi:https://doi.org/10.1016/B978-075067963-3/50012-6

Wu B, Yan DYS, Khan M, Zhang Z, Lo IMC (2017) Application of magnetic hydrogel for anionic pollutants removal from wastewater with adsorbent regeneration and reuse. J Hazard Toxic Radioact Waste 21:04016008. https://doi.org/10.1061/(ASCE)HZ.2153-5515.0000325

Yaser AZ, Safie NN (2020) Sewage treatment in campus for recycling purpose: a review. In: Yaser AZ (ed) Green engineering for campus sustainability. Springer, Singapore, pp 207–243. https://doi.org/10.1007/978-981-13-7260-5_15

Advances in Wastewater Treatment in Food Processing Industries: Sustainable Approach

4

Ankit Paliwal

Abstract

Water containing 0.1% or more solids could be termed as wastewater. The primary objective of the wastewater treatment is just to remove most or all of the solid particles. Sufficient treatment is required to make water suitable for their best or intended use. Usually, treatment of wastewater includes collection of the wastewater to a centralized quarantined area where it is subjected to different treatment methods.

Around more than 66% freshwater is utilized in food production process. Food and beverage processing industry required large chunk of drinking quality water during production process. Few industries require up to 90% water during processing. The amount of water required depends upon different factors like diversity of plant, capacity of plant, applies equipment, their automation level and their cleaning system, raw material and final packaging washing, and various heat transfer operation, like cooling and heating. Due to variations in process and unit operation, different industries produce different quality of the wastewater. It generally contains microorganisms, organic materials, and different chemical elements used in sanitization process and as fertilizers, and heavy metal compounds and other materials.

Keywords

Wastewater · Treatment processes · Sustainable wastewater management

A. Paliwal (✉)
Western Dairy, Nadi, Fiji

M. Thakur et al. (eds.), *Sustainable Food Waste Management*,
https://doi.org/10.1007/978-981-15-8967-6_4

4.1 Typical Wastewater Treatment Processes

As substantial amount of wastewater is involved, treatments are generally done on consistently running wastewater also known as continuous stream or "open" system. But few selected operations are handled as periodic batch operations, for example vacuum filtration, which required batch or semi-batch mode of operation. Other examples include storage of sludge, the addition of chemicals, and removal or disposal of the treated sludge. On the basis of operation wastewater treatment could be categorized into three different stages which basically show different types of operation used in treatment of wastewater (Fig. 4.1). The whole treatment process may consist combination of various treatment techniques.

Pre-treatment processes are dependent on the type of biological treatment to be undertaken as well as the waste material being treated.

The bioreactor used in the treatment stage may be a large vessel, a heap of soil, or, in the case of in situ soil remediation or land treatment, may be the soil itself.

Post-treatment stages may require filtration or sedimentation before the cleaned effluent can be discharged, plus sludge dewatering, solidification, or incineration (Table 4.1).

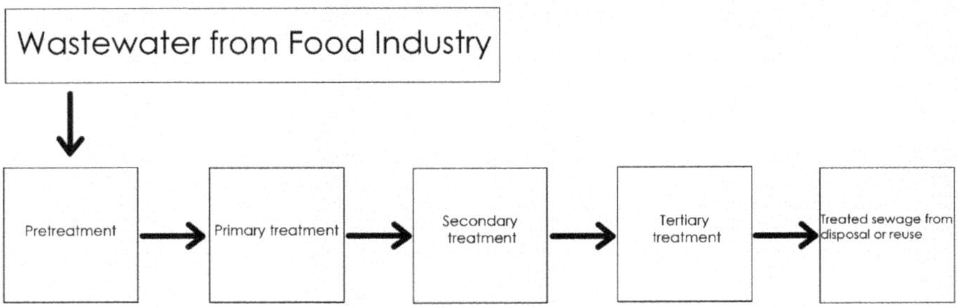

Fig. 4.1 Wastewater treatment processes

Table 4.1 Various levels of wastewater treatment

Treatment	Description
Preliminary	Removal of wastewater constituents Such as floatables, sticks, stems, leaves, grit & grease by screening, filtration, grit remove
Primary	Removal of suspended solids and organic matter by sedimentation
Secondary	Removal of biodegradable organic matter in suspension by aerobic or anaerobic digestion
Tertiary	Removal of residual suspended solids and soluble solids and disinfection by membrane process, ozone process, UV radiation, ion exchange, etc.

Adapted from Metcalf and Eddy (2002)

4.1.1 Preliminary Treatment

The first stage of wastewater treatment is the removal of all the substances which can damage the treatment equipment by clogging or by interfering in their operation. The primary purpose here is to protect and streamline the whole process of the wastewater treatment plant. This stage mainly includes the physical separation of all the substances which can interfere or retard the operation. The main focus of primary treatment includes:

1. Removal of large suspended or floating solids. This could be achieved by filtration or size reduction. It includes all type or organic and inorganic matters
2. Removal of grit, which includes inorganic solids like sand and gravel, metal, and glass particles
3. Removal of lubricative agents like grease and oil

Primary treatment is mainly focused on mechanical techniques which reduces total suspended solid by some 50–60%. Depending upon type of industry this process also reduces BOD of wastewater by 20–40%. The process includes series of mechanical separation process from filtration, to sedimentation to flocculation. While large food waste material could be easily removed by filtration, gravity induced sedimentation is generally used for small sized particles, mostly like sand. Although it is a mechanical process, sometimes few chemicals are also used to accelerate or simplify the operations.

Being the first stage of wastewater treatment process it works in continuous manner, although few steps may include accumulation of water. This accumulation reduces the flow velocity of water which further helps in flocculation and sedimentation of particles. Because of variations in design, operation, and application, settling tanks can be divided into four general groups:

1. Septic tanks
2. Two story tanks—Imhoff and several proprietary or patented units
3. Plain sedimentation tank with mechanical sludge removal
4. Upward flow clarifiers with mechanical sludge removal

When chemicals are used, other auxiliary units are employed. These are:

1. Chemical feed units
2. Mixing devices
3. Flocculators

The results obtained by primary treatment, together with anaerobic sludge digestion as described later, are such that they can be compared with the zone of degradation in stream self-purification.

4.1.2 Secondary Treatment

Secondary treatment is also known as biological treatment as it primarily removes the dissolved organic matter from the wastewater which escaped primary treatment. Biological treatment process involves utilization of microbes which degrade organic matter, consume them as food for their own energy and growth. Secondary treatment removes up to 85–90% of the remaining suspended solid matter and further reduces BOD up to 80%. The biological process is then followed by additional settling tanks ("secondary sedimentation") to remove more of the suspended solids. This incorporates the basic activated sludge system, the series of secondary ponds and wetland systems, trickling filters, and other forms of treatment which use biological activity to convert and separate organic matter from the wastewater stream.

On the basis of their functionality, the devices used in secondary treatment may be divided into following groups:

1. Trickling filters with secondary settling tanks
2. Activated sludge and modifications with final settling tanks
3. Intermittent sand filters
4. Stabilization ponds

Sludge Treatment

Sludge treatment is an important step in both primary and secondary treatment for the removal of solid waste from wastewater. Although purpose and methodology involved during each treatment stage is different, both of them remove the water, and constitute wastewater sludge in the form of settleable solids and sewage sludge. As sludge contains large amount of solid waste, both organic and inorganic in nature, some kind of treatment or conditioning of it is required for final disposal. Main objective of such treatment includes the removal of remaining part of the liquid from the sludge to reduce its volume, and the disintegration of the putrescible organic solids to mineral solids or to relatively stable organic solids. This is accomplished by a combination of two or more of the following methods (Mareddy 2018):

1. Thickening
2. Digestion with or without heat
3. Drying on sand bed—open or covered
4. Conditioning with chemicals
5. Elutriation
6. Vacuum filtration
7. Heat drying
8. Incineration
9. Wet oxidation
10. Centrifuging

4.1.3 Tertiary Treatment

Tertiary treatment is the final stage of wastewater treatment, which is responsible for the removal of more than 99% of the impurities from the effluent water and makes it suitable for discharge. It may include combination of biological and chemical methods with utilization of more advance sophisticated equipment. The related technology requires a well-trained plant operator and a high level of technical know-how. Also it is very expensive and requires a constant energy supply, chemicals, and specific equipment.

This technique uses intermittent sand filters which are required for the removal of suspended solids from the wastewater. Many times tertiary treatment removes plant nutrients, primarily nitrogen and phosphorous from wastewater. Tertiary treatment use has increased because of its improvement and upgrading and also to minimize environmental effects. Advanced treatment is the terms many times used to produce a high-quality effluent water.

4.2 Physical-Chemical Treatment System

The following main technologies are used as physical-chemical system for wastewater remediation:

4.2.1 Gravity Separation

Gravity separation is one of the first step to remove solid and semi-solids from the wastewater. Bar screens alone or along with sedimentation tanks are used to separate pollutants on the basis of their density. Sometimes chemicals are also used as assistant. While oil and grease are lighter in density compared to water, solid particles occupy bottom of the sedimentation tank. Gravity separation reduces BOD by more than 50%. But it fails to remove dissolved compounds.

4.2.2 Evaporation

Evaporation is the concentration process, where dissolved pollutant like heavy metal ions and inorganic salts are condensed while evaporated water could be reutilized. Mechanical evaporator, evaporator tank, and other equipment based upon forced circulation are used for evaporation. Because of its high-power consumption with high maintenance, the application of this process is very limited.

4.2.3 Centrifugation

Various types of centrifugation equipment are used for removal of smaller particles and oil compounds from the wastewater. Type of machine depends upon operation effectiveness and level of contamination in the wastewater.

4.2.4 Filtration and Flotation

While all above mentioned methods come under primary method of water treatment, depending upon filtration process, filtration could be considered as primary treatment and could also be used as tertiary treatment. While basic purpose of primary filtration and flotation is to remove larger particles from the water to make it safe for further treatment, cartridges and membrane filters are used as final polishing steps to clean the water. Resin based and diatomaceous earth based pre-coated filters are also used to remove bad foul smell from the water. Air flotation system is used for removal of oil and grease particles from wastewater by using gas bubble system.

4.2.5 Membrane Technologies

Membrane technology is new emerging technology which is used in filtration. Polymer based materials like polyamides, polycarbonates are used for fabrication of film, which can filter out material from size range 10 mm to nanometres. While larger pour size is used for removal of microbial agglomerations and colloidal compounds from the water, small pour sized member is used for ultrafiltration. Membrane technology is also used for reverse osmosis.

4.2.6 Biological Treatment System

The main objective of biological system is to reduce organic matter present in effluent stream. Biological processes rely on maximizing biological activity by improving the bacterial life that will naturally develop. This may involve a fixed medium (such as a bacterial filter bed) or may rely on an aqueous medium (such as activated sludge), most often with aeration. After the bioreactor stage, subsequent physical processes such as decanting of liquids (sedimentation) will allow separation of the liquid phase (cleaned effluent) and the solids, which must then be appropriately treated. When the conventional techniques are not sufficient to reach the level required by the regulations, it may be necessary to use complementary or finishing treatments. These range from simple lagoons to the use of activated carbon.

4.3 Aerobic Wastewater Treatment

Wastewater treatment processes typically occur in stages, with the final stage, *i.e.* the quality of the discharged water, often being determined by regulations. Because of the need to comply with these, as well as the need to match treatment processes to the waste stream, there are some basic data collection requirements associated with wastewater treatment.

The quality of the wastewater produced by an industrial process depends in part on the operation of the plant, as well as on the process being undertaken. Possible and actual variations in the wastewater composition must be taken into account by the operator of the treatment facility.

4.3.1 Bacterial Filter Beds

Bacterial filter beds use cultivated bacteria on fixed media, which are mostly made from natural elements such as gravel, or plastic membranes. The wastewater being treated trickles through these elements. Cleaned effluent may be 'recycled' through the bed to dilute the charge and to maintain the humidity of the bed.

This is a stable process which is not sensitive to minor changes in the raw material. This technique is particularly suited to use with effluents containing high organic loadings such as those from the agro-nutritional industries (*e.g.* canning industry, sugar industry).

4.3.2 Activated Sludge

This treatment is suitable for municipal wastewater and aqueous hazardous wastes which have less than 1% of suspended solids. During the process, suspended microorganisms are continuously recycled through a bioreactor. The recycling process allows microorganisms to become adapted to the incoming wastewater composition. The bacteria develop freely in the liquid mass and collect in 'flocs', which float according to the aeration currents of the basin. Various techniques are used to ensure an excess of oxygen, which is indispensable for development of the bacterial mass. This process is very flexible, since the charge can be controlled according to the situation (day, night, peak hours, seasons, etc.), but it requires regulating and control apparatus, and is therefore more complex than bacterial filter beds.

Numerous variations of this technique are used, and it is frequently used in refineries, canneries, and pharmaceutical plants. Wastes can be successfully treated providing there is efficient mixing and a high level of dissolved oxygen is maintained throughout the bioreactor. The process produces a relatively cleaner effluent than other biological processes, but requires a considerable amount of energy.

4.3.3 Lagoons

Lagooning describes the process of placing effluents in a shallow impermeable water basin to allow degradation to take place. Artificial aeration may be required for depths greater than a metre. This type of containment is used for wastewater with low organic contents. The cleaning of the effluents takes place as a result of bacterial action, algae, or aquatic vegetation.

The residence time is generally several weeks. This process is often used as a sustainable pre-treatment step. This process is inexpensive and sustainable but it needs a large area and is not suitable for wastes containing mainly chemical components. In addition, when the lagoon is not sufficiently aerated, there is the risk of it becoming anaerobic, which causes odour problems.

4.3.4 Membrane Bioreactor (MBR)

It is the new technology used for biological degradation of soluble organic impurities. Just like conventional activated sludge process, the membrane bioreactor process has mixed liquor solids in suspension in an aeration tank. But in case of membrane bioreactor process, the bio-solids are separated by means of a polymeric membrane based on microfiltration or ultrafiltration unit while in case of activated sludge process the secondary clarifier is based upon the gravity settling process.

4.3.5 Slurry-Phase Treatments

In this treatment process, wastes (which may be solids, sludges, or contaminated soils) are suspended with water or wastewater in a reactor to form slurry. Agitation of the slurry homogenizes its consistency, breaks down solid particles, oxygenates it, and increases contact between the organic waste and the microorganisms. The treatment process can be batch or continuous. This treatment method was originally used for treating old lagoons and is now being extended to reactor systems.

Mixing and particle suspension are critical parts of the process. Keeping large concentrations of sludge in suspension requires high-energy mixers. The design of the mixer, and the reactor, is probably the most crucial aspects of this treatment method. Slurry-phase treatment degrades waste at a faster rate and requires less land area than solid-phase treatment. It is being developed for use with a number of waste types, including wastes from wood preserving with high concentrations of creosote oil and PAHs, petroleum refinery sludges.

4.3.6 Anaerobic Wastewater Treatment

Opposite to aerobic treatment, anaerobic digestion decomposes organic matter, and produces biogas with the help of microorganisms in the absence of oxygen. As microorganisms used in anaerobic process can survive in harsh environment, it is considered as an energy-efficient sustainable method which can be applied directly to industrial wastewaters, which generally contain high amount of bio-decomposable organic matter and have higher temperature compared to require for general aerobic process. Hence anaerobic treatment process is also suitable to utilize sustainably as pre-treatment before discharging wastewater to general wastewater treatment plant or before polishing in an aerobic process.

Due to their simplicity, anaerobic processes are more sustainable compared to aerobic treatment options as they required less chemicals, less energy, less pre-treatments, and lower sludge handling costs. Biogas is mixture of methane and carbon dioxide, which are generated during the anaerobic treatment and could be utilized as renewable energy to replace fossil fuels.

Microorganisms which are used for anaerobic treatment could be classified into three different categories on the basis of reaction they carried out:

1. Fermenting microorganisms: they involve in basic fermentation, converting simple organic molecules into carbon dioxide, alcohol, and ammonia.
2. Methane-producing microorganisms: these microorganisms convert carbon dioxide and molecular hydrogen into methane.
3. Acetic acid bacteria: they are gram-negative bacteria, which can convert organic material into acetic acid.

These microorganisms are involved in different stages of process. On the basis of biological reaction involved anaerobic digestion could be divided into following stages:

1. Hydrolysis
2. Acidogenesis
3. Acetogenesis
4. Methanogenesis

Organic waste is mostly made up of large polymers. These large polymeric compounds are digested into their smaller constituent parts by the bacteria for energy production by the process known as hydrolysis. Organic molecules are converted into simple sugars, amino acids, and fatty acids during the process. It is the first and most important step of anaerobic digestion as it breaks down the complex organic molecules and makes them available for other bacteria.

The second stage of the process is acidogenesis, where fermentative bacteria further digest and breakdown the remaining components. During the process, mainly volatile fatty acids along with carbon dioxide, ammonia, and hydrogen sulphide are generated.

The acetogenesis is the third stage of anaerobic digestion process. During acetogenesis, mainly acetic acid is produced by digestion of simple molecules generated during the acidogenesis stage by acetogens, along with carbon dioxide and hydrogen.

The last stage of anaerobic digestion process is the methanogenesis. Methanogen bacteria convert products from previous stages like acetate, and hydrogen into methane, carbon dioxide, and water. Complex molecules like volatile fatty acids need to be converted into simple molecules before they can be utilized by methanogens. Methane and carbon dioxide make up the large fraction of the biogas generated from the anaerobic digester.

The residual, indigestible materials along with any dead bacterial remains create the digestive sludge. An ideal generic chemical equation for the overall processes outlined above is as follows:

$$C_6H_{12}O_6 \rightarrow 3CO_2 + 3CH_4$$

4.4 Factors Influencing Biological Treatment

The choice of biological waste treatment depends upon numerous factors. Organic wastes are more suitable for biological treatment, especially with adequate moisture content. If the organic waste is present in dry form, it cannot be treated by biological system, until minimum moisture is not present which may be required for supporting microbial activity. Wastes pH should also be within the optimum range, which could be between pH 5 and 10. Due to nature of raw material, nature of reaction, and presence of other matters, the 100% digestion of organic matter is not possible. Further the bacterial activity depends on various factors like nutrients availability, change in pH, generation of toxic compounds or inhibitors, decrease in nutrient levels, etc. Due to this, maximum efficiency achievable is considered to be up to 95% for a biological treatment process.

4.4.1 Process Conditions

Major process conditions which control the biological treatment processes include the optimum temperature range, desired moisture content and water activity, optimum pH along with required acidity or alkalinity level, and the required aeration.

- Temperature range—every bacterial species required an optimum temperature range for their growth and activity. On the basis of their required temperature range, bacteria could be divided into different categories, for example, mesophilic bacteria required temperature between 30 and 40 °C, while thermophilic bacteria favours temperature range from 45 to 55 °C.
- Water activity—a minimum moisture is required to maintain minimum water activity for the growth and activity of bacteria. Although presence of adequate water during wastewater treatment, water activity is not a concern.
- pH range—different bacteria grow and operate at different pH. An optimum pH is required for their action on organic waste throughout the process. As acid is

generally generated during the biological process, a buffer is generally desired to use for maintaining pH of the media at optimum level.

- Aeration—on the basis of process, aeration may be required. An aerobic process required the presence of free oxygen to carry on while during anaerobic process oxygen acts as an inhibitor.

  Inhibitor—various heavy metals like zinc, copper and chemical compounds like pesticides act as inhibitor and can adversely affect the growth of microorganism and hence affect the biological processes in the negative way.

4.4.2 Advantages of Biological Waste Treatment

Biological treatment has a major advantage as it is very tolerant to changes in waste composition, and the ongoing process and its effectiveness once the pre-requisite conditions have been met like temperature humidity, etc.

In biological treatment method the living organisms are able to regulate their activity according to the composition of the medium. Hence for the efficiency of the plant the simple intervention is the re-timing of a recirculation pump and hence changes in the composition of the raw material like dilution may result in a period of inactivity.

4.5 Sustainable Wastewater Treatment Systems

When we talk about sustainable approach towards waste treatment we need to look into economic sustainability, environmental sustainability, societal sustainability, and overall sustainability. While economic indicator looks into capital requirement and operational cost of the process, environmental indicators take account of resource utilized in treatment process and final reusability of water (Muga and Mihelcic 2008).

Lagoon and wetlands, being economical viable and self-sufficient, are considered as traditional sustainable approach towards wastewater treatment process. Lagoons in the form of shallow pond series provide a space for stabilization, during which period duckweed acts as accumulator of heavy metals and ecosystem of bacteria and algae helps in water purification. Although requirement of large land mass for construction of lagoons and wetlands could make them unviable in urban space.

Anaerobic treatment is also considered as sustainable approach being cost effective, requiring less space, and reducing waste volume up to 70%. The by-product of process, biogas, and sludge could also be utilized in better way. Construction and operation of anaerobic treatment plant is also much cheaper and viable.

Soil aquifer treatment is another traditional sustainable approach where partially treated effluent wastewater is used for recharging artificial aquifers. Unsaturated soil layer used here provides an additional layer in natural water filtration and purification method. Such filtered ground water could be utilized in water scare areas. Although such approach could not be utilized for wastewater with high amount of organic waste which could clog the filtration bed or high amount of minerals which cannot be removed by soil filtration (Bdour et al. 2009).

Lots of work have been done in sustainable wastewater treatment approach. Khan et al used organic hydrophobic and hydrophilic ligands for removal of zinc oxide and copper oxide nanoparticles from waste stream using coagulation at different pH. The environmental pH affects the solubility, hence coagulation efficiency (Khan et al. 2018, 2019). The simultaneous effect of charge neutralization and coagulants can enhance removal of metallic nanoparticles, which are more effective in removal of nanoparticles during wastewater treatment. In another study Xu et al. proposed optimum level of chemical oxygen demand to nitrogen and dissolve oxygen for better growth of microorganism in activated sludge treatment system (Xu et al. 2019). This approach could be utilized for better construction of wetlands for sludge treatment. Use of agro-forest waste in wetland construct could further enhance their efficiency (Melián 2020).

4.6 Scope of Treatment Application

Wastewater treatment is the most widely used application of biological treatment in waste management, principally for municipal sewage but also for industrial wastewater in a number of industries such as paper manufacture, food processing, tanneries, and the pharmaceutical industry. After effluent treatment, wastewaters from agriculture and food industries typically leave solid residues (sludges) that must be further processed to be transformed into fertilizers, or treated before land filling where there is no market for the fertilizer.

Food industry wastewater contains different category of waste like organic waste (biodegradable and non-biodegradable), heavy metals, suspended solids, pesticides, and other nutrient matter. Not a single approach could be utilized for treatment of such waste and hence combination of treatment must be applied. Separation of waste from the source could be one of the approaches which can simplify the whole process and make it more sustainable. Use of chemical treatment along with traditional sustainable methods could be a good hybrid approach for future of sustainable wastewater treatment.

References

Bdour AN, Hamdi MR, Tarawneh Z (2009) Perspectives on sustainable wastewater treatment technologies and reuse options in the urban areas of the Mediterranean region, Desalination. Elsevier B.V 237(1–3):162–174. https://doi.org/10.1016/j.desal.2007.12.030

Khan R et al (2018) Influence of organic ligands on the colloidal stability and removal of ZnO nanoparticles from synthetic waters by coagulation processes. MDPI 6(9):170. https://doi.org/10.3390/pr6090170

Khan R et al (2019) Coagulation and dissolution of CuO nanoparticles in the presence of dissolved organic matter under different pH values. Sustainability. MDPI 11(10):2825. https://doi.org/10.3390/su11102825

Mareddy AR (2018) Environmental impact assessment, 1st edn. Elsevier. https://doi.org/10.1016/C2015-0-06055-5
Melián JAH (2020) Sustainable wastewater treatment systems (2018–2019). Sustainability 12 (5):387–400. https://doi.org/10.3390/su12051940
Metcalf and Eddy (2002) Wastewater engineering treatment and reuse. Tata McGraw Hill, New Delhi, pp 10–11
Muga HE, Mihelcic JR (2008) Sustainability of wastewater treatment technologies. J Environ Manage 88(3):437–447. https://doi.org/10.1016/j.jenvman.2007.03.008
Xu J et al (2019) Shifts in the microbial community of activated sludge with different COD/N ratios or dissolved oxygen levels in Tibet, China. Sustainability. MDPI 11(8):1–12. https://doi.org/10.3390/su11082284

Part II

Utilization of Waste from Food Processing Industries

Fruits and Vegetable By-Product Utilization as a Novel Approach for Value Addition

5

Maysam Sarafrazy and Urba Shafiq Sidiqi

Abstract

Among the various food processing industries, fruit and vegetable processing industry is the second prime generator of wastes into the environment only after the domestic sewage. An enormous amount of wastes in form of solids and liquids generated through fruit and vegetable processing industries contains certain reusable substances of high value with huge financial prospective. Processing of fruits and vegetables generates higher amounts of wastes which include skin, seeds, stones, and unused flesh produced during the various steps of the processing chains. Thus leading into various pollution problems due to the reason of not being utilized or disposed-of properly. A disposal of these waste materials frequently signifies a problem that is additionally provoked by different legal restrictions. Howbeit, the waste product, which is discarded into the environment, is loaded with valuable compounds. They are new, innate, and monetary sources of colorants, protein, dietary fiber, flavoring, antimicrobials, and antioxidants, which can be utilized in the food industry as a basis of natural food additives. Therefore, novel aspects concerning to the use of these by-products for advance utilization on the production of food additives or supplements with high nutritional value have gained immense interest because of their characterization among high-value products; therefore, their recovery and improvement may be economically attractive and beneficial.

M. Sarafrazy
Horticultural Research in Urdu Khan Agricultural Research Station, Herat, Afganistan

U. S. Sidiqi (✉)
Islamic University of Science and Technology, Pulwama Kashmir, Jammu & Kashmir, India

Collection and Conservation of Horticultural Germplasm, Urdokhan Agricultural Research Station, Herat, Afganistan

M. Thakur et al. (eds.), *Sustainable Food Waste Management*,
https://doi.org/10.1007/978-981-15-8967-6_5

Keywords

Antimicrobials · Flavorings · Colorants · Fruit and vegetable wastes (FVW)

5.1 Introduction

India is a largely populated country which exists as the major reason for massive waste generation created frequently out of domestic & industrial actions which includes removal of peel followed by cutting of raw fruits and vegetables former to processing, eating, and cooking. FAO revealed that every year, about one-third of all food produced for the purpose of human consumption globally is lost or wasted. This food wastage generation predicts a huge missed opportunity to enhance global food security, but also to alleviate environmental impacts and exhaustive resource use from food chains. Globally, the various quantitative food losses and waste estimations per year are roughly 40–50% for root crops, fruits, and vegetables; 20% for oil seeds, meat and dairy plus; 35% for fish, and 30% for cereals (Schieber et al. 2001).

Referring to the individual supplement, fruit processing industries contribute more than 0.5 billion tons of waste. Thus globally providing the accessibility of this feedstock and its intact potent alien encouraging researchers and other authorities to exercise comprehensive studies on the various value-added potential of fruit processing waste (FPW). On the other hand, vegetables are essential but uneconomical, i.e., they produce ample waste concentration. Around twenty peculiar kinds of plants are generally developed for vegetables in the United States (US). Including this in each State these plants are grown on commercial basis out of which maximum population resides in New York, Texas, California, and Wisconsin. The profits from them are approximately 300 million dollars annually despite of the fact not more than 20–30% of the crops consumed. Out of the total wastes 4 million tons of them are generally leaves. Several wastes are left as such on the soil to be plowed underneath. Few of them are fed, some are discarded in dumps and some are a simple nuisance; a small portion is synthetically dehydrated for feed. The most prominent component of the wastes is water which accounts nearly about 75 to 90% (Willaman and Eskew 1948). On a comparison fruit processing wastes are originated to be selective and concentrated in nature as compared to other biomass derived waste. Besides, the greatest contribution provided is the utilization of peels, pomace, and seed fractions as an excellent feedstock for recovery of bioactive compounds which include flavonoids, lipids, dietary fibers, pectin, etc. (Kowalska et al. 2017; Banerjee et al. 2017). A novel bio-refinery method would aim to manufacture a wider variety of important chemicals from fruit and vegetable processing waste. The wastes from bulk of the withdrawal processes may supplementary be used as recycle sources for creation of biofuels. These all benefits will open up as a scope for future utilization of fruit and vegetable waste for therapeutic and nutraceuticals purpose as well as a great source for value addition of the end products (Table 5.1).

Table 5.1 Production percentage of food wastes and their by-products in fruit and vegetable sector

Production process	% of wastes and by-products
White wine production	20–30
Red wine production	20–30
Fruit and vegetable juice production	30–50
Fruit and vegetable processing and preservation	5–30
Vegetable oil production	40–70
Corn starch production	41–43
Potato starch production	80
Wheat starch production	50
Sugar production from sugar beet	85

Source: Agro food wastes minimization and reduction network, 2004

According to the global trend for fruits and vegetable production, the total sum of residues after processing has been anticipated in millions of tons every year. This demands the use of different forms of energy, water, and other factors providing a by-product potential as the cardinal significance. This comes into being due to the presence of bio-components, which may be utilized for novel food production. This demands an appropriate measure to convert conventional products into value-added ones for the reason of their calculus natural components (Fig. 5.1).

5.2 Functional Characteristics

Presently, consumers are becoming progressively more concerned in maintaining a healthy diet and standard of living (Schieber et al. 2003a, 2001). The by-products of fruit and vegetable processing wastes always give emphasis on the most important functional bioactive constituents such as tocopherols, carotenoids, polyphones etc. Various other components are present which stipulate the producers to generate products which hold a value-added feature, such as dietary fiber or in more modern epoch, phytochemicals. The invention and accumulation of such nutrients can be fairly costly for the manufacturer. In the fruit and vegetable industry, the preparation and dispensation measures can lead to one-third of the product being useless and hence discarded. This can be expensive for the producer and also may have a depressing impact on the surroundings. Researches have revealed that these by-products can have an elevated nutritional significance. It has also been recommended that they may possibly be used as a food component. This is mainly due to their purposeful and practical abilities such as water holding capacity and gelling. Several of their widely pronounced applications are proved as such.

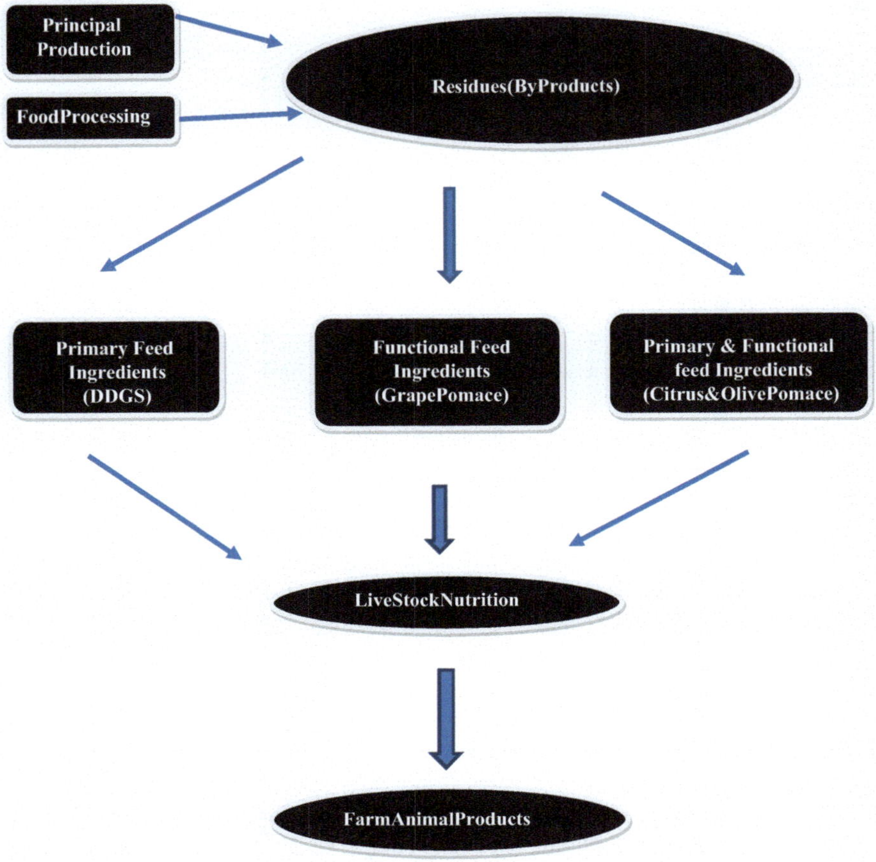

Fig. 5.1 Fruit & Vegetable by-product utilization in Food Processing Chain

5.3 Source of Dietary Fiber

One of the most important nutrients incorporated in diet is dietary fiber. Over the past years, dietary fiber has acknowledged much affirmative concentration with regard to its prospective as a pharma food. This is because of its ability to lessen coronary heart disease, diabetes, cholesterol and it also has ability to ease constipation (Telrandhe et al. 2012). In recent epoch, fiber has also been considered as a constituent with particular and specific functions in food fabrication. Due to the nature of fiber accounting both insoluble as well as soluble properties, it has a wide array of technical and scientific attributes. These properties mainly include water holding capacity, structure building, and gelling. In addition, it can be used as a fat replacer. It was elaborated that the "ideal fiber" is having the certain uniqueness such as exempting any such constituents that are nutritionally odious or increasing its use.

It must be of elevated concentrate in a minute quantity. The ideal fiber should be tasteless and no depressing odor, color, or any such texture properties. It requires maintaining equilibrium among soluble and insoluble fiber with an adequate existence of bioactive compounds. Its accumulation must not alter the food as is being used supplementary. Further, the ideal fiber must also have an extended shelf life. It should work cordially with food processing and it must have a constructive buyer reflection (Larrauri 1999; Kunzek et al. 2002).

5.4 Utilization of Dietary Fiber

Characteristically, fibers such as corn, wheat, and rice have been utilized for food production in the past for their health as well as technical functions. Howbeit, recently novel discovery and utilization of fiber sources came into existence. Among the discovery one of the sources discovered and utilized is the by-product fraction obtained from various types of food processing. In particular, the by-products obtained from any such fruit and vegetable processing unit (e.g., juices, drinks etc.) are seeking great attention as being a novel and economic and cardinal sources of a healthy functional ingredient (Ayala-Zavala et al. 2011). Such by-products can be explained as the remnants or residues left after the manufacturing of fruit and vegetable-based product processing. These residues include pips, peel, stems, skins, and cores. Presently these by-products are disposed of and wasted at a specific cost to the producer for the production of animal feed, landfill, or incineration (Angulo et al. 2012).

5.5 Source of Phytochemicals

Phytochemicals are bioactive component derived from plants (Thakur et al. 2020) and from various by-products also we extract them. The major product produced from apple processing is apple juice. The whole fruit is typically hard-pressed in a cold press to remove the juice as of the fruit. This can result in a large amount of waste, which is termed as Apple pomace. Apple pomace consists of around 25% of fresh apple weight (Angulo et al. 2012). In broad-spectrum, producers typically dispose tones of pomace at a cost to themselves.

On the other hand, it is used as animal feed. Contrasting cereals, there is an elevated proportion of the soluble fiber portion in apple fiber, thus giving rise to the accessibility of the polymer pectin. Pectin has such distinctiveness as it can be used as gelling, thickening stabilizing agent in foods. It is also a health-promoting polymer, considered to lower cholesterol and setback gastric emptying (Hwang et al. 1998; Royer et al. 2006). It was considered that the phytochemicals which was a part of apple pomace produced throughout cider processing (Garcia et al. 2009). The phytochemicals during various researches came into being were polyphenols such as flavonoids, e.g., flavanols and phenolic acids such as caffeic acid, chlorogenic acid, and protocatechuic acid.

Moreover, investigations reported the phenolic compound content of apple pomace and similar findings in their result (Schieber et al. 2003b; Garcia et al. 2009). A number of quercetin glycosides with quercetin 3-galactoside being the principal, flavonol chlorogenic acid, phloridzin were also found. Derivatives of flavanols, i.e., catechins and procyanidin were also obtainable in huge amounts. Phloridzin was revealed to be the most abundant multiple components present, while chlorogenic acid also existed. Procyanidins and flavonol glycosides were either seen in lesser quantities or missing on the whole. The phytochemicals there in apples have been linked with many health-promoting benefits, e.g., cancer cell propagation, decreased lipid oxidation, and lesser cholesterol. In turn these valuable phytochemicals have peripheral dropping effects on chronic diseases there in the western world, e.g., obesity, cancer, and heart disease. Various conclusions highlighted the accessibility of phenolic acids, e.g., chlorogenic acid which has an elevated alkyl peroxyl radical scavenging accomplishment; hence, apples have a self-protective outcome against cancer. Procyanidins have an elevated antioxidant action and restrain low density lipoprotein oxidation (Rice-Evans et al. 1997; Henriquez et al. 2010).

Quercetin is one of the major flavonoids originated in apple, mainly in apple peel. It has been connected with compact incidences of breast cancer and leukemia (Rice-Evans et al. 1997; Henriquez et al. 2010). This information illustrates the accessibility of phytochemicals in apple pomace. It is also suggested that the addition of apple pomace as a bioactive component in food products. This addition could noticeably advance the nutritive properties of such products and conceivably the vigor of the buyer. It has many functional properties like water holding capacity, gelling ability, thickening as well as stabilizing abilities), nutrients and availability of phytochemicals also.

Apple pomace has been utilized by researchers in a wide range of food products such as sausages, jams, and baked goods (Rupasinghe et al. 2008). An apple skin powder (ASP) was supplementary to muffins to develop their phenolic substance. It was brought into being to improve the flavor while rising the phenolic and antioxidant contents accomplished that ASP from the producers could be used as a substitute for wheat flour and in muffins (Rupasinghe et al. 2009; Sudha et al. 2016). The substitution of wheat with 16% (weight basis [w/w]) ASP still acknowledged favorable sensory scores. Apple pomace (AP) was also supplemented to cakes; however, it was originated that as the pomace level augmented ahead of a particular level, the volume of the cake decreased. Additionally, it was exposed that due to the water binding capacity of AP, extra water was requisite to fully hydrate the dough. Also it was perceived that the color of the cake became darker as the quantity of AP decreased. The addition of AP resulted in some favorable qualities such as a fruit aroma and taste, thus permitting the level of sugar added to be concentrated (Masoodi et al. 2002; Ayala-Zavala et al. 2011).

5.6 Antioxidant

Fruit peels such as skins are loaded with nutrients and include many phytochemicals that may be competently used as drugs or as food supplements or replacers (Bobinaite et al. 2016). Antioxidants are excellent compounds as additives in foodstuffs in order to enhance their nutritional value or in manufacture of fruit purees (Chacko and Estherlydia 2014). In addition to their sensory properties they also act as a raw material for the creation of food dyes. Very essential is the prospect of preserved and improved quality as a consequence of avoiding food oxidation (Ayala-Zavala et al. 2011). The probable antioxidative position and bioavailability of by-products from creation of nectar of the pomegranate fruit were examined under a study (Surek and Nilufer-Erdil 2016). Pomegranate seeds and the impulsive obtained by sedimentation of nectar comprise a good supply of anthocyanins. Filter cake, sediment peel contained extra phenolic compounds and were characterized by elevated antioxidant action than those extracted from the pomegranate nectar. In the certain studies the antioxidant activity and overall content of polyphenols of tomato skin were found to be 38.2 and 66.5% elevated, correspondingly, in comparison to the seeds of tomatoes (Sarkar and Kaul 2014).

5.7 Antimicrobial Activity

Recent studies have shown that natural extracts have high antimicrobial prospective. Besides, it has to be taken into description that the fruit and vegetables by-products may supply as probable natural antimicrobial agents. Principally due to the antioxidants radical chelating and scavenging activities can hinder or inhibit the oxidation of DNA and proteins as well as lipids. The components have shown various effects and at the same time playing a cardinal role in food protection against pathogenic agents (Ayala-Zavala et al. 2011; Bobinaite et al. 2010).

Certain studies showed that by-product extracts of raspberry exhibited antibacterial activity adjacent to a range of Gram-positive and Gram-negative bacteria (Bobinaite et al. 2010).

A study carried with the aim of preparation and processing of jams from certain fruits peels such as pineapple, pomegranate, orange, and banana evaluates the antimicrobial properties. Among all pomegranate peel jam was observed to possess the maximum activity against Shigella (Bobinaite et al. 2010). Moreover, avocado fruit seeds are highly rich not only in fiber but also in certain antimicrobials, colorants, flavorings, antioxidants, and thickening agents (Vodnar et al. 2017; Barbosa-Martín et al. 2016). Further, their seeds hold various components such as furanoic acids, flavonoid terpenes, fatty acids, and saponin, phytosterols, proanthocyanidins among which some of them are connected to possess certain antifungal activities, antimicrobial activities, and larvicidal activities (Vodnar et al. 2017; Rodríguez-Carpena et al. 2011). A study showed that quercetin and rutin possess certain antifungal effects against Cryptococcus spp. specifically. This arises due to the polyphenolic components. Cauliflower and mandarin by-product mixture of vegetables particularly in aggregation with process of pasteurization by application of highly hydrostatic pressure (HHP) technology presented a result of strong antimicrobial effect against S. Typhimurium (Oliveira et al. 2015; Sanz-Puig et al. 2017). Consequently, these techniques could be used as a supplementary control measure to assure the food safety.

5.8 By-Products Used as a Global Trend

5.8.1 Innovative Foods

The by-product as the innovative food provides a novel approach in every aspect globally. It further extends its contribution towards the attractiveness and the production of high quality food with every essential predominant characteristic. The innovative foods are highly connected towards the growing environmental concern as the waste from fruit and vegetable processing industry is drastically produced. Thus, the innovation comes into existence when the majority of plant waste is converted and recovered into high value and setting a bench mark of novelty.

However, the process demands more reproducibility and wide use. There exist certain berry fruits which pertain to an essential role in maintaining the healthy diet (Aura et al. 2015). The addition of raspberry extract (2%) to certain mix fruit purees showed a cardinal increase of total phenolic content up to 3 folds. Therefore, this addition provided with an enhanced and improved functional properties of the products (Bobinaite et al. 2010).

Further, pumpkin varieties of Muscovy produce a by-product constituting a source of specific bioactive compounds providing antimicrobial, antioxidant pro-health properties (Saavedra et al. 2015). The addition of oligofructose, apple fiber, and inulin, as the source of prebiotic fiber provided a cardinal increase in the antioxidant characteristics. It also signifies a flourishing preservation of sensory attributes of strawberry juice (Cassani et al. 2016). The novelty of fruit pomace used as an ingredient in the functional food processing commonly known for bakery industry may provide a part effectively to promote health benefits. They impart an essential role against various disorders which include ulcer, diabetes, atherosclerosis, and cancer (Sudha et al. 2016). A marked remark of increased bioactivity has been observed by addition of apple pomace. Blueberry bagasse flour may be added for the formation of fermented beverages (Goldmeyer et al. 2014). The flour imparts very good microbiological stability during storage and enhancing various characteristics such as lipid and ash content, pH and soluble solids, moisture, protein, which makes its use possible for the production of novel products. Flour extracted out of guava skin is used for the fortification of wheat flour which then improves the nutritional quality of products manufactured out of it without affecting the sensory quality of the respective product.

Among fruit and vegetable by-products, waste out of salads reveals high water content. On the other hand, various waste management techniques could be efficaciously implemented. Certain important compounds like fibers, polyphenols, and water are the major components out of salad wastes (Plazzotta et al. 2017). Further, in order to reduce the salad waste, certain traditional techniques like composting and anaerobic digestion may be used in combination with various other novel technologies together with those based ultrasound and pressure. By-products out of apple are thought to be considered as one of the most feasible strategies for pectin extraction. Gluten free formulation can be produced out of apple pomaces (Parra et al. 2015). Because of the increased water binding capacity of all the polysaccharides, apple fiber being polysaccharide can be used for preparation of low calorie products (Sharma et al. 2016).

5.8.2 By-Products as a Source of Flavorings

Because of the importance gained by potential flavorings they are being widely used. They are extensively used in various dishes as seasoning. Besides, most of them are utilized by traditional medicine, e.g., certain essential oils that are rich in terpene compounds are acknowledged for supporting in the treatment and curability of enormous health problems. An important purpose of some of them leads to various activities such as antifungal, antiviral, anti-inflammatory, antimutagenic, antibacterial, vermicide, and anticancer activities (Raut and Karuppayil 2014; Felipe et al. 2017). At the present epoch flavor companies are concerned with the aromas and flavors which are more stable and do not escape openly besides under specifically defined conditions, i.e., they are mostly microscopically encapsulated (Arvanitoyannis and Varzakas 2008).

Sometimes, various factors affect the existence or availability of specific flavoring extracts such as their existence being uneconomical due to high cost or its unavailability. This leads to a great switch towards the commercial flavoring adoption. These trendy flavorings are nothing but an equal or adjacent chemical substitutes or equivalents of the existing natural flavors commonly known as the "nature-identical" (Mantzouridou et al. 2015). The production of the specific enzymatic as well as the whole-cell biocatalysts has gained a cardinal attention as well as a novelty. Their approach has been greatly observed as a substitute for the better development of several esters when compared to any naturally or chemically synthesized enzymes (Zhuang et al. 2015). These extractions of the enzymes from the particular organisms serve various vital benefits which mainly include superior productivity with respect to the elevated catalyst concentration, as well as the simpler product refinement. Some researches provide three main methods exercised for the process of aroma compounds production (Felipe et al. 2017). These processes include the method of chemical synthesis, extraction from natural sources, and some of the biotechnological production processes resulting in the development of the bio-aromas.

However, among all the above-mentioned techniques majority advantages are contributed by the biotechnological process. The importance can be approved or confirmed by its natural product development as a versatile resultant. Furthermore, the most novel advantage offered is the development of the better, suitable, and sustainable preservation approach towards the environment. Bio-aroma production extends its approach towards the most unique renewable processing characteristics. This imparts requirement of very simple operation conditions responsible for non-toxic waste generation, thus sometimes availing the agro-industrial residues. The vital substitute is a by-product of fruit and vegetable waste which is a prospective resource off-flavor production.

5.9 Conclusion

There have been growing remarks and evidences of various value-added food products developed from the by-production of the fruit and vegetable residues. Among which the majority have been significantly implemented for the various cardinal purposes such that their value is unhampered and not being adversely affected. However, their production can itself be a challenge to the manufacturer for the reason of imparting additional cost to the processing value and adding to the economical values or the overall cost. Howbeit, in the meantime these value-added product development may also lead to a remarkable growth in terms of the profit, but the identification, selection, manufacturing itself pertains to a major challenge. This depends on the production practices, utilization on the agro-economic level, and the least threats imparted to the global environment.

References

Angulo J, Mahecha L, Yepesi SA, Yepesi AM, Bustamante G, Jaramillo H (2012) Nutritional evaluation of fruit and vegetable waste as feedstuff for diets of lactating Holstein cows. J Environ Manage 95:S210–S214

Arvanitoyannis IS, Varzakas TH (2008) Vegetable waste treatment, comparison and critical presentation of methodologies. Crit Rev Food Sci Nutr 48(3):205–247

Aura AM, Holopainen Mantila U, Sibakov J, Kössö T, Mokkila M, Kaisa P (2015) Bilberry and bilberry press cake as sources of dietary fibre. J Food Nutr Res 59(1):28367

Ayala-Zavala JF, Vega-Vega V, Rosas-Domínguezi C, Palafox-Carlo H, Villa-Rodriguez JA, Siddiqui MW (2011) Agro-industrial potential of exotic fruit byproducts as a source of food additives. Food Res Int 44(7):1866–1874

Banerjee JN, Singh R, Vijayaraghavan R, MacFarlane D, Patti AF, Arora A (2017) Bioactives from fruit processing wastes: green approaches to valuable chemicals. Food Chem 225:10–22

Barbosa-Martín E, Chel-Guerrero L, Gonzalez-Mondrag-on E, Betancur-Ancona D (2016) Chemical and technological properties of avocado (Persea Americana Mill.) seed fibrous residues. Food Bioprod Process 100:457–463

Bobinaite R, Viskelis P, Buskienee L (2010) Extraction of phenolic compounds from raspberry press cake Scientific Works of the Institute of Horticulture, Lithuanian Research Centre for Agriculture and Forestry and Lithuanian University of Agriculture: SODININKYSTĖ IR DARŽININKYSTĖ:29(2)

Bobinaite R, Viskelis P, Bobinas C, Miezeliene A, Alencikiene G, Venskutonis PV (2016) Raspberry marc extracts increase antioxidative potential, ellagic acid, ellagitannin and anthocyanin concentrations in fruit purees. Food Sci Technol 66:460–467

Cassani L, Tomadoni B, Viacava G, Ponce A, Moreira MR (2016) Enhancing quality attributes of fibre-enriched strawberry juice by application of vanillin or geraniol. LWT Int J Food Sci Technol 72:90–98

Chacko CM, Estherlydia D (2014) Antimicrobial evaluation of jams made from indigenous fruit peels. Int J Curr Adv Res 2(1):202–207

Felipe LO, de Oliveira AM, Bicas JL (2017) Bioaromas—perspectives for sustainable development. Trends Food Sci Technol:141–153

Garcia YD, Valles BS, Lobo AP (2009) Phenolic and antioxidant composition of by-products from the cider industry apple pomace. Food Chem 4:731–738

Goldmeyer B, Pena NG, Melo A, da Rosa CS (2014) Physicochemical characteristics and technological functional properties of fermented blue berry pomace and their flours. Revista Brasileira De Fruti cultura 36(4):980–987

Henriquez C, Speisky H, Chiffelle I, Valenzuela T, Araya M, Simpson R (2010) Development of an ingredient containing apple peel as a source of polyphenols and dietary fiber. J. Food Sci 75(6): H172–H181

Hwang JK, Kim CJ, Kim CJ (1998) Extrusion of apple pomace facilitates pectin extraction. J Food Sci 63(5):841–844

Kowalska H, Czajkowska K, Cichowska J, Lenart A (2017) What's new in biopotential of fruit and vegetable by-products applied in the food processing industry Faculty of Food Sciences, Department of Food Engineering and Process Management. S GGW 159c Nowoursynowska St:02-776 Warsaw Poland

Kunzek H, Müller S, Vetter S, Godeck R (2002) The significance of physico chemical properties of plant cell wall materials for the development of innovative food products. Eur Food Res Technol 214(5):361–376

Larrauri JA (1999) New approaches in the preparation of high dietary fibre powders from fruit by-products. Trends Food Sci Technol 10(1):3–8

Mantzouridou FT, Paraskevopoulou A, Lalou S (2015) Yeast flavour production by solid state fermentation of orange peel waste. Biochem Eng J 101:1–8

Masoodi FA, Sharma B, Chauhan GS (2002) Use of apple pomace as a source of dietary fiber in cakes. Plant Food Hum Nutr 57(2):121–128

Oliveira VM, Carraro E, Auler ME, Khalilc NM (2015) Quercetin and rutin as potential agents antifungal against Cryptococcus spp. Braz J Biol 76(4):1029–1034

Parra AF, Ribotta PD, Ferrero C (2015) Apple pomace in gluten-free formulations: Effect on rheology and product quality. Int J Food Sci Technol 50(3):682–690

Plazzotta S, Manzocco L, Nicoli MC (2017) Fruit and vegetable waste management and the challenge of fresh-cut salad. Trends Food Sci Technol 63:51–59

Raut JS, Karuppayil SM (2014) A status review on the medicinal properties of essential oils. Ind Crops Prod 62:250–264

Rice-Evans C, Miller N, Pagang G (1997) Antioxidant properties of phenolic compounds. Trends Plant Sci 2(4):152–159

Rodríguez-Carpena JG, Morcuende D, Andrade J, Kylli P, Estevez M (2011) Avocado (Persea Americana Mill) phenolics in vitro antioxidant and antimicrobial activities and inhibition of lipid and protein oxidation in porcine patties. J Agric Food Chem 59(10):5625–5635

Royer G, Madieta E, Symoneaux R, Jourjo F (2006) Preliminary study of the production of apple pomace and quince jelly LWT. Int J Food Sci Technol 39(9):1022–1025

Rupasinghe HP, Wang LX, Huber GM, Pitts NL (2008) Effect of baking on dietary fibre and phenolics of muffins incorporated with apple skin powder. Food Chem 107(3):1217–1224

Rupasinghe HPV, Wang LX, Pitts NL, Astatkie T (2009) Baking and sensory characteristics of muffins incorporated with apple skin powder. J. Food Qual 32(6):685–694

Saavedra MJ, Aires A, Dias C, Almeida JA, De Vasconcelos M, Santos P (2015) Evaluation of the potential of squash pumpkin by-products (seeds and shell) as sources of antioxidant and bioactive compounds. J Food Sci Technol Mys 52(2):1008–1015

Sanz-Puig M, Moreno PM, Pina-Perez C, Rodrigo D, Martínez A (2017) Combined effect of high hydrostatic pressure (HHP) and antimicrobial from agro-industrial by-products against S Typhimurium. LWT Int J Food Sci Technol 77:126–133

Sarkar A, Kaul P (2014) Evaluation of tomato processing by-products A comparative study in a pilot scale setup. J. Food Process Eng 37(3):299–307

Schieber A, Stintzing FC, Carle R (2001) By-products of plant food processing as a source of functional compounds — recent developments. Trends Food Sci Technol 12:401–413

Schieber A, Berardini N, Carle R (2003a) Identification of flavonol and xanthone glycosides from mango (Mangifera indica L. Cv. "Tommy Atkins") peels by high-performance liquid chromatography-electrospray ionization mass spectrometry. J Agric Food Chem 51 (17):5006–5011

Schieber A, Hilt P, Streker P, Endress H, Rentschler C, Carle R (2003b) A new process for the combined recovery of pectin and phenolic compounds from apple pomace. Innov Food Sci Emerg 4:99–107

Sharma PC, Gupta A, Issar K (2016) Effect of packaging and storage on dried apple pomace and fiber extracted from pomace. J Food Process Preserv 12913:1745–4549

Sudha ML, Dharmesh SM, Pynam H, Bhimangoude SV, Eipson SW, Somasundaram R (2016) Antioxidant and cyto/DNA protective properties of apple pomace enriched bakery products. Int J Food Sci Technol 53(4):1909–1918

Surek E, Nilufer-Erdil D (2016) Phenolic contents antioxidant activities and potential bio accessibilities of industrial pomegranate nectar processing wastes. Int J Food Sci Technol 51 (1):231–239

Telrandhe UB, Kurmi R, Uplanchiwar V, Mansoori MH, Raj VJ, Jain K (2012) Nutraceuticals — a phenomenal resource in modern medicine. J Pharm Clin 2(1):179–195

Thakur M Singh K, Khedkar R (2020) Phytochemicals: extraction process, safety assessment, toxicological evaluations, and regulatory issues. In: Prakash B (ed) Functional and preservative properties of phytochemicals. Elsevier, pp 34–356

Vodnar DC, Calinoiu LF, Dulf FV, Stefanescu BE, Crisan D, Socaciu D (2017) Identification of the bioactive compounds and antioxidant antimutagenic and antimicrobial activities of thermally processed agro-industrial waste. Food Chem 231:131–140

Willaman JJ, Eskew RK (1948) Uses for vegetable wastes. United States Department of Agriculture, p 739

Zhuang S, Fu J, Powell C, Huang J, Xia Y, Yan R (2015) Production of medium chain volatile flavor esters in *Pichia pastoris* whole-cell biocatalysts with extracellular expression of *Saccharomyces cerevisiae* acyl-CoA: ethanol O-acyltransferase Eht1 or Eeb1. Springer plus https://doi.org/10.1186/s40064-015-1195-0

Phytochemicals from the Fruits and Vegetable Waste: Holistic and Sustainable Approach

6

Alok Mishra and Amrita Poonia

Abstract

Food by-products and wastes generate various environmental and health issues. However, the extraction of phytochemicals and bioactive compounds can be a sustainable and lucrative alternative. Fruit and vegetable wastes (FVWs) account for several million tonnes around the globe whilst FVWs are a potential source to many phytochemicals, notably carotenoids, phenolics, vitamins, enzymes, and essential oils. Intriguingly, phytochemical content in by-products sometimes exceeds the main edible parts. Phytochemicals positively influence human health whether consumed as a food component or as nutraceuticals. Numerous techniques are being tapped ranging from conventional to novel approaches. Recently, the focus on sustainable extraction techniques has been increased. The sustainable approach involves innovations in existing techniques, process intensification by combining methods, and search for alternative and green solvents. These approaches are based on greener concepts of extractions. However, their commercialization depends on various factors, viz. selectivity, nature of food matrix, extraction efficiency, operation costs, and time required. The phytochemical extracts from FVWs are of great economic interest for health and food industry. Besides, valorization of such waste may strengthen the local economy and help mitigate environmental issues.

A. Mishra · A. Poonia (✉)
Department of Dairy Science and Food Technology, Institute of Agricultural Sciences, Banaras Hindu University, Varanasi, India

M. Thakur et al. (eds.), *Sustainable Food Waste Management*,
https://doi.org/10.1007/978-981-15-8967-6_6

6.1 Introduction

India ranks 2nd in the production of fruits and vegetables estimating 284 MT in 2017–2018 according to Ministry of Agriculture & Farmers' Welfare. Fruits and vegetables have high nutritional value and are characterized as the simplest form of functional foods due to various bioactive components present in them (Day et al. 2009). High consumption and increased processing of fruits and vegetables have led to significantly high losses and waste of their usable components as well as by-products. Inedible products from vegetables and some fruits amount to 25% and 30%, respectively (Ajila et al. 2007, 2010). It is to be noted that "losses" are the unintended outcome of food production and supply chain systems, however, "waste" accounts for food that is fit for human consumption instead it is not consumed and discarded thereof (FAO 2014). The fruit and vegetable waste (FVWs) comprise trimmings, stems, bran, seeds, peelings, shells, and residues which are left behind after extraction of valuables like juice, oil, starch, and sugar (Kumar et al. 2017). This horticultural waste which is conventionally regarded as "general waste" makes it an ignored altogether (Banerjee et al. 2017). However such wastes contain various phytochemicals that can be utilized in the production of food processing as functional food ingredients or in therapeutic and pharmacological preparations (Baiano 2014).

Phytochemicals (plant chemicals) are a range of chemicals that naturally occur in plants which serve various functions in them. They may be responsible for some bio-activity when ingested by humans. These phytochemicals often help prevent various body ailments and hence help maintain good health (Saxena et al. 2013; Cao et al. 2017). Phytochemicals frequently referred to as phytonutrients are found in foods like fruits, vegetables, nuts and seeds, legumes, whole grain products, tea, and dark chocolate (Singh and Chaudhuri 2018). Out of numerous phytochemicals, only a few have been isolated from plants and got identified (Woźniak et al. 2016). The phytochemicals most found in food include polyphenols, flavonoids, carotenoids, isoflavones, lignans, coumarins, indoles, organosulphur, phenolic acids, catechins, saponins, procyanidins, anthraquinones, ginsenosides, phenylpropanoids, stilbenoids, isothiocyanates, and so on (Kennedy and Wightman 2011; Zhao et al. 2018).

FVWs not only accounts for the waste of food but also considers a waste of critical resources like arable land, water, manure and fertilizers, labour, and energy (Venkat 2011). In the environment, they get degraded due to inappropriate waste management practices and lead to the production of pollutants and foul gases. It is to be understood that waste reduction leads to increased profit, reduced liabilities, less usage of water, and saves energy. Thus, some of the waste may be unavoidable but proper utilization of FVWs should be established to ensure the mitigation of environmental problems and to promote enriching the foods with phytochemicals from these wastes toward improved human health (Sagar et al. 2018). The recovery of phytochemicals from FVWs is a practical approach that will not only reduce the problem of waste disposal but pave a new way for functional food and nutraceutical production.

6.2 Phytochemicals in Fruit and Vegetable Waste

A varied range of secondary metabolites is synthesized by plants to serve hormonal functions in addition to natural defence mechanisms against foreign infections imparting them specific aroma, flavour, and colour. Besides, these phytochemicals can also act as inhibitors, cofactors, scavengers, enhancers of nutrients absorption, fermentation substrates, etc. (Zhang et al. 2015; Martillanes et al. 2018). Studies have revealed that these phytochemicals are beneficial in reducing risks associated with several diseases such as cancer and cardiovascular diseases (Joshi et al. 2012). In recent years, various phytochemicals have been investigated and categorized depending on their structural characteristics (Table 6.1).

6.2.1 Terpenoids

Terpenes (isoprenoids) are the leading set of phytonutrients which are present in grains, soy plants, and green foods (Rao and Rathod 2017). Terpenoids are like terpenes having been made of five-carbon isoprene units which are assembled and modified in different ways and are based on isopentane skeleton (Dillard and German 2000). These are biosynthesized in plants serving various primary functions. Terpenoids of interest in food are discussed below:

Monoterpenoids and sesquiterpenoids are chief constituents of various essential oils, viz. limonene, geraniol, eugenol, myrcene, ocimene, carvacrol, linalool, citronellal, citral, carvone, etc. These are obtained from tissues and sap of certain plants (Singh 2007). Citrus processing residues are chief by-product source along with clove and ginger (Martillanes et al. 2018). Carotenoids are fat-soluble colour producing pigments ranging from yellow to deep red possessing antioxidant and immunostimulatory properties (Krinsky and Yeum 2003). Chemically, they are differentiated into two main groups: carotenes, hydrocarbon in nature, and xanthophylls, oxygenated derivatives of carotenes. α, β, γ-carotenes and lycopene are among carotenes which can be found in reddish and orange sources like tomato pomace and carrot waste. Cryptoxanthin, zeaxanthin, fucoxanthin, and lutein are among xanthophylls which can be obtained from spinach residues (Badui Dergal 2006; Rao and Rathod 2017).

Terpenoids are sought for their aromatic qualities and are significantly used in ethnomedicinal and herbal preparations. In animals, these compounds are utilized for various hormonal as well as growth regulatory functions such as vitamin A synthesis. These molecules when present in animal tissues protect from certain diseases related to growth dysregulation and chronic damages (Dorta et al. 2012). Terpenoids react with free radicals by splitting themselves into fatty membranes owing to their elongated carbon side chain (Krinsky and Yeum 2003).

Table 6.1 Major categories of phytochemicals and their acquisition sources from fruit and vegetable waste

Category	Sub-category	FVWs sources	Bio-activity	Utilization
Terpenoids	Mono- and sesqui-terpenoid	Citric peels soy, chestnuts, seeds	Aroma, digestion, antiseptic, antioxidant, cardiovascular health	Various formulations
	Saponins	Seeds	Antibacterial	Antibacterial formulations
	Carotenoids	Carrot, watermelon, tomato and spinach wastes, pumpkin seeds	Colour, provitamin-A, prostate cancer	Anticancer, anti-ageing
	Xanthophylls	Green vegetables	Age-related macular degeneration	Anti-ageing
Phenolic compounds	Flavonoids: Flavanols, anthocyanins, flavanones, isoflavones	Onion skin waste, citric peels, grape skins, soursop peel, olive leaf, soybean flour	Antioxidant, colour, astringent, enzyme inhibition, phyto-oestrogen	Natural preservatives, anticancer agents, cardioprotective agents
	Lignans and xyloglucans	Sesame cake, watermelon rinds	Phyto-oestrogen, dietary fibre, food additives	Fat substituents, anti-obesity
	Stilbenes	Wine pomace	Heart disease	Cardioprotective agents
	Polyphenols	Pupunha peel, cocoa peel, passion fruit, coffee peel, and spent coffee grounds	Antioxidant	Natural preservatives, anticancer agents
	Phenolic acids	Pomegranate and grape by-products, rice bran, papaya peel	Antioxidant	Natural preservatives, anticancer agents
	Coumarins	Apiaceous waste	Photosensitization	Chronic infections
Organosulphur compounds	Thiosulfinates	*Allium* waste	Aroma, antiseptic	Food preservation
	Glucosinolates	Cauliflower and broccoli by-product	Aroma/burning taste, thyroid hypertrophy, inhibition of *Helicobacter pylori*	Treatment of some serious diseases
Nitrogen alkaloids	Betalains	Caryophyllaceae	Colour	Natural colour
	Capsaicinoids	Capsicum waste	Burning taste, analgesic	Pharma

Pham (2017), Banerjee et al. (2017), Renard (2018), Martillanes et al. (2018), Torres-Valenzuela et al. (2020)

6.2.2 Polyphenols

Phenolic compounds are the largest class of phytochemicals offering diverse biological functions. These are secondary metabolites which act by reducing activity, directly by free radical scavenging, and indirectly by chelation of prooxidant metal ions (Nair 2015). Among dietary phenolics are the phenolic acids, polyphenols, and flavonoids which are of great interest (Rao and Rathod 2017). The basic structure of phenols has one or more aromatic rings in addition to one or more hydroxyl groups (Balasundram et al. 2006). These are present in the peel, rind, and seeds of FVWs. Phenolic compounds are good antioxidants and are well known to possess free-radical scavenging activity (Martillanes et al. 2018). Due to their ability to scavenge free radicals, they can be used in food preservation. Of all the phenolics, flavonoids are studied much having about 4000 structure known (Xiao 2017). Among flavonoids are the flavanols, flavanones, isoflavones, anthocyanidins, and flavones. They are antioxidant, provide relief from oxidative stress, and help in modulating living-body activities (Nair 2015; Terahara 2015). Phenols are present in by-products obtained during the processing of different vegetable oils, coffee, wine, juices, and cereals (Banerjee et al. 2017). Some common phenolics found in FVWs are quercetin in apples, catechins in tea and cocoa, hesperidin in citrus fruits, curcumins in turmeric, etc. (Rao and Rathod 2017).

6.2.3 Nitrogen-Containing Alkaloids

Nitrogen alkaloids are water-soluble having one or more atom of nitrogen and possess biological activity. In the human body, they interact with neurotransmitters and produce various physiological as well as psychological responses (Aniszewski 2015). Piperine (pepper), caffeine (coffee), and theobromine (cocoa) are some alkaloids of interest present in foods (Dillard and German 2000).

6.2.4 Organosulphur Compounds

The organosulphur compounds are thiols having sulphur in their basic structure that include the glucosinolates and allyl sulphides. Glucosinolates, a class of thioglucosides, are the metabolites characteristically present in the cruciferous vegetables, mustard, broccoli, and radish. These are derived from amino acids and glucose (Dillard and German 2000; Devi and Thangam 2010; Pažitná et al. 2014). Glucosinolates activate liver detoxification enzymes and are anti-cancerous (Verkerk et al. 2009).

Allyl sulphides are present in waste of garlic, onions, leeks, and chives. They possess antioxidant and anti-cancerous properties (Williams et al. 2013; Martillanes et al. 2018).

6.3 Extraction: Opportunities and Challenges

The extraction is of renewed interest to obtain functional or bioactive components for food uses or as food supplements. Phytochemicals are extracted to enhance or improve colour, antioxidant, texture, aroma, or taste of the food product (Renard 2018). Though there are several methods investigated for extraction to obtain phytochemical compounds, their selection depends on the type of food matrix (solid or liquid), nature of molecule to be extracted, and the utilization of the extract so obtained (Carciochi et al. 2017). Other factors considered during the extraction process are temperature, plant part, pressure, and type of solvent. The compounds are characterized after getting identified from stem, leaves, flower, and fruits (Hernández et al. 2009).

The extraction methods can be classified into two main categories: conventional and novel techniques. The conventional methods are disadvantageous over novel ones because of longer extraction times, high energy consumption, high solvent consumption, and risk of thermal degradation of heat-labile components (Kadam et al. 2013). The conventional methods are difficult to scale up due to the perishability and bulkiness of the FVWs. The selective nature of extraction methods makes the recovery of phytochemicals from FVWs a challenging task. The extraction rate and performance remain a major concern among the various factors affecting the method selection process (Banerjee et al. 2017).

Sample preparation is also a key factor in the determination of the type and extent of phytochemical compounds to be extracted (Sagar et al. 2018). If not processed immediately once collected, raw materials should be stored at a temperature below -18 ± 2 °C to reduce oxidation besides degradation of compounds due to undesirable conditions; for instance, high temperature, enzymes, or UV-radiation from sunlight. In sample preparation, the drying process is crucial as the enzymes responsible for the degradation of compounds of interest get degraded. The microbial growth rate also gets decreased due to drying (Dorta et al. 2012). Likewise, drying helps to lessen the cost of transportation and preservation by decreasing moisture content along with volume as well as weight (Pham et al. 2015). A flowchart of general steps of phytochemical extraction from FVWs is given in Fig. 6.1. Today, several methods of drying are being applied in the preparation of dried materials: sun-drying, low-temperature air drying, hot air drying, infrared drying, microwave drying, vacuum drying, microwave vacuum drying, and freeze-drying (Pham 2017).

6.4 Conventional Techniques of Extraction

Conventional methods of extraction are in service for a long time and hence are also known as classical methods. The basis of these extraction techniques is the solvent assisted extraction and the application of heat and/or their combination (Khoddami et al. 2013). Some improvements have been made in these methods such as temperature is increased for better dissolution as well as to lower solvent viscosity. Besides,

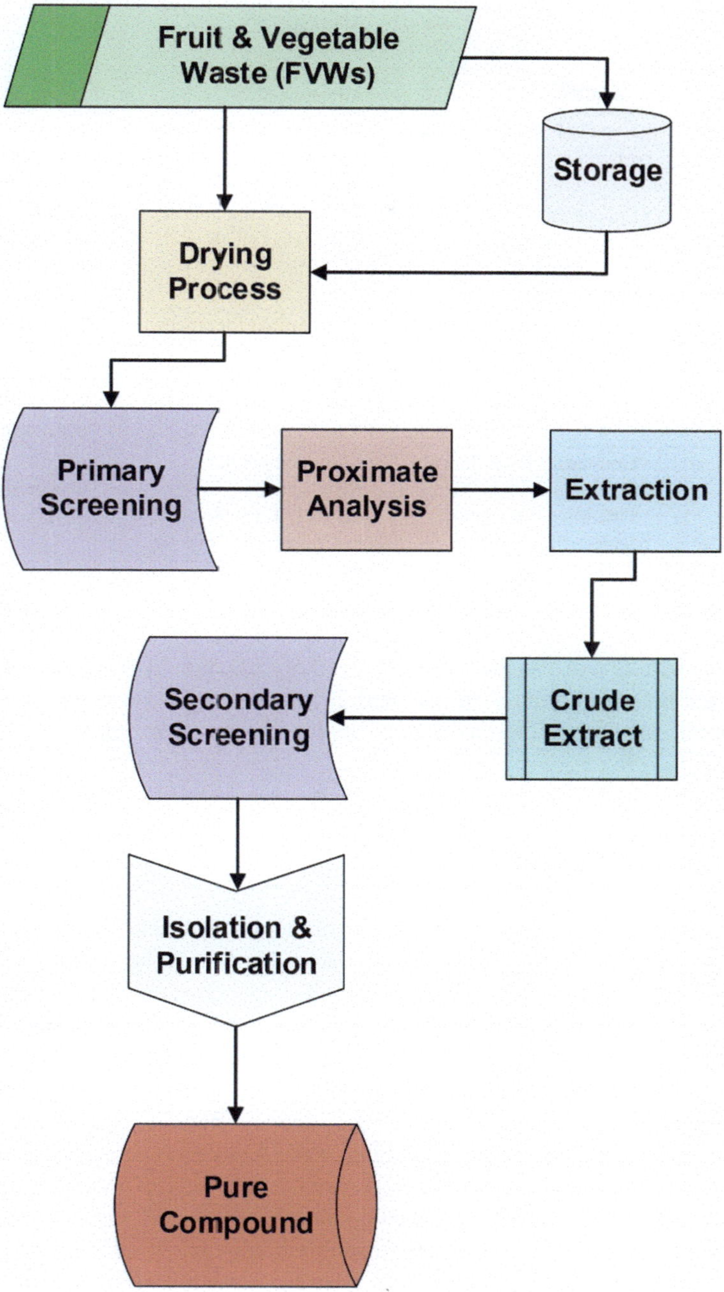

Fig. 6.1 General process for extraction of phytochemicals from fruit and vegetable wastes

Table 6.2 Conventional techniques used for the extraction of phytochemicals from fruit and vegetable waste

Extraction technique	Applications	Advantages	Limitations
Maceration	• Essential oils from plant parts.	• Low investment cost; • Modulation of selectivity by solvent choice.	• Longer processing time; • Low recovery.
Heating reflux, Soxhlet	• Epicatechin from seed-coat of Thai tamarind. • Phenolics from pistachio hull.	• Low investment cost; • Increased yields.	• High temperature; • Toxicity of solvent.
Hydro-distillation	• The oldest and simplest technique for extracting essential oils.	• Best suited for small-scale industry; • Different options of operations.	• Not suited for heat-labile compounds; • Time-consuming .
Supercritical fluid extraction	• Coumarins from the peel of Citrus maxima. • Phenolics from grape bagasse.	• Low temperature; • High yields; • Mostly for molecules of low polarity but can be modulated.	• High investment costs.

Kumar et al. (2017), Pham (2017), Renard (2018), Sagar et al. (2018)

decreasing particle size also increases the speed of extraction by facilitating the mass transfer and thus increasing yield (Renard 2018). A comparison between different conventional methods along with their application-based examples is given in Table 6.2.

6.4.1 Solvent Extraction

Solvent extraction is a very popular and one of the widely used classical techniques of extraction used in the food industry. Extraction is done in heated reflux (having advantage of concentration equilibria) or a Soxhlet apparatus (provides an advantage by separating the soluble fractions from the insoluble ones). Water, hexane, chloroform, ethanol, acetone, etc., are commonly used solvents. Essential oils and fatty acids besides other important polar or non-polar phytochemicals are extracted using this technique (Kaufmann and Christen 2002; Rao and Rathod 2017). In the solvent extraction method, the raw material is suitably sized and exposed to various organic solvents, which takes up solvable components of the interest (Vyas et al. 2009, 2014). Samples are centrifuged followed by filtration to remove any solid residue so that the extract could be utilized as a food additive, supplement, or in the preparation of functional foods (Kumar et al. 2017). Solvent extraction offers low processing cost and ease of operation making it very beneficial when compared to other methods. The disadvantage of this method is the use of toxic solvents which require steps of evaporation and concentration for recovery. Other disadvantages are the use of large amounts of solvent and long periods to carry out extraction. Additionally, there is a huge possibility of thermal degradation of volatile components that cannot be ignored because of high temperatures during the long intervals of extraction (Szentmihályi et al. 2002).

6.4.2 Maceration

Most of the classical extraction methods of phytochemicals rely on maceration, employing intense stirring, using a solvent of suitable polarity (Galanakis 2012). Maceration is a popular technique for low-cost extraction of phytochemicals and comprised of several steps (Joscelyne 2009). However, the process is often time-consuming and may take hours to many days (Margeretha et al. 2012). Sample preparation starts with complete grinding of FVW into tiny particle size to facilitate appropriate mixing of solvent. Subsequently, an apt quantity of the solvent (termed as menstruum) is emptied into a closed container. The liquid is then discarded, attaining the prepared solution in a large amount by pressing the solid residue obtained from this extraction process. Finally, filtration is used to separate the pressed liquid to remove impurities. At times, maceration is accompanied with stirring to surge the extraction in two possible ways: (a) by increasing the diffusion and (b) disposing of the concentrated solution from the surface and adding new solvent to the menstruum (Sagar et al. 2018).

6.4.3 Hydro-Distillation

Hydro-distillation, also one of the classical techniques, is used before dehydrating the sample. There are three different types of hydro-distillation, namely, water distillation, water and steam distillation, and direct steam distillation (Vankar 2004). The sample is packed in a static compartment and then boiled by adding a sufficient amount of water or steam as an alternative. Hot-water and steam, both work as the effective agents for the recovery of phytochemicals from the food matrix. The indirect water-cooling method is used for condensation of vapour mixture which then passes through a separator where the phytochemical compounds split automatically from the water (Silva et al. 2005). Hydro-distillation, thus, includes three significant physicochemical processes: hydro-diffusion, hydrolysis, and thermal decomposition. The disadvantage of this method is heat-labile compounds are prone to be degraded or lost at high extraction temperatures and thus is not suited for such compounds (Sagar et al. 2018).

6.5 Novel Approaches to Extraction

Limitations of conventional techniques compelled researches to look for newer techniques that offer easy extraction with high purity, inexpensive and non-toxic solvents, shorter time of extraction, high extraction selectivity, and protection of

Table 6.3 Different novel approaches of phytochemical extraction from fruits and vegetable waste

Extraction technique	Applications	Advantages	Limitations
Supercritical fluid extraction	• Coumarins from the peel of *Citrus maxima*. • Phenolics from grape bagasse.	• Low temperature; • High yields; mostly for molecules of low polarity but can be modulated.	High investment costs
Microwave-assisted extraction (MAE)	• Polyphenols from apple pomace. • Antioxidants from grape seeds.	• Reduced processing time. • Reduced solvent use.	• Locally high temperatures; • Toxic polar solvents.
Ultrasound-assisted extraction (UAE)	• Anthocyanins from blueberry pomace. • Naringin and hesperidin from orange peel.	• Reduced processing time. • Low temperature.	• Swelling of plant material.
Pressurized liquid extraction (PLE)	• Carotenoids from pressed palm fibre. • Antimicrobials from sea buckthorn seed.	• Reduced processing time. • Reduced solvent use.	• Investment costs; • Temperature; • Low throughput.
Pulsed electric fields (PEF)	• Anthocyanins from grape pomace. • Polyphenols from vine shoots.	• Reduced processing time. • Reduced solvent use.	• Requires conductivity; • Enzyme activity hinders the process.
Enzyme-assisted extraction (EAE)	• Lycopene from tomato processing waste.	• Facilitated extraction from a plant tissue.	• Additional long operation in wet conditions.
High voltage electrical discharges (HVED)	• Anthocyanins from grape pomace. • Polyphenols from grape seeds.	• Low energy. • Less time and solvent use. • Low diffusion temperature.	• Less selective compared with PEF.

Kumar et al. (2017), Pham (2017), Renard (2018), and Sagar et al. (2018)

heat-labile compounds from elevated temperatures during extraction (Putnik et al. 2019a). Numerous emerging techniques have been developed considering these aspects which are discussed here (Table 6.3).

6.5.1 Supercritical Fluid Extraction (SFE)

Supercritical fluid extraction (SFE) is an extraction technique widely used for compounds such as plants, algae, or food by-products. This technique uses solvent in its supercritical condition, i.e., the intermediate stage between a gas and a liquid which is obtained by subjecting a substance at a temperature and pressure higher than the critical point (Martillanes et al. 2018). At such stage, they are known as supercritical fluids possessing gas like properties of diffusion, viscosity, and surface tension along with liquid-like properties of solvation power and the density. Due to such properties, SFE is a suitable technique for extracting of phytochemicals rapidly and with maximum output (Azmir et al. 2013). In most of the cases, CO_2 is used as a supercritical fluid due to its GRAS (generally recognized as safe) status. Supercritical carbon dioxide offers an attractive alternative to the organic solvents being non-toxic, non-explosive, and inexpensive (Wang and Weller 2006). CO_2 is an inert gas and allows extraction of compounds at near room temperature (31 °C), which is appropriate for thermolabile compounds and low critical pressure (74 bar). Use of a co-solvent, such as n-hexane, isopropanol, ethanol, methanol, dichloromethane, or acetonitrile, is often necessitated to increase the polarity of solvent mixture (CO_2/co-solvent) while extracting polar compounds. It is due to the non-polar nature of CO_2 (Azmir et al. 2013). The factors affecting the extraction efficiency are extraction temperature, extraction time, pressure, solvent flow rate, co-solvent and its flow rate, porosity, particle size, density, bed diameter and its height (Bimakr et al. 2012).

The extraction process consists of 4-stages: pressurization stage, adjustment of temperature, extraction step, and separation stage (Carciochi et al. 2017). Initially, the raw material is placed in an extraction container equipped with temperature and pressure controllers to maintain the required conditions. After this, the pressure is applied to the extraction container with the fluid by a pump. The fluid and dissolved compounds are then transported to separators and the yield is collected using a tap which is located in the lower portion of the separators. In the end, the fluid is restored and recycled or released (Chemat et al. 2020). Advantages of SFE are easy separation of solvent and the matrix avoiding temperature rise to remove solvent and use of non-toxic solvent and its reusability. Disadvantages of SFE are the high initial costs.

SFE extraction has been applied in the extraction of lycopene from tomato peels and β-carotene from carrot peels (Kehili et al. 2017; de Andrade Lima et al. 2018). SFE extraction has also been tested for recovery of phenol from apple pomace, mango by-products, cacao pod-husk, or sour cherry pomace (Meneses et al. 2015; Woźniak et al. 2016; Valadez-Carmona et al. 2018; Ferrentino et al. 2018).

6.5.2 Microwave-Assisted Extraction (MAE)

Microwave-assisted extraction (MAE), a relatively newer extraction technique, combines traditional solvent extraction with microwave irradiation (Pimentel-Moral et al. 2018; Pettinato et al. 2019; Wen et al. 2019). The extraction of compounds soluble in specific fluids (liquids or gases) is improved using microwave energy because of alterations in the cellular structure due to electromagnetic waves resulting in an improved rate of mass transfer (Kaderides et al. 2019). The microwaves electromagnetic field ranges from 300 MHz to 300 GHz. Out of which the two frequencies (915 MHz) and (2450 MHz) are commonly used for heating in industrial and domestic settings (Wen et al. 2019). They are composed of two perpendicular fields: electric field and magnetic field. The heating principle is based upon the immediate effects of microwaves on polar materials (Azmir et al. 2013). The energy of microwaves is transformed into heat energy through ionic conduction and dipole rotation mechanisms (Jain et al. 2009). The generation of heat occurs due to the resistance offered by the medium during the ionic flow or conduction and the collision among molecule created by a random change in the direction of the side-ions aligned towards the direction of the field. The MAE technique comprises three subsequent steps (Alupului et al. 2012). The solute molecules are split and separated from the sample matrix due to the increase in pressure and temperature, which is followed by solvent diffusion together with the sample matrix. Subsequently, the solutes are released into the solvent from the sample (Sagar et al. 2018).

The parameters that affect extraction yield are microwave power, extraction time, irradiation time, solvent concentration, the ratio of solvent to sample, and particle size. The use of microwave is limited in laboratory scale because of high initial expenditure, maintenance costs, and safety aspects (Ciriminna et al. 2016). Also, the thermolabile phytochemicals are prone to thermal degradation, further limiting the use of MAE. Advantages of MAE include reduced equipment size, large quantities of extract, high-temperature gradient, and momentary heating of sample (Cravotto et al. 2008).

6.5.3 Pulsed Electric Field (PEF)

Pulsed electric field (PEF) is a promising technique for recovery of compounds of interest from food waste and by-products (Deng et al. 2011). Principle phenomenon behind PEF is electroporation that breaks the structure of cell membranes by splitting the dipolar molecules based on charges (Vorobiev et al. 2004; Vorobiev and Lebovka 2006). The sample material is placed between two electrodes and PEF is applied. Pulse amplitude inside the PEF equipment ranges from 100–300 V/cm to 20–80 kV/cm (Koubaa et al. 2015). PEF treatment is done at ambient temperature (or slightly higher) and the treatment time is less than 1 s (μs to ms) (Barbosa-C et al. 2000). Thus, PEF minimizes the degradation of heat-labile compounds (Ade-Omowaye et al. 2001). PEF treatment is either applied in a batch mode or a continuous mode, which depends on the design of the treatment chamber (Puértolas et al. 2012). The effectiveness of the PEF relies on electric field strength, treatment temperature, specific energy input, and material properties (Heinz et al. 2003). The PEF technique is beneficial in improving the process of drying, diffusion, extraction, and pressing (Wen et al. 2019).

6.5.4 Enzyme-Assisted Extraction (EAE)

Enzymes are widely used for the recovery of bound metabolites from food waste as a novel pre-treatment method. EAE is a mild extraction technology which is eco-friendly and sustainable (Nadar et al. 2018; Zhu et al. 2018). Cell walls of fruits and vegetables contain polysaccharides like cellulose, hemicellulose, and pectin which are barriers to the intracellular substances. Enzymes such as pectinase, cellulase, β-glucanase, β-glucosidase, and xylanase help to degrade the structure of cell wall and depolymerize polysaccharides which facilitates the release of linked compounds (Moore et al. 2006; Singh et al. 2016). This prior treatment of the food matrix using the corresponding enzyme is followed by a process of solvent extraction (Özkan and Bilek 2015). Enzyme treatment, thus, enhances the performance of the extraction process. Enzyme-assisted extraction (EAE) is basically of two types: (a) enzyme-assisted cold pressing (EACP); (b) enzyme-assisted aqueous extraction (EAAE) (Latif and Anwar 2009). The key factors affecting the extraction process are extraction temperature, enzyme concentration, types of enzymes, the molecular size of plant materials, water proportion, and the hydrolysis time (Niranjan and Hanmoungjai 2004). Common food-grade enzymes are used in the EAE method making it a low-cost technique. Aqueous extraction of essential oils and other phytochemical compounds in the EAE method makes it an eco-friendly technique (Puri et al. 2012). EAE has been used for improved extractions method of carotenoids from tomato peels (Prokopov et al. 2017). The use of hydrolytic enzymes has been tested to improve the recovery of phenols (bound and free) from pomegranate peels (Esclapez et al. 2011). EAE has also been efficaciously applied for the polyphenol's extraction from underutilized watermelon rind, citrus peel, and ginger, cauliflower outer leaves winemaking by-products, among others (Martillanes et al. 2018).

6.5.5 Ultrasound-Assisted Extraction (UAE)

Ultrasound-assisted extraction (UAE) is considered as a simple, cheap, and more effective technique when compared to traditional extraction techniques. The UAE technique uses ultrasonic waves that agitate a sample immersed in an organic solvent. Ultrasound is a mechanical wave of frequency more than the audible range (>20,000 Hz) (Rostagno and Prado 2013). Ultrasound treatment creates, enlarges, and collapses gas bubbles which are dissolved in the solvent. Such physical processes generate acoustic cavitation which creates microbubbles bubbles which further act as new cavitation nuclei or get simply dissolved. As a result of this, a "shockwave" is generated which passes through the solvent and increases the mixing resulting in the disruption of the cell membrane. Ultrasound treatment also accelerates infiltration of the solvent into the food matrix which results in the increasing interaction between the superficial area of solid and liquid phase (Esclapez et al. 2011; Tiwari 2015; Putnik et al. 2019b). Following extraction, compounds with the solvent matrix are separated by centrifugation or filtration.

The factors affecting extraction yield along with the composition of the extract are power and time, temperature, the ratio of solvent to sample, and solvent type. At the laboratory scale, the commonly used UAE systems are Ultrasonic baths or probes; which either work in batch or flow mode (Martillanes et al. 2018).

The greatest advantage of UAE technique is the short time interval of the procedure as compared to conventional methods, plus reduced energy consumption and pollution. This lowers the production cost and the operational cost as compared to the cost of conventional procedures and gives the final product with high purity. Also, sonication can be used for extracting thermolabile compounds as the ultimate temperature is not very high. The disadvantage of UAE technique is a high initial investment at industrial levels (Romdhane and Gourdon 2002; Wang et al. 2008; Putnik et al. 2019b).

6.5.6 Pressurized Liquid Extraction (PLE)

Pressurized liquid extraction (PLE) is based on accelerated extraction using solvent by the application of high pressure. PLE combines solvent extraction at high pressures (1500–2000 psi) and temperatures (50–200 °C) for the quick and efficient extraction of compounds from solid matrices (Kaufmann and Christen 2002). Due to high pressure, the temperatures are raised above solvents atmospheric boiling point while it is still in a liquid state. The rise in temperature increases the solubility and mass transfer properties, and, hence, speeds up the extraction kinetics. Increasing the temperature decreases solvent viscosity and the surface tension leading to a greater penetration within the material matrix (Kha and Nguyen 2014). It also weakens the interactions between metabolites and the matrix and increases the solubility of compounds in the solvents. However, the PLE method can cause degradation of thermolabile compounds (Lozano-Sánchez et al. 2014).

PLE technique can be carried out in two modes: static and dynamic. Static extraction offers greater penetration of solvent into the sample spaces and hence is considered more efficient (Carciochi et al. 2017). The parameters affecting the process are extraction temperature, extraction time, pressure, solvent type, flow rate, the ratio of solid to solvent, and material particle size. The advantages of PLE include the shorter extraction times, the increased yields, and the reduced solvent usage. When organic solvents are replaced with water to be used for extraction using PLE, the process is called pressurized hot-water extraction, also known as, subcritical water extraction (Tunchaiyaphum et al. 2013).

6.5.7 High Voltage Electrical Discharges (HVED)

High voltage electrical discharges (HVED) technique is used to recover high added-value compounds from various food materials by the rupture of tissues due to the electrical breakdown (Boussetta and Vorobiev 2014). In an intense electrical field (up to 40 kV and 10 kA), the avalanche of electrons is turned to an initial point of streamer propagation via high voltage needle electrode to the grounded one. Various secondary phenomena get accompanied by the electrical breakdown, such as high-amplitude pressure shockwaves, creation of liquid turbulence, bubbles cavitation, and production of reactive species. All these lead to fragmentation of particles and the cell structure is damaged, further facilitating the release of the intracellular compounds. Air bubbles, initially present in water or formed due to local heating, will also be involved in accelerating the extraction process (Barba et al. 2016).

6.6 Sustainable Approaches to Extraction

There has been an interest in the search for alternatives to present-day extraction methods, procedure, and solvents which are environmentally friendly and are sustainable. These sustainable approaches which are often termed as "green extraction" are based on the concept of "green chemistry" and "green engineering" (Chemat et al. 2012). Chemat et al. (2012) describe green extraction as the extraction process which are innovated and designed in such a way that they will reduce consumption of energy, utilize alternative (green) solvents, harness renewable natural resources, are safer to use, yield high quality, and safe extract. The need of sustainable approaches arises to replace or eliminate inefficient processes, non-toxic reagents, unsustainable raw materials, and unsafe and low-quality extract/product (Wen et al. 2019; Chemat et al. 2019, 2020). The development of such efficient and sustainable extraction methods is also necessary to curb pollution as well as to reduce the operating costs (Wang et al. 2016; Wu et al. 2017).

Chemat et al. (2020) suggest that the processes innovated to be termed as sustainable should carry some specific characteristic features. These are enlisted below:

- Time consumption should be low.
- Reproducibility of the process should be high.
- Solvents should be "green" and minimally spent.
- The final product should be safe and of high quality.
- Consumption of energy should be considered very low when compared to conventional methods.

To achieve this goal, a set of 12 standardized principles each of green chemistry and green engineering have been accepted by the scientific community (Anastas and Warner 1998; Anastas and Zimmerman 2003). Based on these principles, six principles have been established for green extraction purposes (Chemat and Strube 2015; Chemat et al. 2019). These six principles are a key to sustainable extraction of phytochemicals from natural products including FVWs. The most acceptable approaches of sustainable extraction can be classified into three categories: (1) Innovative techniques; (2) Combination of techniques; and (3) Alternative solvents.

6.7 Innovative Techniques

Innovative techniques may be referred to as those techniques which confer "green engineering" concepts (Anastas and Zimmerman 2003). These innovative techniques can be utilized for pre-treatment processes or can be the extraction process itself. The ""novel technologies of extraction" discussed above fall into this category which may be utilized as such or with modification in the protocol, solvents, or unit operations to meet the criteria of "green extraction" (Chemat et al. 2019).

6.8 Combination of Techniques

Apart from developing new techniques, combined with others are microwave (MAE), ultrasound (UAE), and enzyme (EAE) assisted intensification of techniques is yet another approach for environmentally friendly and inexpensive extraction of phytochemicals (Chemat et al. 2019). Studies have been performed to combine the existing novel techniques with other conventional or non-conventional techniques to lower down the cost and maximize yield having good quality and non-toxic. Such efficient combinations can open new horizons of phytochemical extraction as no single technique alone can act as an ideal extraction method. Thus, ways need to be explored to maintain a balance between production costs, product quality, and the use of solvents (Wen et al. 2019). Some of the techniques commonly combined with others are microwave (MAE), ultrasound (UAE) and enzyme (EAE) assisted extraction methods. These techniques can be utilized in raw material preparation, pre-treatment step, enrichment steps, or enhancement of processing capacities such as improvement of mass transfer, enzymatic activity, quick heating up to the cold point, etc. Application-based examples of such combined techniques are given in Table 6.4.

6.9 Alternative Solvents

Typically, the organic solvents like hexane, ethanol, toluene, diethyl ether, and their aqueous solutions have been the foremost solvent phases for the phytochemical extraction since a very long time (Cvjetko Bubalo et al. 2015; Byrne et al. 2016). Though organic solvents offer many recognized gains, identification of their alternatives is of utmost importance for both human and environmental health. Besides being toxic, the volatility of organic solvents adds to risks of fire as well as explosion. Also, they can act as pollutants as they easily vaporize to the atmosphere. Thus, organic solvents are unsustainable affecting health, safety, and environment negatively (Vian et al. 2014, 2017).

Various alternative solvents have been identified and studied to serve the purpose of sustainable extraction. These solvents should neither be toxic nor volatile, can be

Table 6.4 Combination of techniques and their application for sustainable extraction of phytochemicals

Combined techniques	Phytochemical	FVW	Specification	References
US, enzyme	Lycopene	Tomato peel	Cellulase: 3%, 20 min; 50 °C; pH 5.0 US: 10 W Yield: 50%	Ladole et al. (2018)
US, enzyme	Lycopene	Tomato pomace	Pectinase: (9 μl/g), 60 min, 45 °C; pH 4.5 US: 50 W Yield: 38.73% (> UAE, EAE)	Amiri-Rigi et al. (2016)
MW, enzyme	Polyphenols	Waste peanuts shells	Cellulase: 0.81%, 2 h, 66 °C, pH 5.5 MW: 960, 2.6 min Yield: 1.75% (> UAE, EAE)	Zhang et al. (2013)
MW, enzyme, US	Antioxidants (phenols, flavonoids, and anthocyanin)	Berry juice by-product	Cellulase: 0.6%, 43 min, 66 °C, pH 4.5 MW: 420 W US: 800 W Yield: Higher than UAE, MAE, US-enzyme, MW-enzyme	Wu et al. (2015)
US, MW	Anthraquinones	*Heterophyllaea pustulata* Hook	UAE: 60 min, 50 W with benzene; as a pre-treatment MAE: 25 min, 900 W with ethylacetate	Vázquez et al. (2014)

US Ultrasound, MW Microwave

recycled, biologically degraded, and produced without involving high energy costs (Das et al. 2017).

This alternative solvent approach includes:

(a) Eco-solvents, e.g. glycerol, limonene;
(b) Natural solvents, e.g. vegetable oils, alcohols;
(c) Supercritical fluids, e.g. CO_2.
(d) Green solvents, e.g. ionic solvents, deep eutectic solvents, water;
(e) Solvent-free extraction.

The application of such alternative solvents for phytochemical extraction from FVWs has been tabulated in Table 6.5.

Table 6.5 Application of alternative green solvents for sustainable extraction of phytochemicals

Alternative solvent	Phytochemical	FVW	Specification	References
Ionic liquids (IL)	Levulinic acid	Rice husk	IL: $[C_4(mim)_2]$ $[(2HSO_4)H_2SO_4)_2]$ EC: 110 °C, 60 min EP: 47.52%	Khan et al. (2018)
	Oleanolic acid	Olive tree leaves	IL: $[C_{12}mim]Cl$ EC: 80 °C, 2 h EP: 47.52%	Cláudio et al. (2018)
Super critical fluids (SCF)	Phenolic compounds	Grape seeds	SCF: CO_2 EC: 40 °C, 30 MPa, 1.5 ml/min EP: 25 mg GAE/g	Pérez et al. (2015)
	Limonene	Orange peel	SCF: CO_2 EC: 50 °C, 40 MPa, 1.6 mL/min EP: 0.25%	Xhaxhiu and Wenclawiak (2015)
Subcritical water	Flavonoids	Apple by-products	Water EC: 125 °C, 10.3 MPa, 3 min EP: 1.8 μmol GAE/g	Plaza et al. (2013)
Deep eutectic solvents (DES)	Phenolic compounds	Lemon peels	DES: Glycerol: Choline chloride (ChCl) EC: 90 min, 80 °C, UAE with DES and 10% water EP: 53.76 mg GAE/g	Mouratoglou et al. (2016)
	Flavonoids	Grape skins	DES: ChCl:Proline: Malic acid EC: 50 min, 65 °C, UAE with DES and 25% water EP: 25 mg/g	Cvjetko Bubalo et al. (2016)
Bio-based eco-solvents (BES)	Carotenoids	Tomato waste (skin and seeds)	BES: Ethyl lactate EC: 30 min, 70 °C EP: 243 mg/kg	Strati and Oreopoulou (2011)
	Ellagic acid	Pomegranate peel	BES: Ethyl acetate EC: 6 h, Soxhlet extraction EP: 37.7–63.6 mg/g	Masci et al. (2016)
Supra-molecular solvents (SMS)	Betaine	Beet molasses	SMS: Triton X-114 EC: Surfactant 0.5% (w/v), molasses 27.5% (w/v), incubation 20 min, pH 6.1, extraction 30 min EP: 80%	Mohammadzadeh et al. (2018)

(continued)

Table 6.5 (continued)

Alternative solvent	Phytochemical	FVW	Specification	References
		Aloe peel	SMS: Triton X-114 EC: Surfactant 10% (w/v), NaCl 2.0% (w/v), 40 °C, pH 3.0, 30 min EP: 96.9%	Tan et al. (2012)

SSR Sample Solvent Ratio, EC Extraction Conditions, EP Extraction Performance, GAE Gallic Acid Equivalent, mim 1-alkyl-3-methylimidazolium cation

6.10 Applications of Phytochemical Residues

Phytochemicals are found in the foods originating from plants. The food industry is involved in the extraction and utilization of phytochemical compounds from waste generated during fruit and vegetable processing. This offers an economic advantage to them (da Silva et al. 2014). The extravagant properties of phytochemicals open a wide range of possibilities. Being health-related compounds, phytochemicals are known to lower the risk of developing many diseases such as cancer, cataracts, Alzheimer, and Parkinson, etc. These beneficial effects have been attributed mainly to their antioxidant and radical scavenging activities which can delay or inhibit the oxidation of DNA, proteins, and lipids (Martillanes et al. 2018).

The application of phytochemicals in different foods like fish and meat, vegetable oils, or packed food has been tested and antioxidant values obtained were like synthetic antioxidant, particularly of the flavonoids and hydroxycinnamic acids (Martínez-Valverde et al. 2002). These phytochemicals from fruit and vegetable by-products can be added in the diet directly, e.g. the consumption of green tea leaves (flavonoids), or indirectly, e.g. a food additive or supplement (Martillanes et al. 2018).

Alternatively, phytochemicals can be used in food processing by creating active packaging systems. Various films and coatings (proteins, polysaccharides, etc.) are being incorporated with a wide variety of plant extracts or essential oils to develop active packaging having antioxidant/antimicrobial properties which help in the improvement of food preservation method as in the preservation of cut-fruits, meat, and fish (Martillanes et al. 2018).

6.11 Conclusion

Fruit and vegetable wastes are a threat to the environment causing pollution. Instead, they can be harnessed to yield several phytochemicals that strengthen many biological activities. This chapter presents the major classes of phytochemicals obtained from fruit and vegetable wastes, their chemical nature, and health benefits associated with them. The extraction techniques, both conventional (solvent extraction, maceration, hydro-distillation, and SFE) and novel (MAE, UAE, EAE, PLE, PEF, and HVED), have been discussed. Novel techniques are considered sustainable and greener because of reduced or no energy requirements, less solvent, shorter extraction time, improved extraction performance. However, the selection of optimum technique for a particular raw depends on its physical characteristics and the chemical composition. Thus, phytochemical extraction from fruit and vegetable waste is essential for their valorization and help mitigate the risk of environmental pollution. These phytochemicals find their application in the food industry as an additive, supplement, or ingredient to produce nutraceuticals or functional foods.

Glossary

Additive A substance added to something in small quantities to improve or preserve it.

Cavitation The formation of bubbles in a liquid, typically by the movement of a propeller through it.

Distillation The action of purifying a liquid by a process of heating and cooling.

Electroporation The action or process of using a pulse of electricity to open the pores in the cell membranes briefly.

Green Extraction Processes with reduced or no energy consumption and using alternative solvents producing safe and good quality extract.

Hydrolysis The chemical breakdown of a compound due to reaction with water.

Maceration The extraction of a compound by allowing it to stand in contact with a solvent.

Neurotransmitters Any of a group of chemical agents released by neurons to stimulate neighbouring neurons or muscle or gland cells.

Phytochemical Any of various biologically active compounds found in plants.

Supercritical Above a critical threshold.

Supplement A thing added to something else to complete or enhance it.

Viscosity The state of being thick, sticky, and semi-fluid inconsistency due to internal friction.

References

Ade-Omowaye BIO, Angersbach A, Taiwo KA, Knorr D (2001) Use of pulsed electric field pre-treatment to improve dehydration characteristics of plant based foods. Trends Food Sci Technol 12:285–295

Ajila CM, Bhat SG, Rao UP (2007) Valuable components of raw and ripe peels from two Indian mango varieties. Food Chem 102:1006–1011

Ajila CM, Aalami M, Leelavathi K, Rao UP (2010) Mango peel powder: a potential source of antioxidant and dietary fiber in macaroni preparations. Innovative Food Sci Emerg Technol 11:219–224

Alupului A, Calinescu I, Lavric V (2012) Microwave extraction of active principles from medicinal plants. UPB Sci Bull Series B 74:129–142

Amiri-Rigi A, Abbasi S, Scanlon MG (2016) Enhanced lycopene extraction from tomato industrial waste using microemulsion technique: optimization of enzymatic and ultrasound pre-treatments. Innovative Food Sci Emerg Technol 35:160–167

Anastas PT, Warner JC (1998) Green chemistry: theory and practice. Oxford University Press, New York

Anastas PT, Zimmerman JB (2003) Principles of green engineering. Environ Sci Technol 37:94–101

Aniszewski T (2015) Alkaloids: chemistry, biology, ecology, and applications, 2nd edn. Elsevier

Azmir J, Zaidul ISM, Rahman MM, Sharif KM, Mohamed A, Sahena F, Jahurul MHA, Ghafoor K, Norulaini NAN, Omar AKM (2013) Techniques for extraction of bioactive compounds from plant materials: a review. J Food Eng 117:426–436

Badui Dergal S (2006) Química de los alimentos, 4th edn. Pearson Education

Baiano A (2014) Recovery of biomolecules from food wastes-a review. Molecules 19:14821–14842

Balasundram N, Sundram K, Samman S (2006) Phenolic compounds in plants and Agri-industrial by-products: antioxidant activity, occurrence, and potential uses. Food Chem 99:191–203

Banerjee J, Singh R, Vijayaraghavan R, MacFarlane D, Patti AF, Arora A (2017) Bioactives from fruit processing wastes: green approaches to valuable chemicals. Food Chem 225:10–22

Barba FJ, Zhu Z, Koubaa M, Sant'Ana AS, Orlien V (2016) Green alternative methods for the extraction of antioxidant bioactive compounds from winery wastes and by-products: a review. Trends Food Sci Technol 49:96–109

Barbosa-C AGV, Pierson MD, Zhang QH, Schaffner DW (2000) Pulsed electric fields. J Food Sci 65(S8):65–79

Bimakr M, Rahman RA, Ganjloo A, Taip FS, Salleh ML, Sarker MZI (2012) Optimization of supercritical carbon dioxide extraction of bioactive flavonoid compounds from spearmint (Mentha spicata L.) leaves by using response surface methodology. Food Bioprocess Technol 5:912–920

Boussetta N, Vorobiev E (2014) Extraction of valuable biocompounds assisted by high voltage electrical discharges: a review. C R Chim 17(3):197–203

Byrne FP, Jin S, Paggiola G, Petchey THM, Clark JH, Farmer TJ, Hunt AJ, Robert McElroy C, Sherwood J (2016) Tools and techniques for solvent selection: green solvent selection guides. Sustain Chem Process 4:7–24

Cao H, Chai TT, Wang X, Morais-Braga MFB, Yang JH, Wong FC, Wang R, Yao H, Cao J, Cornara L, Burlando B (2017) Phytochemicals from fern species: potential for medicine applications. Phytochem Rev 16(3):379–440

Carciochi RA, D'Alessandro LG, Vauchel P, Rodriguez MM, Nolasco SM, Dimitrov K (2017) Valorization of agrifood by-products by extracting valuable bioactive compounds using green processes. In: Grumezescu AM, Holban AM (eds) Ingredients extraction by physicochemical methods in food, vol 4. Academic Press, pp 191–228

Chemat F, Strube J (eds) (2015) Green extraction of natural products: theory and practice. John Wiley & Sons, Weinheim

Chemat F, Vian MA, Cravotto G (2012) Green extraction of natural products: concept and principles. Int J Mol Sci 13(7):8615–8627

Chemat F, Abert-Vian M, Fabiano-Tixier AS, Strube J, Uhlenbrock L, Gunjevic V, Cravotto G (2019) Green extraction of natural products. Origins, current status, and future challenges. TrAC Trends Anal Chem 118:248–263

Chemat F, Vian MA, Fabiano-Tixier AS, Nutrizio M, Jambrak AR, Munekata PE, Lorenzo JM, Barba FJ, Binello A, Cravotto G (2020) A review of sustainable and intensified techniques for extraction of food and natural products. Green Chem 22:2325–2353

Ciriminna R, Carnaroglio D, Delisi R, Arvati S, Tamburino A, Pagliaro M (2016) Industrial feasibility of natural products extraction with microwave technology. Chem Select 1 (3):549–555

Cláudio AFM, Cognigni A, de Faria ELP, Silvestre AJD, Zirbs R, Freire MG, Bica K (2018) Valorization of olive tree leaves: extraction of oleanolic acid using aqueous solutions of surface-active ionic liquids. Sep Purif Technol 204:30–37

Cravotto G, Boffa L, Mantegna S, Perego P, Avogadro M, Cintas P (2008) Improved extraction of vegetable oils under high-intensity ultrasound and/or microwaves. Ultrason Sonochem 15 (5):898–902

Cvjetko Bubalo M, Vidović S, Radojčić Redovniković I, Jokić S (2015) Green solvents for green technologies. J Chem Technol Biotechnol 90(9):1631–1639

Cvjetko Bubalo M, Ćurko N, Tomašević M, Kovačević Ganić K, Radojčić Redovniković I (2016) Green extraction of grape skin phenolics by using deep eutectic solvents. Food Chem 200:159–166

da Silva LMR, De Figueiredo EAT, Ricardo NMPS, Vieira IGP, De Figueiredo RW, Brasil IM, Gomes CL (2014) Quantification of bioactive compounds in pulps and by-products of tropical fruits from Brazil. Food Chem 143:398–404

Das S, Mondal A, Balasubramanian S (2017) Recent advances in modeling green solvents. Curr Opin Green Sustain Chem 5:37–43

Day L, Seymour RB, Pitts KF, Konczak I, Lundin L (2009) Incorporation of functional ingredients into foods. Trends Food Sci Technol 20(9):388–395

de Andrade Lima M, Charalampopoulos D, Chatzifragkou A (2018) Optimisation and modelling of supercritical CO2 extraction process of carotenoids from carrot peels. J Supercrit Fluids 133:94–102

Deng Q, Penner MH, Zhao Y (2011) Chemical composition of dietary fiber and polyphenols of five different varieties of wine grape pomace skins. Food Res Int 44(9):2712–2720

Devi JR, Thangam EB (2010) Extraction and separation of glucosinolates from brassica oleraceae var rubra. Adv Biol Res 4:309–313

Dillard CJ, German JB (2000) Phytochemicals: nutraceuticals and human health. J Sci Food Agric 80(12):1744–1756

Dorta E, Lobo MG, González M (2012) Using drying treatments to stabilise mango peel and seed: effect on antioxidant activity. LWT Food Sci Technol 45(2):261–268

Esclapez MD, García-Pérez JV, Mulet A, Cárcel JA (2011) Ultrasound-assisted extraction of natural products. Food Eng Rev 3(2):108

FAO (2014) Global initiative on food loss and waste reduction. In: Definitional framework of food losses and waste. FAO, Rome

Ferrentino G, Morozova K, Mosibo OK, Ramezani M, Scampicchio M (2018) Biorecovery of antioxidants from apple pomace by supercritical fluid extraction. J Clean Prod 186:253–261

Galanakis CM (2012) Recovery of high added-value components from food wastes: conventional, emerging technologies and commercialized applications. Trends Food Sci Technol 26(2):68–87

Heinz V, Toepfl S, Knorr D (2003) Impact of temperature on lethality and energy efficiency of apple juice pasteurization by pulsed electric fields treatment. Innovative Food Sci Emerg Technol 4(2):167–175

Hernández Y, Lobo MG, González M (2009) Factors affecting sample extraction in the liquid chromatographic determination of organic acids in papaya and pineapple. Food Chem 114 (2):734–741

Jain T, Jain V, Pandey R, Vyas A, Shukla SS (2009) Microwave assisted extraction for phytoconstituents–an overview. Asian J Res Chem 2(1):19–25

Joscelyne V (2009) Consequences of extended maceration for red wine colour and phenolics. PhD Thesis. University of Adelaide, Adelaide

Joshi VK, Kumar A, Kumar V (2012) Antimicrobial, antioxidant and phyto-chemicals from fruit and vegetable wastes: a review. Int J Food Ferment Technol 2(2):123–136

Kadam SU, Tiwari BK, Donnell PC (2013) Application of novel extraction technologies for bioactives from marine algae. J Agric Food Chem 61:4667–4675

Kaderides K, Papaoikonomou L, Serafim M, Goula AM (2019) Microwave-assisted extraction of phenolics from pomegranate peels: optimization, kinetics, and comparison with ultrasounds extraction. Chem Eng Process Process Intensif 137:1–11

Kaufmann B, Christen P (2002) Recent extraction techniques for natural products: microwave-assisted extraction and pressurised solvent extraction. Phytochem Anal 13:105–113

Kehili M, Kammlott M, Choura S, Zammel A, Zetzl C, Smirnova I, Allouche N, Sayadi S (2017) Supercritical CO2 extraction and antioxidant activity of lycopene and β-carotene-enriched oleoresin from tomato (Lycopersicum esculentum L.) peels by-product of a Tunisian industry. Food Bioprod Process 102:340–349

Kennedy DO, Wightman EL (2011) Herbal extracts and phytochemicals: plant secondary metabolites and the enhancement of human brain function. Adv Nutr 2:32–50

Kha TC, Nguyen MH (eds) (2014) Plant bioactive compounds for pancreatic cancer prevention and treatment. Nova Science Publishers, New York

Khan AS, Man Z, Bustam MA, Nasrullah A, Ullah Z, Sarwono A, Shah FU, Muhammad N (2018) Efficient conversion of lignocellulosic biomass to levulinic acid using acidic ionic liquids. Carbohydr Polym 181:208–214

Khoddami A, Wilkes M, Roberts T (2013) Techniques for analysis of plant phenolic compounds. Molecules 18(2):2328–2375

Koubaa M, Rosello-Soto E, Šic Žlabur J, ReŽek Jambrak A, Brncic M, Grimi N, Boussetta N, Barba FJ (2015) Current and new insights in the sustainable and green recovery of nutritionally valuable compounds from Stevia rebaudiana Bertoni. J Agric Food Chem 63(31):6835–6846

Krinsky NI, Yeum KJ (2003) Carotenoid-radical interactions. Biochem Biophys Res Commun 305:754–760

Kumar K, Yadav AN, Kumar V, Vyas P, Dhaliwal HS (2017) Food waste: a potential bioresource for extraction of nutraceuticals and bioactive compounds. Bioresour Bioprocess 4:1–14

Ladole MR, Nair RR, Bhutada YD, Amritkar VD, Pandit AB (2018) Synergistic effect of ultrasonication and co-immobilized enzymes on tomato peels for lycopene extraction. Ultrason Sonochem 48:453–462

Latif S, Anwar F (2009) Physicochemical studies of hemp (Cannabis sativa) seed oil using enzyme-assisted cold-pressing. Eur J Lipid Sci Technol 111(10):1042–1048

Lozano-Sánchez J, Castro-Puyana M, Mendiola JA, Segura-Carretero A, Cifuentes A, Ibáez E (2014) Recovering bioactive compounds from olive oil filter cake by advanced extraction techniques. Int J Mol Sci 15(9):16270–16283

Margeretha I, Suniarti DF, Herda E, Masud Z (2012) Optimization and comparative study of different extraction methods of biologically active components of Indonesian propolis Trigona spp. J Nat Prod 5:233–242

Martillanes S, Rocha-Pimienta J, Delgado-Adámez J (2018) Agrifood by-products as a source of phytochemical compounds. In: Díaz AV, García-Gimeno RM (eds) Descriptive food science. IntechOpen, pp 43–58

Martínez-Valverde I, Periago MJ, Provan G, Chesson A (2002) Phenolic compounds, lycopene and antioxidant activity in commercial varieties of tomato (Lycopersicum esculentum). J Sci Food Agric 82(3):323–330

Masci A, Coccia A, Lendaro E, Mosca L, Paolicelli P, Cesa S (2016) Evaluation of different extraction methods from pomegranate whole fruit or peels and the antioxidant and ant proliferative activity of the polyphenolic fraction. Food Chem 202:59–69
Meneses MA, Caputo G, Scognamiglio M, Reverchon E, Adami R (2015) Antioxidant phenolic compounds recovery from Mangifera indica L. by-products by supercritical antisolvent extraction. J Food Eng 163:45–53
Mohammadzadeh M, Honarvar M, Zarei AR, Mashhadi Akbar Boojar M, Bakhoda H (2018) A new approach for separation and recovery of betaine from beet molasses based on cloud point extraction technique. J Food Sci Technol 55:1215–1223
Moore J, Cheng Z, Su L, Yu L (2006) Effects of solid-state enzymatic treatments on the antioxidant properties of wheat bran. J Agric Food Chem 54(24):9032–9045
Mouratoglou E, Malliou V, Makris DP (2016) Novel glycerol based natural eutectic mixtures and their efficiency in the ultrasound-assisted extraction of antioxidant polyphenols from Agri-food waste biomass. Waste Biomass Valor 7:1377–1387
Nadar SS, Rao P, Rathod VK (2018) Enzyme assisted extraction of biomolecules as an approach to novel extraction technology: a review. Food Res Int 108:309–330
Nair DG (2015) Use of phytochemicals as functional food: an overview. Ind J Res 4:1–3
Niranjan K, Hanmoungjai P (2004) Nutritionally enhanced edible oil processing. AOCS Publishing
Özkan G, Bilek SE (2015) Enzyme-assisted extraction of stabilized chlorophyll from spinach. Food Chem 176:152–157
Pažitná A, Džúrová J, Špánik I (2014) Enantiomer distribution of major chiral volatile organic compounds in selected types of herbal honeys. Chirality 26(10):670–674
Pérez C, Ruiz del Castillo ML, Gil C, Blanch GP, Flores G (2015) Supercritical fluid extraction of grape seeds: extract chemical composition, antioxidant activity and inhibition of nitrite production in LPS-stimulated raw 264.7 cells. Food Funct 6:2607–2613
Pettinato M, Casazza AA, Perego P (2019) The role of heating step in microwave-assisted extraction of polyphenols from spent coffee grounds. Food Bioprod Process 114:227–234
Pham HN, Nguyen VT, Vuong QV, Bowyer MC, Scarlett CJ (2015) Effect of extraction solvents and drying methods on the physicochemical and antioxidant properties of Helicteres hirsuta Lour. Leaves. Technologies 3(4):285–301
Pham HNT (2017) Recovering bioactive compounds from fruit and vegetable wastes. In: Nguyen VT (ed) Recovering bioactive compounds from agricultural wastes. Wiley Publishers, pp 81–99
Pimentel-Moral S, Borrás-Linares I, Lozano-Sánchez J, Arráez-Román D, Martínez-Férez A, Segura-Carretero A (2018) Microwave-assisted extraction for Hibiscus sabdariffa bioactive compounds. J Pharm Biomed Anal 156:313–322
Plaza M, Abrahamsson V, Turner C (2013) Extraction and neoformation of antioxidant compounds by pressurized hot water extraction from apple byproducts. J Agric Food Chem 61:5500–5510
Prokopov T, Nikolova M, Dobrev G, Taneva D (2017) Enzyme-assisted extraction of carotenoids from Bulgarian tomato peels. Acta Aliment 46(1):84–91
Puértolas E, Luengo E, Álvarez I, Raso J (2012) Improving mass transfer to soften tissues by pulsed electric fields: fundamentals and applications. Annu Rev Food Sci Technol 3:263–282
Puri M, Sharma D, Barrow CJ (2012) Enzyme-assisted extraction of bioactives from plants. Trends Biotechnol 30(1):37–44
Putnik P, Barba FJ, Lucini L, Rocchetti G, Montesano D (2019a) Conventional, non-conventional extraction techniques and new strategies for the recovery of bioactive compounds from plant material for human nutrition. Food Res Int 123:516–517
Putnik P, Kresoja Ž, Bosiljkov T, Jambrak AR, Barba FJ, Lorenzo JM, Roohinejad S, Granato D, Žuntar I, Kovačević DB (2019b) Comparing the effects of thermal and non-thermal technologies on pomegranate juice quality: a review. Food Chem 279:150–161
Rao P, Rathod V (2017) Phytochemicals: an insight to modern extraction technologies and their applications. In: Grumezescu AM, Holban AM (eds) . Academic Press, Ingredients extraction by physicochemical methods in food, pp 495–521

Renard CM (2018) Extraction of bioactives from fruit and vegetables: state of the art and perspectives. LWT 93:390–395

Romdhane M, Gourdon C (2002) Investigation in solid–liquid extraction: influence of ultrasound. Chem Eng J 87(1):11–19

Rostagno MA, Prado JM (eds) (2013) Natural product extraction: principles and applications. Royal Society of Chemistry

Sagar NA, Pareek S, Sharma S, Yahia EM, Lobo MG (2018) Fruit and vegetable waste: bioactive compounds, their extraction, and possible utilization. Compr Rev Food Sci Food Saf 17 (3):512–531

Saxena M, Saxena J, Nema R, Singh D, Gupta A (2013) Phytochemistry of medicinal plants. J Pharmacogn Phytochem 1:168–182

Silva LV, Nelson DL, Drummond MF, Dufossé L, Glória MB (2005) Comparison of hydrodistillation methods for the deodorization of turmeric. Food Res Int 38(8–9):1087–1096

Singh D, Chaudhuri PK (2018) A review on phytochemical and pharmacological properties of Holy basil (Ocimum sanctum L.). Ind Crop Prod 118:367–382

Singh G (2007) Terpenoids. In: Arora MP (ed) Chemistry of Terpenoids and carotenoids. Discovery Publishing House, New Delhi

Singh G, Verma AK, Kumar V (2016) Catalytic properties, functional attributes and industrial applications of β-glucosidases. 3 Biotech 6:3

Strati IF, Oreopoulou V (2011) Effect of extraction parameters on the carotenoid recovery from tomato waste. Int J Food Sci Technol 46:23–29

Szentmihályi K, Vinkler P, Lakatos B, Illés V, Then M (2002) Rose hip (Rosa canine L.) oil obtained from waste hip seeds by different extraction methods. Bioresour Technol 82 (2):195–201

Tan Z-J, Li F-F, Xing J-M (2012) Cloud point extraction of aloe anthraquinones based on non-ionic surfactant aqueous two-phase system. Nat Prod Res 26:1423–1432

Terahara N (2015) Flavonoids in foods: a review. Nat Prod Commun 10(3):521–528

Tiwari BK (2015) Ultrasound: a clean, green extraction technology. TrAC Trends Anal Chem 71:100–109

Torres-Valenzuela LS, Ballesteros-Gómez A, Rubio S (2020) Green solvents for the extraction of high added-value compounds from Agri-food waste. Food Eng Rev 12(1):83–100

Tunchaiyaphum S, Eshtiaghi MN, Yoswathana N (2013) Extraction of bioactive compounds from mango peels using green technology. Int J Chem Eng Appl 4(4):194–198

Valadez-Carmona L, Ortiz-Moreno A, Ceballos-Reyes G, Mendiola JA, Ibáñez E (2018) Valorization of cacao pod husk through supercritical fluid extraction of phenolic compounds. J Supercrit Fluids 131:99–105

Vankar PS (2004) Essential oils and fragrances from natural sources. Resonance 9(4):30–41

Vázquez MB, Comini LR, Martini RE, Montoya SN, Bottini S, Cabrera JL (2014) Comparisons between conventional, ultrasound-assisted and microwave-assisted methods for extraction of anthraquinones from Heterophyllaea pustulata Hook f.(Rubiaceae). Ultrason Sonochem 21 (2):478–484

Venkat K (2011) The climate change and economic impacts of food waste in the United States. Int J Food Syst Dyn 2(4):431–446

Verkerk R, Schreiner M, Krumbein A, Ciska E, Holst B, Rowland I, De Schrijver R, Hansen M, Gerhäuser C, Mithen R, Dekker M (2009) Glucosinolates in Brassica vegetables: the influence of the food supply chain on intake, bioavailability and human health. Mol Nutr Food Res 53 (52):S219

Vian M, Breil C, Vernes L, Chaabani E, Chemat F (2017) Green solvents for sample preparation in analytical chemistry. Curr Opin Green Sustain Chem 5:44–48

Vian MA, Allaf T, Vorobiev E, Chemat F (2014) Solvent-free extraction: myth or reality? In: Chemat F, Vian M (eds) Alternative solvents for natural products extraction. Green chemistry and sustainable technology. Springer, Berlin, pp 25–38

Vorobiev E, Lebovka NI (2006) Extraction of intercellular components by pulsed electric fields. In: Raso J, Heinz V (eds) Pulsed electric fields technology for the food industry. Springer, Boston, pp 153–193

Vorobiev E, Jemai AB, Bouzrara H, Lebovka N, Bazhal M (2004) Pulsed electric field-assisted extraction of juice from food plants. In: Barbosa-Canovas GV, Tapia MS, Cano MP (ed) Novel food processing technologies. CRC Press, p 127–152

Vyas P, Chaudhary B, Mukhopadhyay K, Bandopadhyay R (2009) Anthocyanins: looking beyond colors. In: Bhowmik PK, Basu SK, Goyal A (eds) Advances in biotechnology. Bentham Science Publishers Ltd, pp 152–184

Vyas P, Haque I, Kumar M, Mukhopadhyay K (2014) Photocontrol of differential gene expression and alterations in foliar anthocyanin accumulation: a comparative study using red and green forma Ocimum tenuiflorum. Acta Physiol Plant 36(8):2091–2102

Wang J, Sun B, Cao Y, Tian Y, Li X (2008) Optimisation of ultrasound-assisted extraction of phenolic compounds from wheat bran. Food Chem 106(2):804–810

Wang L, Weller CL (2006) Recent advances in extraction of nutraceuticals from plants. Trends Food Sci Technol 17(6):300–312

Wang M, Bi W, Huang X, Chen DD (2016) Ball mill assisted rapid mechanochemical extraction method for natural products from plants. J Chromatogr A 1449:8–16

Wen L, Zhang Z, Sun DW, Sivagnanam SP, Tiwari BK (2019) Combination of emerging technologies for the extraction of bioactive compounds. Crit Rev Food Sci Nutr 11:1–6

Williams DJ, Edwards D, Hamernig I, Jian L, James AP, Johnson SK, Tapsell LC (2013) Vegetables containing phytochemicals with potential anti-obesity properties: a review. Food Res Int 52(1):323–333

Woźniak Ł, Marszałek K, Skąpska S (2016) Extraction of phenolic compounds from sour cherry pomace with supercritical carbon dioxide: impact of process parameters on the composition and antioxidant properties of extracts. Sep Sci Technol 51(9):1472–1479

Wu D, Gao T, Yang H, Du Y, Li C, Wei L, Zhou T, Lu J, Bi H (2015) Simultaneous microwave/ultrasonic-assisted enzymatic extraction of antioxidant ingredients from Nitraria tangutorun Bobr. Juice by-products. Ind Crop Prod 66:229–238

Wu K, Ju T, Deng Y, Xi J (2017) Mechanochemical assisted extraction: a novel, efficient, eco-friendly technology. Trends Food Sci Technol 66:166–175

Xhaxhiu K, Wenclawiak B (2015) Comparison of supercritical CO2 and ultrasonic extraction of orange peel essential oil from Albanian Moro cultivars. J Essent Oil Bear Plants 18:289–299

Xiao J (2017) Dietary flavonoid aglycones and their glycosides: which show better biological significance? Crit Rev Food Sci Nutr 57(9):1874–1905

Zhang G, Hu M, He L, Fu P, Wang L, Zhou J (2013) Optimization of microwave-assisted enzymatic extraction of polyphenols from waste peanut shells and evaluation of its antioxidant and antibacterial activities in vitro. Food Bioprod Process 91(2):158–168

Zhang YJ, Gan RY, Li S, Zhou Y, Li AN, Xu DP, Li HB (2015) Antioxidant phytochemicals for the prevention and treatment of chronic diseases. Molecules 20:21138–21156

Zhao C, Yang C, Liu B, Lin L, Sarker SD, Nahar L, Yu H, Cao H, Xiao J (2018) Bioactive compounds from marine macroalgae and their hypoglycemic benefits. Trends Food Sci Technol 72:1–12

Zhu Z, Li S, He J, Thirumdas R, Montesano D, Barba FJ (2018) Enzyme-assisted extraction of polyphenol from edible lotus (Nelumbo nucifera) rhizome knot: ultra-filtration performance and HPLC-MS2 profile. Food Res Int 111:291–298

Fruit Peels: A Sustainable Agro Waste Utilization Approach

7

Simple Kumar, Girish N. Mathad, and Rongsenyangla Ozukum

Abstract

Fruit peel contains components which are extracted and utilized in different products. One of the components is pectin. Among other fruits citrus peel is one of the most important sources of commercial pectin. Pectin is a natural, biocompatible, biodegradable, and reusable polysaccharide classified as an emulsifier, gelling agent, glazing agent, stabilizer, and/or thickener in commercial applications of food processing industries. The utilization of fruit peels as edible coating material can also be the promising aspect of food packaging industries. Pectin based coatings and the food products developed from the by-product of fruits, i.e., peel, are an important material innovation and sustainable food waste management approach because it not only reduces the amount of by-products obtained from the food processing industries but it also decreases dependence on petroleum while still yielding a product that provides similar benefits of traditional plastics where the quality of fruits and vegetable's shelf life are increased.

Keywords

Fruit peels · Pectin · Sustainable approach · Edible coatings · Waste utilization

S. Kumar (✉)
Amity International Centre For Post-Harvest Technology and Cold Chain Management, Amity University, Noida, Uttar Pradesh, India

G. N. Mathad · R. Ozukum
Amity Institute of Horticulture Studies and Research, Amity University, Noida, Uttar Pradesh, India

M. Thakur et al. (eds.), *Sustainable Food Waste Management*,
https://doi.org/10.1007/978-981-15-8967-6_7

7.1 Introduction

The problem challenging the world's food industry is how to utilize the waste materials. Waste products such as fruit peels and other residues from 60% to 65% of bulk fruit after processing such residues are usually dumped as waste during processing (Phatak et al. 1988). Fifty percent of the fresh weight of fruits like citruses, mango, apple is lost in seed, skin, and peel processing (Gupta and Joshi 2000; Macagnan et al. 2015; Chau and Huang 2003). One of these components is pectin. Hence, fruit peels are one of the significant sources of commercial pectin.

Pectin is present in almost all the plants, but is most concentrated in citrus fruits (oranges, lemons, grapefruits) and apples. Commercial pectin is produced from almost entire citrus peel or apple pomace, which are by-products from the processing of juice. Citrus varieties like lime, grapefruit, sweet orange, lemon, and mandarins are grown in Sudan. Apple pomace contains 10–15% and citrus peel contains 20–30% of pectin on a dry matter basis (Virk and Sogi 2004). The most widely citrus species grown commercially is sour orange (*Citrus aurantium*, L.). Commercial production of citrus currently spreads to many parts of the world.

The wastes generated from fruit processing industries are pomace, seeds, and peels. Orange waste contains 16.9 g 100 g′1 soluble sugar, 9.21 g 100 g″1 cellulose, 10.5 g 100 g″1 hemicellulose, and 42.5 g 100 g 4 pectin. Suitable methods have to be adopted to utilize them by converting into value-added product. Orange peel, which constitutes approximately 45% of the total bulk of the processed citrus fruit, is a good source for the extraction of pectin.

On a dry matter basis, it contains about 30% pectin. Consequently, amounts of orange peels are available as by-product. While present in most plant cell walls, apple pomace and peels of orange are the two primary sources of commercial pectin.

7.1.1 Pectin

Pectin is a complex mixture of polysaccharides that contain methylated ester of poly (galacturonic acid), which consists of chains of 300–1000 galacturonic acid unit. It also contains non-sugar substituents, essentially methanol, acetic acid, phenolic acids, and occasionally amide groups. The presence of acetyl groups inhibits the formation of gel with calcium ions which gives emulsion-stabilizing properties to the pectin. It is a white to light brown powder, primarily obtained from citrus fruits, used as a gelling agent especially in jams and jellies, medicine, sweet. Pectin derived from by-products contains significant quantities of anti-oxidants, colorants, proteins, and other bio-active substances (Vania et al. 2020). They are also expected to have additional health benefits and contribute to the dietary fibre's nutraceutical properties (Chantaro et al. 2000; Xu et al. 2018).

The primary application of pectin in food is as a gelling agent, thickening agent, and stabilizer. This implementation gives jams or marmalades the jelly like consistency which otherwise would be sweet juices. It is an ingredient for household use in gelling sugar where it is mixed with sugar and some citric acid to change pH to the appropriate concentration. It is also used for home jam-making in certain countries as a paste or concentrate, or as a mixed powder. At the other side, by removing other ingredient from the compost, an increase in the waste value of citrus fruit peel as a source of pectin may be accomplished. With respect to food processing, the common trend in this country is to study the use of pectin (different sources) in jam processing. The study's aims included the extraction of pectin from orange and lemon fruit peels as well as the use of extracted pectin in the manufacture of jams and the evaluation of the chemical and sensory consistency of the product.

Not only in overall production but also in economic importance citrus fruits are at the top. Oranges are fully peeled and eaten or processed to acquire orange juice and fragrance. These essentials oils are a combination of volatile compounds such as terpenes and oxygenated derivates, as citrus aldehydes, alcohols, and esters. Due to their fragrance and functional properties they are of great commercial importance and are used in many products of the food industry which make them excellent additives.

Pectin use has been found to decrease cholesterol levels in the body. Microorganism degrades pectin in the large intestine and colon, and releases short-chain fatty acids that have beneficial health effects. A literature survey shows that different authors have used different methods to extract pectin but no unified or well-established technology has been reported yet. The most trending developments in food packaging research aimed at enhancing the consistency and health of products by increasing the use of environmentally sustainable materials, preferably those that can be derived from bio-based commodities and present biodegradable properties. Edible films represent a key development area of modern multifunctional materials because of their character and properties to protect food effectively without waste production. Application of edible films can be used as a safe and stylish alternative to waste management issues in packaging materials. It has been identified as one of the key raw materials for obtaining edible films. Citrus peels (30–35% of pectin) obtained in fruit juice industries are the main raw materials used for the pectin extraction. The use of edible films is the latest innovation in food packaging industries with special focus on the use of pectin as base material for edible coatings. It can be described as a thin layer of material that covers the food surface and can be consumed as part of the whole product. The formulation of edible coatings must also be in compliance with the law specific to the food product concerned. Through isolating the coated commodity from the environment, they are able to create a modified atmosphere on the coated fruits. Gas selectively permeable coatings are capable of reducing the interaction of oxygen and carbon dioxide between coated fruit and the atmosphere. The important advantage of edible coating is the reduction of synthetic packaging, because these coatings are composed of biodegradable raw material.

High CO_2 concentration in fruit tissues often slows the maturing cycle by decreasing the production of ethylene, an important hormone for maturation. The important advantage of edible coating is the reduction of synthetic packaging, because these coatings are composed of biodegradable raw material (Fig. 7.1).

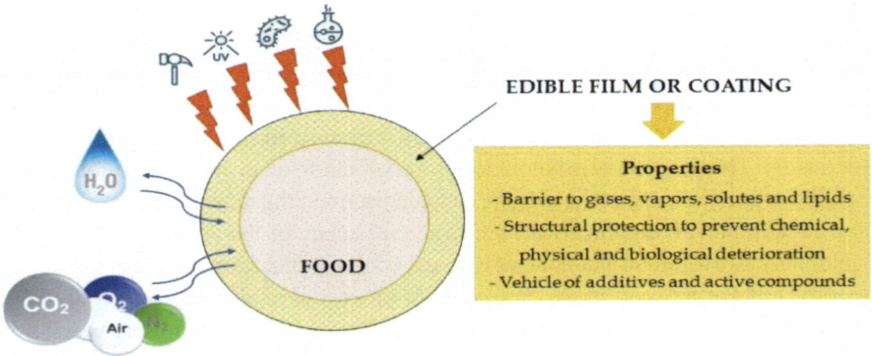

Fig. 7.1 Edible film properties

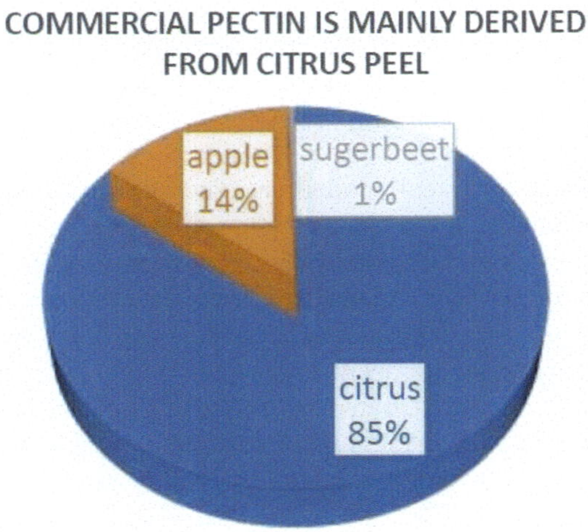

Fig. 7.2 Sources of commercial pectin

Edible film and coating are classified into three groups according to their ingredient's hydrocolloids, protein, and polysaccharides, while waxes, acylglycerols, and fatty acids are found in the lipids. Composites comprise all elements of hydrocolloids and lipids. Their existence and concentration decide material's barrier properties in terms of water vapor, oxygen, CO_2, and transfer of lipids in the food systems (Figs. 7.2 and 7.3).

7.1.2 Pectin: Chemical and Physical Properties

Pectin is a structural polysaccharide, usually called as nature's glue as it helps in maintaining the integrity and stability of plant cell walls. The cross-linking of cellulose and hemicellulose fibers which is primary biological function of pectin provides rigidity to the cell wall. This is also an essential part of the middle lamella,

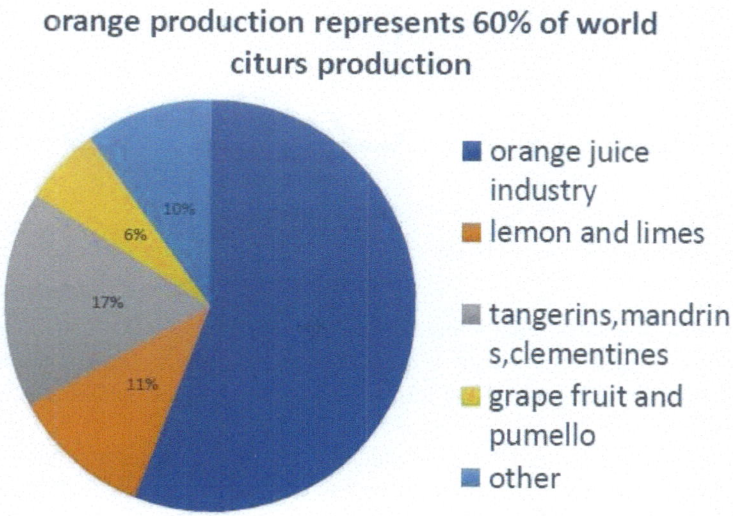

Fig. 7.3 Citrus overall production

Fig. 7.4 Molecular structure of pectin

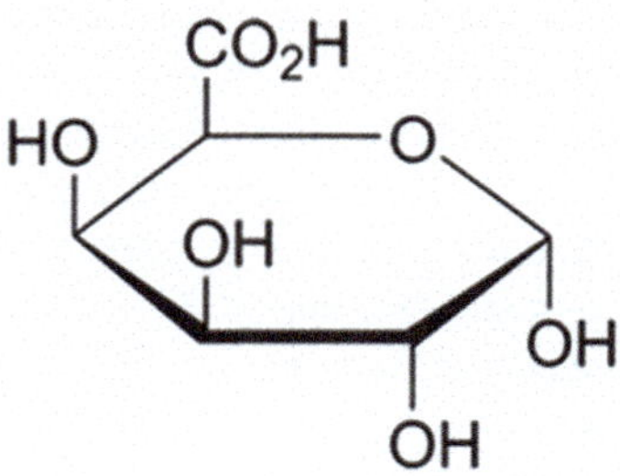

where it serves to connect cells. Galacturonic acid is the pectin's key ingredient, a sugar acid, an oxidized source of D-galactose, and it occurs as polymer polygalacturonic acid (Fig. 7.4).

Pectin comprises a base of polysaccharide backbone of a D-galacturonic acid connected to a-(1-->4). Within the natural product the acid group in the chain is mostly esterified with methoxy group. Acetyl groups can be found on the free hydroxy groups (Fig. 7.5).

7.1.3 Pectin's Gelling Properties

A pectin gel is produced when the ingredients are heated and the pectin dissolves. A gel starts to form while cooling of dissolved pectin below gelling temperature. When gel forming is too solid, liquid part extraction syneresis occurs probably causing a granular texture. Poor gelling results in incredibly fragile gels. Parameters such as content of the sugar, pH, and calcium ions define pectin's gelling properties.

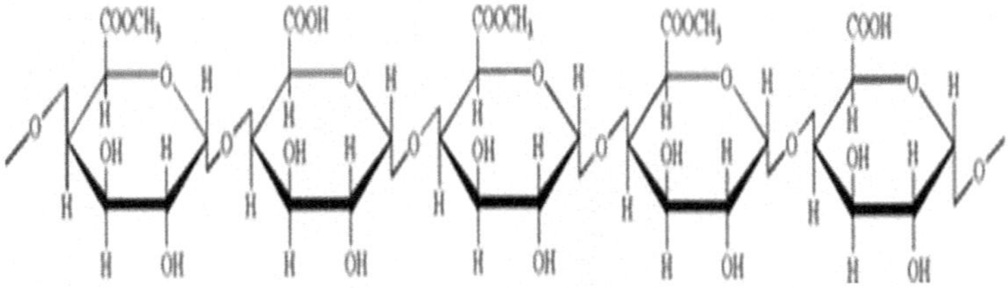

Fig. 7.5 Cell wall matrix-polysaccharide

The individual pectin chains bound together because of hydrogen bonds and hydrophobic interactions. As water is bound by sugar, it leads to the formation of the linkages which further leads to the formation of a three-dimensional molecular net that forms gel.

Ionic bridges are formed between calcium ions and galacturonic acid in case ionized carboxy groups of low ester pectin and calcium are required to shape a gel. Most of the pectins are extracted from peels of the citrus and apple pomace (in which pectin content ranges from20 to 40 g/100 g on dry weight basis at high temperatures by hydrolyzing protopectin).

Cho and Hwang (2000) extracted pectin from apple pomace using conventional method of heating. Wang et al. extracted pectin from the pomaces of apple using conventional method of heating. Wang et al. applied microwave assisted extraction of pectin from orange peel. Zhao Hui et al. optimized the conditions for pectin extraction from orange peels. Fishman et al. (2006) extracted pectin from lime peel. Singh and Dhillon (2007) used 0.05 N hydrochloric acid at 90 °C to remove pectin from dried kinnow peels. This research aimed to extract the pectin from the peels of kinnow.

7.2 Extraction Methods

7.2.1 Conventional Extraction Method

Conventional extraction method is used for extracting pectin from pulp of sugar beet by the use of hydrochloric acid by variation of time extraction, pH, and temperature (pH to 1.5, extraction time, and temperature being 4 h and 80 °C, respectively). The resulting pectin yield was 19.53% on dry basis at these extraction conditions.

Pectin was extracted from orange peel by the hydrochloric acid extraction and the ethanol deposition method. The result showed that the pectin extraction rate was higher when the ratio of solution/material was 10–15, pH of solution 2–3, time 45 min, and temperature 80 °C. The feasible technological conditions are solution solid ratio of 10, pH 2.0, extraction temp 80 °C, and extraction time 45 min for extraction yield of 70.68%.

Pectin from dried citrus fruit peel is extracted under pH 2, ethanol ratios (ER) 1:1, and extraction period 120 min, using hot water boiling method; at this condition highest yield was obtained 18.21%. Pectin.

Pectin was extracted using conventional method in distilled water solutions of different pH values 1, 1.5, 2, 3, 4, and 5. Pectin from dried kinnow peel and pomace using water acidified with HN03 at varying temperature (40–80 °C), time (10–90 min), and pH of solution (1.25–2.25) was extracted (Cho and Hwang 2000).

Pectin yield was found to be more in peel than in pomace irrespective of the extraction conditions. Temperature, time, and pH extraction greatly impacted pectin yield. Yield of pectin increased with increase in temperature up to 60 °C followed by slight decrease at 80 °C.

7.2.2 Microwave Assisted Extraction

It is used to extract pectin from young sugar palm. MAE for pectin extraction from the dried orange peel. They used Box–Behnken response surface design to optimize the effects of processing variables microwave power (160–320–480 W), extraction time (60–120–180 s), pH (1–1.5–2), and solid–liquid ratio (10–20–30 mL) on the yield of pectin. With increase in microwave power, the volume of pectin extracted increased, although it decreased as the time, pH, and solid–liquid ratio increased and optimum conditions were dependent on each individual and combinations of all independent variables were: microwave power of 422 W, irradiation time of 169 s, pH of 1.4, and solid–liquid ratio of 1:16.9 g/ml for maximum pectin yield of 19.24%.

Pectin from waste *Citrullus lanatus* fruit rinds using microwave energy. Microwave assisted extraction parameters employed in this study were microwave power (160–480 W), irradiation time (60–180 s), pH (1–2), and solid–liquid ratio (1:10–1:30 g/ml) and using a four-factor three-level Box–Behnken reaction surface design (BBD) combined with desirability feature methodology, they were designed. The results revealed that all method factors had a major impact on pectin's extraction yield. Optimum MAE conditions for pectin at its maximum residue yield C. Lanatus fruit rinds (25/79%) with a microwave capacity of 477 W, 128 s irradiation time, 1.52 PH, 1:@0.3 g/ml solid–liquid ratio, respectively.

7.2.3 Ultrasonic Assisted Extraction Method

This method is used for sisal waste. The experimental yield (29.32%) was obtained under the optimal condition ultrasonic power of 61 W, temp of 50 °C, time of 26 min, and SL ratio of 1:28 g/ml. Response surface methodology (RSM) based on a three-level four-factor optimizes the extraction conditions (Xu et al. 2018).

7.2.4 Enzymatic Extraction

Enzyme assisted method of extraction can be used to produce pectin from fruit peel. For this purpose, a microorganism that produces a proto pectin-solubilizing enzyme was isolated and classified as *Trichosporon penicillatum*. Citrus (*Citrus umhiu*) peel was suspended in water (1:2, wt./vol), applied to the organism and fermentation proceeded at 30 °C 15 to 20 h. During fermentation 20–25 g of pectin/kg of peel was produced using this process.

Using microwave to process orange peels and then adding complex enzyme to extract pectin; the yield and purity of pectin can be increased. Orange peel pectin contained a relatively huge amount of carbohydrates, but no protein has been shown by infrared spectrum. The extract technology was optimized by orthogonal experiments at 250 W microwave power, 50 °C microwave heating temperature, 10 min microwave processing time, 1 h. single extraction time, buffer pH 4.5, extraction temperature of 50 °C enzyme, and enzyme dosage of 0.9%. Extraction of the enzyme has been carried out 3 times. The average yield of orange peel pectin is 22.12%, which under the same conditions was much higher than yield of pectin extraction using traditional water bath process. (Yu and Sun 2013).

The enzymes are used efficiently to remove, alter, and improve the functional properties of lime pectins. Enzyme preparation was Laminex C2K derived from Penicillium fimiculosum which, during 4 h treatment at pH 3.5, 50 °C, released pectin with similar yield (23% w/w), molecular weight (69 kDa), and functional properties, e.g., gelling, stabilization of acidified milk drinks, and viscosity as the classically acid-extracted pectin (8 h treatment at 70 °C, pH < 2). Carbohydrate microarray analysis showed that enzymatically extracted pectin mainly contained highly methylated pectin (chemical compositional analysis indicated degree of esterification up to 82%). The Laminex CK2 extracted pectin polymers were not sensitive to the presence of Ca^{+2} ions; they formed a gel at low pH in the presence of sugar and were able to stabilize acidified milk drinks. Further modification of the pectin derived with Laminex C2 K by enzymatic de esterification increased its calcium sensitivity and the ability to stabilize acidified milk beverages (Figs. 7.6, 7.7, and 7.8 and Table 7.1).

7.3 Sustainable Agro Waste Management Approach and Food Industries

It is a promising sustainable food waste management approach as fruit peels as well as their active components like pectin because of its non-toxic, safe, and eco-friendly behavior it can be efficiently utilized in food packaging and processing industries. To minimize the product quality loss and to improve the marketability of pectin based coating material, the kinetics of quality changes of the products under different storage conditions has also been studied (Maftoonazad and Ramaswamy 2019). Thus, in order to secure the sustainable development for exponentially growing population, the increasing demands for the light-weighted high performance

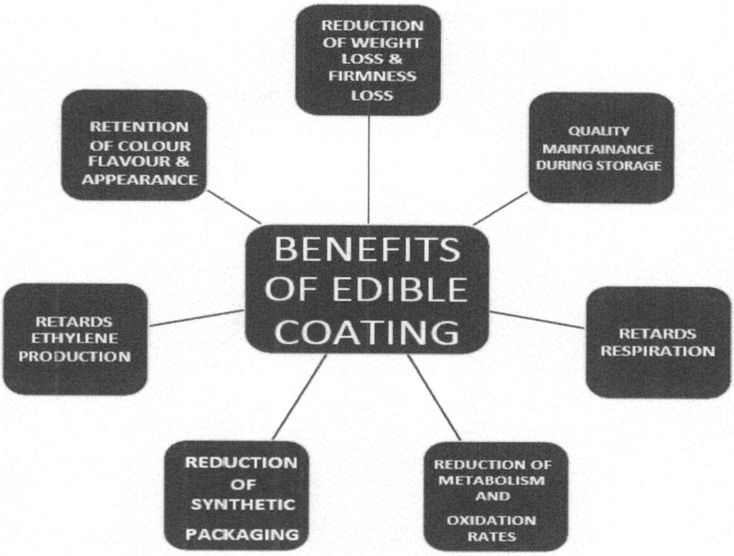

Fig. 7.6 Role of edible coating

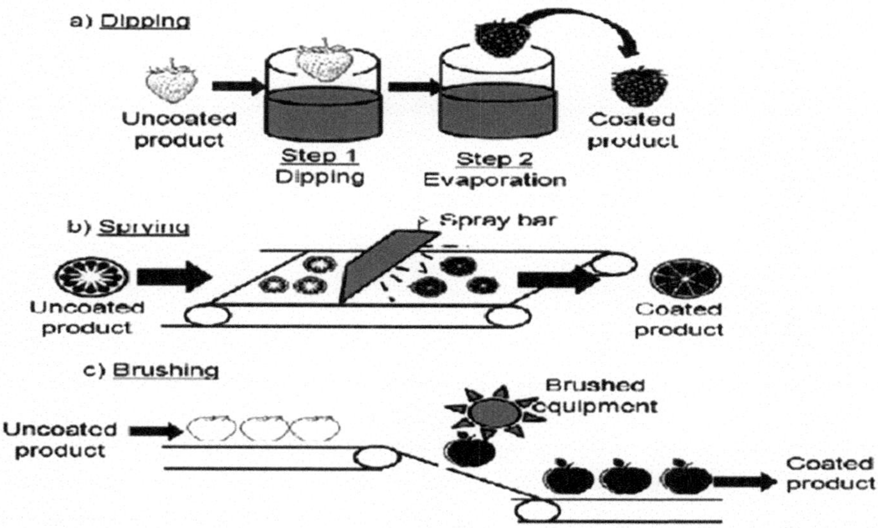

Fig. 7.7 Application of pectin as edible coating

materials and growing concerns over environmental impact of the materials have enforced academic and industrial researchers to develop new materials from biodegradable natural resources (Hui Suan et al. 2020). To successfully produce this key technology needed to meet the following profiles:

Fig. 7.8 Recent trends in utilization of fruit peel in processing and packaging industry

- Transparency
- Sealing ability
- Printability
- Flexibility
- Storage ability
- Shelf life

7.4 Conclusion

Pectin based coatings and the food products developed from the by-product of fruits, i.e., peel, are an important material innovation and sustainable food waste management approach because it will not only reduces the amount of by-products obtained from the food processing industries but it also decreases dependence on petroleum while still yielding a product that provides similar benefits of traditional plastics to increase the shelf life and quality of fruits and vegetables. Pectin based edible coating has the potential for improving the food quality and shelf life. They are safe for the consumer and to the environment. The utilization of fruit peels for the extraction of active component like pectin because of its excellent gelling property can be the promising aspect of food processing industries to develop new products like jam, jellies, etc. There is an urgent need to overcome the challenges for making this sustainable approach as successful technology. There is still a large amount of work to be performed since some of the most remarkable improvements are not yet

Table 7.1 Pectin based coatings to extend shelf life of fresh fruits and vegetables

S. no	Coating	Active Agent	Food Matrix	Effect	Reference
1	Pectin(3%), sorbitol, TIC gums		Lime fruits (*Citrus aurantifolium)*	Kinetics of quality change in stored limes fruits(10–25 °C, 22 days)	
2	Multi-layered+C6: C10 chitosan (2%)-	Trans-cinnamaldehyde encapsulated in β-Cyclodextrins	Papaya fruits (*Carica papaya* L. cv Maradol)	Antimicrobial activity maintaining quality sensory attributes (4 °C, 15 days)	
3	Pectin (3%), sorbitol, beeswax	–	Mangoes (cv. Ataulfo)	Reduction of physiological changes, respiration rates, and temperature)	Moalemiyan et al. 2012
4	Pectin (3%), sorbitol, beeswax		Avocardo	Reduction of firmness, color, respiration rates, and moisture loss (10 °C, 1 month)	Maftoonazad and Ramaswamy (2008)
5	Pectin (3%) + glycerol	Cinnamon leaf essential oil	Peach (*Prunus persica*)	Antimicrobial, antioxidant activity. Odor acceptability upto 10 days (5 °C)	
6	Pectin (2%), glycerol, CaCl2	N-acetylcysteine glutathione	Pears (*Pyrus communis* L.)	Anti-browning, antimicrobial, antioxidant maintaining sensory attributes (4 °C, 14 days)	Oms-Oliu et al. (2008)

experimentally attained or reproducible at large scale. In general terms, there is still a need for a better understanding of the composition structure processing–properties relationships in edible films based on pectin for food packaging, both at the laboratory and industrial scale. Moreover, since many of the studies related to this issue have been carried out using some basic pectin already in the market, there is still a lot of room for variation and maturation in the development of edible coatings for application in food packaging.

References

Chantaro P, Devahastin S, Chiewchan N (2000) Production of antioxidant high dietary fiber powder from carrot peels. LWT Food Sci Technol 41:1987–1994

Chau CF, Huang YL (2003) Comparison of the chemical composition and physicochemical properties of different fibers prepared from the peel of *Citrus sinensis* L. Cv. Liucheng. J Agric Food Chem 51:2615–2618

Cho YJ, Hwang JK (2000) Modeling the yield & intrinsic viscosity of pectin in acidic solubilization of apple pomace. J Food Eng 4:85–89

Fishman ML, Chau ELK, Hoagl PD, Hotchkis AT (2006) Microwave-assisted extraction of lime pectin. Food Hydrocoll 20:1170–1177

Gupta K, Joshi VK (2000) Fermentative utilization of waste from food processing industry. In: Verma LR, Joshi VK (eds) Postharvest technology of fruits and vegetables: handling processing fermentation and waste management, vol 2. Indus Pub, New Delhi, p 693

Hui Suan N et al (2020) Recent advances on the sustainable approaches for conversion and reutilization of food wastes to valuable bioproducts. Bioresour Technol 2:302

Macagnan FT, dos Santos LR, Roberto BS, de Moura FA, Bizzani M, da Silva LP (2015) Biological properties of apple pomace, orange bagasse and passion fruit peel as alternative sources of dietary fibre. Bioact Carbohydr Diet Fibre 6:1–6

Maftoonazad N, Ramaswamy HS (2008) Effect of pectin-based coating on the kinetics of quality change associated with stored avocados. J Food Process Preserv 32:621–643

Maftoonazad N, Ramaswamy HS (2019) Application and evaluation of a pectin-based edible coating process for quality change kinetics and shelf-life extension of lime fruit (*Citrus aurantifolium*). Coatings 9:285

Moalemiyan M, Ramaswamy HS, Maftoonazad N (2012) Pectin-based edible coating for shelf-life extension of ataulfo mango. J Food Process Eng 35:572–600

Oms-Oliu G, Soliva-Fortuny R, Martín-Belloso O (2008) Edible coatings with antibrowning agents to maintain sensory quality and antioxidant properties of fresh-cut pears. Postharvest Biol Technol 50:87–94

Phatak L, Chang KC, Brown G (1988) Isolation & characterization of pectin in sugar-beet pulp. J Food Sci 53:830–833

Singh M, Dhillon SS (2007) Extraction of pectin from kinnow peels. Int J Environ Stud 64:287

Vania GZ et al (2020) Towards a green and sustainable fruit waste valorization model in Brazil: optimization of homogenizer-assisted extraction of bioactive compounds from mango waste using a response surface methodology. Pure Appl Chem 92(4):617–629. https://doi.org/10.1515/pac-2019-1001

Virk S, Sogi DS (2004) Extraction and characterization of pectin from apple (*Malus pumila* cv Amri) peel waste. Int J Food Sci 7:693–703

Xu SY, Liu JP, Huang X, Du LP, Shi FL, Dong R et al (2018) Ultrasonic-microwave assisted extraction, characterization and biological activity of pectin from jackfruit peel. LWT Food Sci Technol 90:577–582

Yu XC, Sun DN (2013) Microwave and enzymatic extraction of orange peel pectin. Asian J Chem 25:5333–5336

Waste from Dairy Processing Industries and its Sustainable Utilization

8

Falguni Patra and Raj Kumar Duary

Abstract

Mostly, the waste materials from dairy processing industries are the waste water and generally considered to be the largest source of food processing wastewater in many countries. Dairy industries throughout the world varies in their sizes and the types of manufactured products due to which, it is hard to give any general characteristics. The dairy industry can be divided into several production divisions. Each division produces wastewater of a characteristic composition, depending on the kind of product that is produced (liquid milk product, cheese, butter, ice cream, powdered dairy product, concentrated milk like condensed milk, evaporated milk, etc.). Waste water in the dairy processing mainly arise from heating and cooling processes, the cleaning of equipment, spillage of milk and milk products, whey, pressing and brining, Clean-In-Place (CIP), and resulting from equipment malfunctions and even operational errors. Waste waters from dairy plants generally have a high organic load due to the presence of diluted milk and milk products; significant quantities of cleaning compounds and sanitizers and are high in sodium content (use of caustic soda for cleaning). The pH of the waste water varies widely due to the use of acidic and caustic cleaning agents. There are also large variations in the characteristics, volume, flow rate, and composition of the effluent generated on an hourly, daily, and seasonal basis. Like any other industries, the effluent from the dairy industry poses environmental problems like water and soil pollution due to the high amounts of nutrients and

F. Patra
Mansinhbhai Institute of Dairy and Food Technology (MIDFT), Mehsana, Gujarat, India
e-mail: falguni@midft.com

R. K. Duary (✉)
Department of Food Engineering and Technology, Tezpur University, Tezpur, Assam, India
e-mail: duary@tezu.ernet.in

M. Thakur et al. (eds.), *Sustainable Food Waste Management*,
https://doi.org/10.1007/978-981-15-8967-6_8

organic matter. Currently, sustainable innovational technologies have been targeted towards utilization of the waste for generation of compounds which have application in food, chemical, plastic, fuel, pharmaceutical, and other industries simultaneously solving pollution problem.

Keywords

Dairy waste · Whey · Waste utilization · Waste minimization · Waste characteristics · Waste water

Abbreviations

BOD_5	Biological Oxygen Demand 5 days
COD	Chemical oxygen demand
FOG	Fats, oil, and grease
NH^{4+}-N	Ammonium nitrogen
TDS	Total dissolved solids
TKN	Total Kjeldahl nitrogen
TN	Total nitrogen
TP	Total phosphorus
TS	Total solids
TSS	Total suspended solids
VSS	Volatile suspended solids

8.1 Introduction

Food industry is one of the major sectors that consumes large quantity of water and generates a huge amount of waste water per unit production. This industry is also responsible for generating a giant amount of sludge in biological treatment. Among food processing industries, the dairy industry is the source of largest volume of food processing wastewater. India is the world's highest milk producer, reported a production of 176.3 million tonnes in 2017–18 which is approximately 6.6% higher than the previous year (https://www.nddb.coop/information/stats/milkprodindia. Accessed 17 June 2019). Currently, 35% of milk produced in India is processed in organized sector. In general, wastes generated from any industries can be in the form of solid waste, wastewater, and volatile compounds. Dairies do not produce solid waste, except for packing material, etc., and sludge from waste water treatment plants (Klemes et al. 2008). In general, 0.5 kg sludge is produced per kg removed COD in aerobic process and in anaerobic process 0.1 kg sludge is generated/kg removed COD. Inorganic wastes may include plastics, metals, cardboard, wood, paper, etc., from product packaging and maintenance, the majority of which can be recycled. Approximately 5.8 tonnes (t) of solid waste is generated by dairy processing industry per million litre of raw milk (Rad and Lewis 2014). In dairy

Table 8.1 BOD and COD levels of milk and milk products (Britz et al. 2006)

Product	BOD g/L	COD g/L
Whole milk	104–120	183–190
Skim milk	67–90	120–147
Cream	400	750–860
Yoghurt	91	–
Butter milk	61–75	110–134
Evaporated milk	208–271	378
Whey	34–45	65–80
Ice cream	292	–

industry, gasses like CO_2, CO, NO_x, and SO_2 are discharged mainly due to the need for energy. Chlorofluorocarbons (CFC's) and NH_3 may be emitted into the air due to the leakage and stripping of chilling machines when not in use.

Most of the wastes generated by the dairy processing industry are the waste water. Wastes from dairy processing industries are almost entirely composed of organic material in solution or colloidal suspension. Also, some larger suspended solids may be present in wastewater from cheese or casein manufacturing plants. Lower amount of sand and other foreign material may also be present from floor or truck washes. Approximately 3–5% of the milk received in a dairy plant may be lost in waste discharges from the plant (Wendorff 2001).

Dairy plant wastewater can be categorized into three groups (Britz et al. 2006):

1. Processing waters, such as water used in the cooling and heating processes.
2. Cleaning wastewaters mainly from the cleaning of equipment that has been in contact with milk or milk products, spillage of milk and milk products, whey, pressings and brines, CIP cleaning options, and waters resulting from equipment malfunctions and even operational errors. This wastewater stream may contain anything from milk, cheese, whey, cream, separator and clarifier dairy waters, to dilute yoghurt, starter culture, dilute fruit and stabilizing compounds, sodium hydroxide, phosphoric acid, nitric acid, sodium hypochlorite, etc.
3. Sanitary wastewater.

There is a wide variation in volume, concentration, and composition of the waste water generated in dairy industry which typically depends on the product profile of the plant, the production programme, operating methods, design of the plant, the water management programme, and subsequently the amount of water being conserved (Britz et al. 2006). This variation occurs from one industry to another dairy industry, from season to season, and hour to hour. Broadly, dairy wastes are high in biological oxygen demand (BOD), chemical oxygen demand (COD), fats, oils, and grease (FOG), proteins, carbohydrates, lipids, nitrogen, phosphate, and suspended solids. The key parameters either BOD or COD are measured to determine the organic wastewater strength. Typical dairy process wastewater has a (BOD_5) of about 2.0 g/L and a dissolved solids concentration of 1.8 g/L. The BOD and COD values of some dairy product are presented in Table 8.1.

Table 8.2 Characteristics of the effluent from dairy plant

pH	–	–	4.7–11	7.10	6.5–8	10.07
BOD_5	1.6	8.239	0.65–6.24	0.442	1.294	0.16
COD	2.8	18.045	0.43–15.2	8.96	2.143	0.257
TS	–	–	–	0.797	–	–
TSS	–	7.175	0.25–2.75	0.253	0.68	0.099
TDS	–	–	–	0.543	–	0.997
VSS	–	–	0.2–1.89	–	–	–
Fat, oil, grease	–	4.89	0.16–1.76	–	–	26.13 ppm
TN	0.14	0.329	0.014–0.09	–	–	
TP	0.03	–	–	–	–	
Alkalinity as $CaCO_3$	–	–	–	–	–	0.186
Chloride	–	0.593	–	0.186	–	0.377
References	Schwarzenbeck et al. (2005)	Arbeli et al. 2006	Passeggi et al. (2009)	Qasim and Mane (2013)	Kadu et al. (2013)	Al-Shammari et al. (2015)

Values are in g/L, except pH

As acid and alkaline detergents are used for cleaning operations in dairy, the pH of dairy effluent varies between 2.0 and 12.0 (Tetrapak 1995). Waste water with a pH of more than 10.0 or below 6.5 must not be discharged to the sewage system. The total suspended solids (TSS) content can range from 0.3 to 5.0 g/L and is often between 0.5 and 2.0 g/L (Liu and Haynes 2011). Suspended solids originate mainly from coagulated milk, cheese curd, and flavouring ingredients (e.g. fruit and nuts), and are predominantly organic in nature. About, 70–75% of TSS is present as volatile suspended solids (VSS; i.e. those lost upon heating to 500 °C). The total fats, oils, and greases (FOG) content is commonly between 0.1 and 0.3 g/L (Liu and Haynes 2011). Wastewater from dairy may also contain a large microbiological load as well as pathogenic viruses and bacteria. The characteristics of effluent of dairy plant are presented in Table 8.2.

As the demand of milk is increasing, the size of the dairy industry is also growing and resulting in increasing volume and nature of the waste generation which are the major concern for the industry in terms of adverse environmental effect. Wastewater with high levels of organic pollutants creates major environmental problems when directly discharged to the surface water or land. The dairy effluents when are discharged directly in the surface waters, the streams, and waterways are contaminated and cause deoxygenation. Also, excessive concentration of nitrogen and phosphorus in surface and subsurface waters contributes to excessive growth of plants and algae blooms which makes downstream water unsuitable for domestic,

Table 8.3 Standard for direct discharge of dairy effluent

Parameter	Maximum permissible value mg/L	
	CPCB, India	World Bank report
pH	6.5–8.5	6–9
BOD	100	50
COD	–	250
TSS	150	50
Oil and grease	10	10
Total nitrogen	–	10
Total phosphorus	–	2
Temperature increase	–	≤3 °C
Coliform bacteria (MPN/100 mL)	–	400

agricultural, and industrial uses. Land degradation and damage to pastures and crops, and long-term damage of soil productivity may appear from direct land discharge. Over-application of effluent to land also results in contaminated groundwater. To protect the environment from pollution, Central Pollution Control Board (CPCB), India has imposed regulation for the direct discharge of dairy effluent (Table 8.3).

To control pollution from these huge amounts of byproducts and waste water generated from the dairy processing, biotechnological intervention has been done to reduce organic load of these waste and to generate industrially important compounds like bioplastics, bioenergy, biofuel, biosurfactant, biofertilizers, organic acids, bioactive peptides, etc. This chapter describes characteristics of dairy processing waster, their minimization and utilization to generate industrially important compounds.

8.2 Dairy Processing Waste Water and Their Minimization

Effluents arising out of dairy plants include rinse and wash water from milk cans, equipment, bottles, and floors. They also include portions of spilled milk, spoiled or sour milk, skimmed milk, whey, butter milk, and entrainment during evaporation. Various sources for generation of waste in the dairy processing have been generalized in Table 8.4. The quantity of effluents generated from different dairy industries varies from 6 to 10 L per litre of milk processed depending upon the processes used, products made, care taken in the use of water, and the quantity of water available. Within a particular dairy, the rate of discharge of the effluent varies considerably from hour to hour. Milk powder, cheese, and/or liquid processing produce 0.4–60.0 L wastewater/kg product, cheese plant generates 0.7–60.0 L/kg product, market milk and cultured product manufacture generates 0.9–25.0 L/kg of product, and ice cream plant generates 2.7–7.8 L/kg product (Rad and Lewis 2014). Sources of dairy wastewater from different sections of a dairy processing plant have been enlisted in Table 8.4.

In the following section, process-wise waste generation in dairy plant has been discussed.

Table 8.4 Sources of dairy waste water in dairy processes (Victoria Environmental Protection Authority 1997)

Dairy processes	Sources of waste
Preparation stages	
Milk receiving/storage	Poor drainage of tankers, spills and leaks from hoses and pipes, spills from storage silos/tanks, foaming, cleaning operations
Pasteurization/UHT	Liquid losses/leaks, recovery of downgraded product, cleaning operations, foaming, deposits on surfaces of pasteurization and heating equipment
Homogenization	Liquid losses/leaks, cleaning operations
Separation/clarification (centrifuge, reverse osmosis)	Foaming, cleaning operations, pipe leaks
Product processing	
Market milk	Foaming, product washing, cleaning operations, overfilling, poor drainage, sludge removal from clarifiers/separators, leaks, damaged milk packages, cleaning of filling machinery
Cheese making	Overfilling vats, incomplete separation of whey from curd using salt in cheese making, spills and leaks, cleaning operations
Butter manufacture	Vacreation and salt use, butter washing, cleaning operations
Condensed/evaporated milk	Evaporator entrainment losses, boil over cleaning, cleaning processes
Powder manufacture	Spills of powder handling, start-up and shut-down losses, plant malfunction, stack losses, cleaning of evaporators and driers, bagging losses
Ice cream manufacture	Spills and leaks, product change over, filling machines jams Broken packages, CIP of mixing vats, freezer, pipe lines
Dahi/yoghurt manufacture	HTST/batch pasteurization start-up and shut-down losses, filling machines jams, broken packages, over fill, cleaning operations (starter vat, yoghurt/dahi vats, filling machines, pipelines, etc.)

8.2.1 Waste from Milk Reception and Storage Areas

Although the product profile of dairy industries varies widely, every dairy plant has a section where milk is received and stored. Milk is either received by tanker or milk can. The empty cans are drained so as to allow residual milk in the cans to drip into a receptacle after emptying. The empty can are then transferred to the cleaning station, where the cans are washed with water and detergent to remove all traces of milk. Tankers are also cleaned every day. Tankers cleaning can be done by connecting the tanker to a cleaning system in the reception area or by driving it to a special cleaning station. The type of wastage in milk reception and storage section could be due to poor drainage of milk transport trucks and storage tanks before CIP starts, overflows, improper hose connections, and also pipeline leaks. Effluent generated in this section contains mostly milk fat, milk protein, and lactose as well as cleaning solution from CIP (Hale et al. 2003). Briao and Granhen Tavares (2007) reported that effluent generated from milk reception had pH of 10.06 ± 1.60, COD 1.794 ± 0.98 g/L, TN 0.0453 ± 0.0246 g/L, TP 0.0252 ± 0.0142 g/L, and oil and grease 0.2533 ± 0.1052 g/L. Later, Janczukowicz et al. (2008) also characterized effluent from milk reception point and reported that the effluent had pH of 7.18, BOD_5 0.8 g/L, COD 2.54 g/L, BOD_5/COD ratio of 0.31 with TSS 0.65 g/L and fats 1.06 g/L. They also reported that the effluent from pumping station had pH of 8.35, BOD_5 1.75 g/L, COD 4.44 g/L, BOD_5/COD 0.39, TSS 1.07 g/L, and fats 0.57 g/L respectively.

Waste water loads in the reception and storage areas can be minimized by the complete drainage of storage tanks and tanker trucks. To prevent overflows, storage tanks should have level controls and self-draining pipes should be installed at a slight angle. To prevent leakage, hoses should be properly connected and well-stitched. CIP processes can also be used for the cleaning of tankers and storage tanks so that final rinse water can be reused for the initial rinse of the next cleaning process (Hale et al. 2003; Tetra Pak 2003).

8.2.2 Waste from Market Milk

The manufacturing steps for processing of liquid milks includes treatments like raw milk reception, filtration/clarification, separation (cream/skim milk) standardization, homogenization, pasteurization/sterilization/UHT, cooling, and packaging/aseptic packaging (UHT milk). After packaging, the product should be stored and transported at 4 °C, except for sterilized/UHT milk. Wastewaters from this section are generally due to cleaning operations of heat exchangers, where most of the organic loading of cleaning water occurs at the product and water interfaces. Generally, the heat exchanger is purged with milk at the starting of the process, and the milk is purged with water for cleaning at the end of the process to avoid accidental adulteration of milk with water. Effluent generated from liquid milk section contains mostly whole milk components such as milk fat, milk protein, and lactose. Briao and Granhen Tavares (2007) reported that effluent generated from fluid dairy product had pH of 9.62 ± 3.69, COD 2.27 ± 0.797 g/L, N 0.0712 ± 0.0387 g/L, P 0.0421 ± 0.0212 g/L, and oil and grease 0.5235 ± 0.3452 g/L.

Organic loads from the liquid dairy product section can be minimized mainly by managing heat-treatment processes such as reducing the number of process interruptions and changeovers. Also, any disturbance during packaging/filling should be managed by adequately sized storage tanks. The interface milk/water fraction should not be wasted, rather can be used as animal feed or recycled into other products (Hale et al. 2003).

8.2.3 Waste from Butter Section

Commercially for manufacturing of butter, cream is the main raw material. Cream is churned and converted to butter and buttermilk (as byproduct). Butter can be

Table 8.5 Characteristics of the effluent from butter plant

Parameter	Butter and ghee	Butter/ milk powder	Butter/ milk powder	Butter/ Comte cheese plant	Butter
pH	7.1	5.8	10–11	5–7	12.08
BOD_5	1.377	–	1.5	1.250	2.42
COD	3.218	1.908	–	2.52	8.93
BOD_5/ COD	–	–	–	–	0.27
TS	3.4	1.720	–	–	–
TSS	2.24	–	–	–	5.07
Fat	1.32	–	0.4	–	2.88
Alkalinity as $CaCO_3$	–	0.532	–	–	–
References	https://law.resource.org/pub/in/bis/S02/is.8682.1977.html	Strydom et al. (1997)	Donkin (1997)	Torrijos et al. (2001)	Janczukowicz et al. (2008)

Values are in g/L, except pH, BOD/COD

manufactured from cultured (soured/ripened) cream or sweet cream and butter with or without added salt, colour, respectively. In India and USA, ripened cream butter is not much popular, although in Europe and Canada, ripened cream butter is an important product. Use of cultured cream for butter manufacture generates sour butter milk which is considered as a waste. The buttermilk from sweet cream can be powdered for further use, also mixed with milk or discharged as waste. Manufacturing of butter using continuous butter making machines have advantages over batch churning, since use of continuous butter making machines generates less waste water, as the washing step after churning is avoided.

Major quantity of the effluent is generated when the residual product is flushed from equipment and floors and when equipment is hosed down at the end of production, as butter wash water or during CIP of the butter churner. Wastewater generated in this section contains mostly cream, butter, buttermilk, high concentrations of milk fat, starter (for ripened cream butter), salt, and colour. Butter sections (plants) generally generate strongly alkaline wastewater (Singh et al. 2014). The physicochemical characteristics of the wastewater generated from butter plant are presented in Table 8.5.

Waste load of butter section can be minimized by manually scraping of the production surfaces wherever possible to reduce the quantity of cream and butter in waste water streams before CIP process (Hale et al. 2003; Tetra Pak 2003). Waste water volume can also be reduced by optimizing the timing of equipment rinsing. Spills of butter milk should be avoided using sufficiently large vat to hold all the buttermilks discharged. Alternatively, buttermilk can be dried or used as animal feed and solids recovered from butter wash water can also be sold as stock feed to minimize waste.

8.2.4 Waste from Cheese Manufacture

Worldwide, Cheese is considered to be one of the major dairy products. The European Union tops the list of cheese production and consumption, followed by the United States (Carvalho et al. 2013). Cheese manufacturing plant generates effluents that pose significant environmental impact irrespective of the type of cheese (Cheddar, Parmesan, Mozzarella, Gouda, Danish blue, Brie, Camembert, Feta, Serpa, etc.) (Lee et al. 2003). Regardless of the cheese type, the production of cheese has general procedures such as pasteurization of milk, cooling of milk at optimum temperature for starter addition, addition of starter, and addition of rennet. Then formed curd is cut and cooked to remove whey followed by whey drainage and compressing of the curd. Other ingredients are now added and the cheese blocks are cut and packaged for sale after ripening (avoided for non-ripened one). Whey obtained from hard cheese productions are also used for cottage cheese or curd cheese manufacturing. Whey generated from cottage cheese production is also known as second cheese whey (SCW). Three main types of effluent generated during cheese production are cheese whey, second cheese whey, and the washing water of vats, storage, tanks, and pipelines which is also known as cheese whey wastewater. Major portion of wastewater from cheese plant which typically contain milk, milk fat, brine, whey, and cheese fines are generated when equipment is flushed at the end of a processing run, as well as during CIP operations. A significant amount of organic load also comes from whey and pieces of curd during cheese-making operations and spillages of starter milk.

Cheese plant effluent generally have high COD and BOD values in the range of 0.8–102.0 g/L and 0.6–60.0 g/L (Carvalho et al. 2013), respectively, and lactose and fat are the major contributor to the organic load. In general, cheese plant effluent has a low pH, although various researchers have reported the pH in the wide interval of 3.3–9.0. Suspended solids, TKN, and total phosphorus content can vary widely in the intervals of 0.1–22.0 g/L, 0.01–1.7 g/L, and 0.006–0.5 g/L, respectively. Cheese effluents pose a considerable risk of eutrophication in receiving waters particularly in lakes and slow moving rivers as it contains high amount of nutrients like total nitrogen and phosphorus (Prazeres et al. 2012). In addition to this, the ammonium nitrogen (NH^{4+}-N) which ranges from 0.06 to 0.27 g/L can also cause toxic effects to aquatic life (Farizoglu et al. 2007). Characteristics of effluent generated in the cheese processing are summarized in Table 8.6.

Waste generation in the cheese section could be reduced by using correct sized cheese vats, also taking care that it should not be filled up to the rim with milk. Whey should be collected sparingly and used in commercial applications. Curd spillages should be collected before the floors are washed and should be treated as solid waste (Hale et al. 2003; Tetra Pak 2003). Organic load of the waste water can also be reduced by completely removing whey and curds from vats before rinsing, segregating all whey drained from cheese, sweeping up pressings (particles), and screening all the liquid streams to collect fines.

Table 8.6 Characteristics of the effluent from cheese plant

Parameter	Cheese plant	Cheese factory	Cheese processing industry	Cheese factory	Cheese section	Cheese whey waste water	Cheese plant
pH	6.70	5.22	3.38	4.99	7.90	–	5.00
BOD_5	2.15	–	–	–	3.46	14.5 ± 1.5	0.36
COD	3.188	5.34	63.3	2.83	11.75	18.5 ± 1.4	0.54
BOD_5/COD	–	–	–	–	0.29	0.8	–
TS	2.3	4.21	53.2	–	–	7.7 ± 0.6	0.8
TSS	0.6	–	12.5	–	0.94	–	0.2
TDS	–	–	–	–		–	0.6
Fat	0.52	–	–	–	0.33	–	–
FOG	–	–	2.6	–	–	–	–
Alkalinity as $CaCO_3$	0.49	0.335	–	–	–	–	–
TP	–	–	–	–	–	–	0.32
Calcium	–	–	–	–	–	–	0.04
Chloride	–	–	–	–	–	2.41 ± 0.25	0.06
References	https://law.resource.org/pub/in/bis/S02/is.8682.1977.html	Strydom et al. (1997)	Hwang and Hansen (1998)	Sparling et al. (2001)	Janczukowicz et al. (2008)	Rivas et al. (2011)	Kannahi and Sangeetha (2014)

Values are in g/L, except pH, BOD/COD

8.2.5 Waste from Concentrated Milk Products

The production of concentrated milk such as condensed milk and evaporated milk starts with standardization of fat and dry solids content after which it is pasteurized, concentrated in an evaporator, homogenized, packaged, sterilized (evaporated milk), and cooled for storage. During manufacture of sweetened condensed milk, sugar is added (18–20% on milk basis) in the evaporation stage and an additional cooling and crystallization step is done to get optimum texture of the product. The organic loads of wastewaters generated during production of concentrated milks are majorly contributed by organic and inorganic compound deposits that formed as a result of the heating processes during milk evaporation, from equipment flushes at the start and end of production, sugars (in SCM production) and product spillages during packaging (Hale et al. 2003). The organic loads of waste water from concentrated milk production can be minimized by reducing the deposits at the product heating surfaces, which can be achieved by running evaporators at the lowest possible temperature (Tetra Pak 2003).

8.2.6 Waste from Powdered Dairy Products

Manufacturing of milk powder and other powdered dairy products like whey powder is a two-step process that involves the concentration of liquid pasteurized milk or whey by evaporation, after that the concentrated products are spray-dried. For production of whey powder, casein fines and milk fat are recovered; the whey is thermized, concentrated, and dried. Waste water from milk powder section includes mostly milk protein, lactose, and milk fat, while whey proteins and lactose, as well as casein fines and milk fat losses are generally found in waste water from whey powder section. Major source of waste water generation is at the start-up and shut-down operations of evaporation and drying processes as well as during cleaning operations. Effluents of milk powder tend to be alkaline in nature. Donkin (1997) characterized the waste water from butter/milk powder plant and found that the pH, BOD, and fat content of the effluent were 10–11, 1.5 g/L, and 0.4 g/L, respectively. Briao and Granhen Tavares (2007) reported that effluent generated from powder milk and powdered whey had pH of 10.43 ± 2.87, COD of 2.391 ± 1.928 g/L, N content of 0.0882 ± 0.0729 g/L, P content of 0.055 ± 0.0389 g/L, and oil and grease content 0.2966 ± 0.1663 g/L, respectively.

To minimize waste from powder section, it is suggested that liquid level should be maintained low enough to stop product boil-over in the evaporator, also avoid very long runs with higher than specified running rates which ultimately lead to blocked tubes that are difficult and time-consuming to clean and also causes higher pollution. It is better to use effluent entrainment separators to avoid carry-over of milk droplets during condensation of evaporated water. Condensate generated during evaporation processes should be used in cooling processes or for feeding the boiler. Also low concentration milk or other feed-stock should be recirculated until it reaches the required concentration. Rinse with 7% or more solids should be processed before scheduled shutdowns rather than discharging to the sewer (Victoria Environmental Protection Authority 1997). Uses of fabric filters or wet scrubbers are also suggested to minimize air emissions. The evaporator should be properly drained before CIP starts as well as between each CIP step, which will ultimately reduce the organic load of wastewaters generated. Dry powder from cleaning of spray towers and spills should be treated as solid waste (Hale et al. 2003; Tetra Pak 2003). In milk powder plant, water consumption can be further reduced to below 0.5 L/L, because the condensate produced during evaporation can be collected and used within the factory (e.g. for CIP systems, cheese curd wash, cooling towers, boilers) (Liu and Haynes 2011).

8.2.7 Waste from Dahi, Yoghurt Manufacture

Production of dahi/yoghurt typically involves standardization in terms of fat content and fortification with milk solids, addition of sugar (in case of sweetened variety) and stabilizers, heated to 60 °C, homogenization, heated again to about 95 °C for 3–5 min, and cooling to 30–45 °C and inoculation with a starter culture. For set dahi/yoghurts, after addition of starter to milk, the content is packed directly and containers are incubated for the desired period, after which they are cooled and distributed. For stirred yoghurts/lassi, after addition of starter to milk, the milk is incubated in bulk after which it is cooled and packaged, and then distributed. Major quantity of the wastewater is generated during cleaning equipment such as starter vat, fermentation vats, heat exchangers, and ingredients' storage vessels. The wastewaters from dahi/yoghurt plant contain residue starter, dilute yoghurt, milk fat, heat deposits from heat exchangers, pieces of fruit and diluted fruit preserve, diluted sugars (including lactose), and stabilizers as well as flavour compounds (Tamime and Robinson 1999; Hale et al. 2003). As the product is highly viscous, it tends to stick to the equipment, pipelines surfaces, thus contributing higher organic load to the cleaning waters. Spillages of fruit concentrates which has COD levels of $\geq$500 g/mL can also increase the organic load (Hale et al. 2003). Another important issue is the generation of acid whey as a byproduct in the manufacture of Greek or strained yoghurt. Greek or strained yoghurt is a popular product in countries like USA, UK, middle-east countries, etc. In contrast to sweet whey, acid whey is almost regarded as waste, has a BOD of over 35,000 ppm and pH of acid whey tends to be lower than 5.1, and frequently even lower, 4.5 or less. The waste water sample from a milk/yoghurt plant had pH 6.92, COD 4.656 g/L, TS content 2.75 g/L, and alkalinity ($CaCO_3$) 0.546 g/L (Strydom et al. 1997). Koyuncu et al. (2000) also reported that waste water from yoghurt and butter milk processing had COD of 1.5 g/L, BOD 1 g/L, pH 7.2, TSS 0.191 g/L, and TN 0.063 g/L.

To reduce the organic load of waste water from dahi, in yoghurt plant some measures have been adopted, e.g. spilled dry ingredient was collected and treated as a solid waste, drainage times of dahi/yoghurt vats and pipes were extended. Also heat deposition during the heating of sweetened yoghurt was reduced by more gradual heating. Adoption of proper CIP management like use of burst rinsing rather than continuous rinsing for cleaning yoghurt/dahi vats and reuse of final rinse water also reduces further generation of waste water (Tamime and Robinson 1999).

8.2.8 Waste from Ice Cream Processing

Ice cream processing starts with preparation of ice cream mix by combining raw materials such as milk, cream, butter, milk powder, whey powders, etc., and then the mix is homogenized, pasteurized, and transferred to a vat for ageing, after which flavourings, colourings, and fruit are added before freezing. During primary freezing the mixture is partially frozen and air is incorporated to obtain the required texture, often regarded as softy. Then the product is packaged, transferred to hardening chamber, stored, and distributed. Majority of the waste generated is due to clean up operations of plant starting with equipment, pipelines, pumps, etc. Also a significant quantity of wastewater is generated from the backflushes whenever there is major changes in flavour is to be made. Wastewater from the ice cream industry is mainly composed of organic matter and suspended solids from the raw materials (milk, cream, milk powder, fruit, nuts, emulsifier, stabilizers, colour, flavour, etc.), having high concentrations of carbohydrates, proteins, and fats. Some of the reports regarding characteristics of effluent from ice cream plant are presented in Table 8.7.

Load of waste water in the ice cream section can be reduced by completely emptying the ice cream storage tanks before cleaning. Flush out water should also be

Table 8.7 Characteristics of the effluent from ice cream plant

pH	6.96 ± 0.33	5.2	8.46	6.25 ± 0.15	4.4	6.96–7.95
BOD_5	–	2.45	–	0.523	–	1.8
COD	4.94 ± 1.16	5.2	13.6	11.9	5.902	1.6–3.2
BOD_5/COD	–	0.47	–	–	–	1.52–1.84
TS	–	3.9	0.7	0.664	3.269	3.8
TSS	1.04 ± 0.40	3.1	–	0.265	0.0025	1.158–1.183
TDS	–	–	1.275	0.399	0.028	–
TVS	–	–	–	–	2.739	–
TKN	–	0.06	–	0.088	–	–
N-$NH4^+$	–	0.015	–	–	–	–
TP	–	0.014	–	–	–	–
K	–		–	0.157	–	–
Chloride	–		0.8	0.256	–	–
Sodium	–		–	0.098	–	–
References	Hawkes et al. (1995)	Borja and Banks (1995)	Patel et al. (2013)	Qasim and Mane (2013)	Ana et al. (2014)	Dubey and Joshi (2015)

Values are in g/L, except pH, BOD/COD

conveyed to the mixing tank rather than wasting. Automatic steam/water mixing valves should be installed to reduce water wasting when a hose is dropped and left to run. To reduce the quantity of residual mix in the flavour tanks closer mix formulation control should be started.

8.2.9 Waste from Casein Manufacture

Effluents from this plant primarily contain whey of milk, mineral acids when used for precipitation, and washings. This unit generally functions on a batch basis and hence discharge of effluent is intermittent. Effluent from a casein plant with 3 tonnes casein per day capacity reported to have pH 7.7, alkalinity (as $CaCO_3$) 0.49 g/L, TS 0.68 g/L, TSS 0.16 g/L, BOD 5 0.2 g/L, COD 0.372 g/L, and oil and grease contents are in nil amount (https://law.resource.org/pub/in/bis/S02/is.8682.1977.html).

8.2.10 Wastewater from Associated Processes

Most of the water consumed in a dairy processing plant is used in associated processes such as the cleaning and washing of floors, bottles, crates, and vehicles, and the cleaning-in-place (CIP) of factory equipment and tanks as well as the inside of tankers. Most CIP systems consist of three steps: a pre-rinse step to remove any loose raw material or product remains, a hot caustic wash to clean equipment surfaces, and a cold final rinse to remove any remaining traces of caustic. Significant quantity of waste water also generated from washing of bottles and crates. As these plants use detergents and/or caustic or washing soda in solution to wash bottles and crates, the effluent are generally alkaline due to the use of caustic or washing soda. Scharnagl et al. (2000) reported that waste water from bottles washing plant had COD of 210 ppm, pH of 11.8, and total organic carbon content of 92 ppm. Characteristics of waste water from water softening plants depend upon the process used. The effluent could be mildly acidic if both cation and anion exchange resins are used or highly acidic or saline if only cation exchange is used. Waste water from boiler blow-down generally contains suspended solids of mineral origin, but the amounts of waste water generated in these sections are low in quantity. The chilled water used for pasteurization of milk, cream is generally not contaminated and can be reused after cooling. Although some cooling water is usually removed to prevent build up of inorganics and consequent scaling.

8.3 Treatment of Dairy Waste

Wastes from dairy processing plant are generally high in dissolved organic matter, contain nearly 1000–2000 ppm and are nearly neutral in pH. Since these wastes are mainly composed of organic matter, they are suitable to treatment by biological method. Aerobic processes are most suitable but the final selection of treatment method depends on location and size of the plants. Before selecting any treatment method, a complete process evaluation should be done along with economic analysis. This should include the wastewater composition, concentrations, volumes generated, and treatment susceptibility, as well as the environmental impact of the solution to be adopted. The most effective conventional methods are aeration, trickling filtration, activated sludge, irrigation, lagooning, and anaerobic digestion. As there is wide variation in the strength and flow rate of the dairy waste, holding and equalization are desirable for uniform waste treatment. A provision of grease trap is also needed as a pretreatment to remove fats and other greasy substances from the waste. Both, the high rate trickling filter and activated sludge plants can be effective for the complete treatment of the dairy wastes, with limitations of more maintenance, involvement of skilled personal, and requirement of special type of equipment. Whereas oxidation pond, aerated lagoon, waste stabilization pond can be used with simpler equipment, less maintenance, thus lower cost. In general, biological methods are the most cost-effective for the removal of organics from the waste water, although aerobic methods are easier to control, but anaerobic methods have advantages such as lower energy requirements and lower sludge production rates. As no single process for treatment of dairy wastewater is competent by itself of complying with the minimum effluent discharge requirements, it is better to use combined process especially designed to treat a specific dairy wastewater.

Waste water generation can be reduced in the range of 0.5–1.0 L/L using advanced equipment (e.g., CIP systems that re-circulate water) along with awareness and water conservation practices by employees and management setup (Crothers 2007). Generation of waste water load can also be reduced by some advanced CIP systems with microfiltration, ultrafiltration, and nanofiltration and regenerate spent cleaning solutions, particularly NaOH (Durham and Hourigan 2007).

8.4 Utilization of Dairy Waste: Sustainable Approach

The waste water from dairy processing contains higher organic load compared to other food industry waste. Thus, a lot of efforts have been directed towards utilization of the enormous waste water for generation of industrially important compounds like biofuel, bioenergy, bioplastics, biosurfactants, biofertilizers, polysaccharide, organic acids, enzymes, and others which will also reduce the pollution problem arising from these wastes. Ghee residue, a byproduct of ghee plant is currently partially utilized as animal food, some portion is converted into food products like confectionery, bakery, sweets and a considerable amount of ghee residue is wasted after fat separation. Whey, a byproduct of paneer, cheese industry is further processed into whey protein concentrate, whey protein isolate, whey powder, delactosed whey, etc. And also lactose, lactoperoxidase, lactoferrin, α-lactalbumin, β-lactoglobulin have been separated from whey and incorporated into development of functional foods like baby food, diet food, sports drink, and other beverages. Still a large amount of whey remains with higher organic load and researchers have used the whey as a substrate for production of SCP, enzymes, vitamins, penicillin, alcohol, lactic acids, etc. Table 8.8 shows some of the current reports regarding application of biotechnology for dairy waste utilization.

8.4.1 Biofertilizer Production

Biofertilizers are substances composed of living cells of different types of microorganisms which promoted the plant growth and protect the plants from diseases and infestations. They convert nutritionally important nitrogen, phosphorous from unavailable to available form by nitrogen fixation, rock mineral solubilization when applied to soil, plant surface, and to seed. Nitrogen (N) fixing, potassium (K) solubilizing, and phosphorus (P) solubilizing microorganisms are commonly used as a biofertilizer. *Rhizobium*, *Azotobacter*, *Azospirillum*, and blue green algae (BGA) have been used as a biofertilizer since long. As such sludge generated after dairy waste water treatment have been used as a biofertilizer. A few studies have been carried out for direct conversion of waste water into biofertilizer; however, in some cases it have been used for producing biomass to be used as biofertilizer.

Yadavalli and Heggers (2013) treated dairy effluent in two stage, i.e. first in a photobioreactor using immobilized *Chlorella pyrenoidosa* and the second stage with two column sand bed filtration technique and reported that two-stage treatment was a very effective way of removing BOD, COD, nitrate, and phosphate contents. The biomass of *C. pyrenoidosa* separated after treatment was used as biofertilizer for rice seedlings and there was 35% increase in growth in rice plant. Chaudhuri (2018) developed tailor made microbial consortium to covert dairy effluent to phosphate rich biofertilizer which increased the production of mung bean about 1.4-folds when compared with chemical fertilizer. Similarly Halder et al. (2020) also converted dairy waste water into biofertilizer with tailor made microbial consortium (5 isolates from activated dairy sludge and one from marine coastal waters) and the biofertilizer improved seed yield of mung bean by 2.6-fold, it also protected mung bean from aphid (*Aphis craccivora* Koch) infestation.

8.4.2 Single-Cell Protein (SCP) Production

Single-cell protein (SCP) refers to crude or refined protein of algal, bacterial, mould, or yeast origin which is used either as animal feed or human food. The production and utilization of microbial biomass as a source of food proteins gained particular interest as an alternative source for proteins of agricultural origin due to its high content of protein. In addition to proteins, SCP contains other nutrients such as lipids and vitamins. Dairy waste mainly whey and waste milk has been used as a medium for production of SCP.

Sisman et al. (2013) used whey based agar medium for cultivation of *Trichoderma harzianum* to produce SCP. Kebbouche-Gana and Gana (2014) produced SCP with cheese whey for cultivation of *Candida kefyr* and maximum 19 g/L dry bio mass production was achieved in a batch submerged fermentation method. Paneer whey was also used for production of SCP with *Kluyveromyces marxianus* and highest cell biomass and crude protein content achieved was 4.5 g/L cell dry weight and 48.1%, respectively (Babu et al. 2014). Yadav et al. (2014) utilized whey

Table 8.8 Utilization of dairy waste for generation of industrially important compounds

Product	Waste	Process	Ref
Biofertilizers (treated water)	Dairy waste water	Treatment of waste water with tailor made microbial consortia or consortium (5 isolates from activated dairy sludge and one from marine coastal waters).	Halder et al. (2020)
Biofertilizers (biomass)	Dairy waste water	Two-stage effluent treatment with *Chlorella pyrenoidosa* in photobioreactor and sand bed filtration	Yadavalli and Heggers (2013)
SCP *Kluyveromyces lactis* (TY-98) and *Rhodotorula graminis* (TY-99)	Waste milk	Fermentation with the yeast, biomass separation and biomass yield was 43.8 g/L (dry cell weight)	Myint et al. (2020)
Bioplastics	Dairy waste water	Bioplastic films from casein isolated from dairy wastewater using a dissolved air flotation method and combining this DAF-casein with κ-carrageenan, sodium carboxymethylcellulose and gelatin.	Ryder et al. (2020)
PHA	Fermented cheese whey	Fermentation of cheese whey medium with mixed microbial culture (enriched from activated sewage sludge) PHA production 60–70 g/kg cheese whey total solid	Colombo et al. (2016)
Biosurfactant	Milk whey along with barley	Fermentation by *Penicillium sclerotiorum* Ucp 1361 Surface tension reduction from 72 to 27 mN/m	Truan et al. (2020)
Biosurfactant (glycolipopeptide)	Whey based medium	Fermentation by *Lactococcus lactis* CECT-4434 Glycolipopeptide production—8.9 mg/L	Vera et al. (2018)
H_2 and CH_4	75% dairy waste water	Fermentation with *Enterobacter aerogenes* and methanogenic bacteria of cow dung H_2 production—0.562 L/L CH_4/production—0.59 L/L	Kothari et al. (2017)
Biodiesel (biomass)	Dairy waste water	*Rhodococcus opacus* was grown in dairy waste water and biomass was used for biodiesel production	Kumar et al. (2015)
Ethanol	Cheese whey permeate	Fermentation of cheese whey permeate was studied using *Kluyveromyces lactis* CBS2359 Ethanol production—15.0 g/L, yield 0.47 g/g lactose	Sampaio et al. (2019)

(continued)

Table 8.8 (continued)

Product	Waste	Process	Ref
Polysaccharide	Salt whey from cheese plant and milk permeate	Fermentation by *Alteribacillus bidgolensis* and *Bacillus licheniformis*; Polysaccharide production—52 g/L, and 42 g/L in whey and 43 g/L and 36 g/L in milk permeate	Hegazy et al. (2019)
EPS	Milk whey	Fermentation by *Lactobacillus delbrueckii* subsp. *bulgaricus* EPS production—69.60 ± 9.48 mg/L	Malaka et al. (2020)
Lactic acid	Cheese whey powder	Fermentation with *Lactobacillus bulgaricus* in fed batch fermentation LA production—113.18 g/L and productivity—2.36 g/L/h	Liu et al. (2018)
Acetic acid	Cheese whey	Fermentation Productivity—96 g/L/h	Pal and Nayak (2016)
Propionic acid	Skimmed whey	*Fermentation with Lactobacillus helveticus* and *Propionibacterium freudenreichii* Propionic acid production—3.78 g/L	Ngome et al. (2017)

for SCP production by mono and mixed culture of *K. marxianus* and *Candida krusei* and to remove organic load of whey. The author demonstrated that mixed culture showed better COD removal (8.8% more), higher biomass production (19% more) than monoculture. The same research group used whey for production of food graded SCP for cultivation of mono and mixed culture of *K. marxianus* and *Saccharomyces cerevisiae* (Yadav et al. 2016). Whey was used as a low-cost raw material for production of SCP from *K. marxianus* and after 24 h of fermentation the biomass yield was 36 mg/mL with crude protein content of 83.33% (Nayeem et al. 2017). In a recent study, SCP and carotenoid production were carried out with mixed culture (1:10 ratio of a mixed TY-98: TY-99) of flower yeast *Kluyveromyces lactis* (TY-98) and *Rhodotorula graminis* (TY-99) from waste milk and under optimum condition of 30 °C, initial pH 6 maximum biomass yield 43.8 g/L (dry cell weight), and carotenoid production of 1.8 mg/L was achieved (Myint et al. 2020).

8.4.3 Bioplastics Production

Bioplastics are the families of polymer materials which include bio-based or biodegradable, or both. Various bacteria and archaea are known to produce bioplastics like polyhydroxyalcanoates (PHAs), poly-β-hydroxybutyrate (PHB) from various carbon sources, and these bioplolymers are accumulated in cells under the form of granules as carbon and energy storage. Dairy waste, whey, fermented whey have been used as a substrate along with other nutrients and other waste material for production of PHA and PHB.

Polyhydroxyalcanoates (PHAs) production was studied using milk whey and dairy wastewater activated sludge to an optimum C/N ratio (Bosco and Chiampo 2010). The researchers evaluated the effect of different treatment of whey to remove protein and effect of pH on PHAs production. Treatment of whey to autoclaving at 121 °C for 15 min, cooling, centrifugation at 8000 g for 15 min at 4 °C, and then filtration with a glass fibre filter, C/N ratio 50 without pH correction resulted in highest PHA production of 13.82% (g/g biomass). Pandian et al. (2010) utilized dairy waste along with rice bran and sea water for PHB production by *Bacillus megaterium* and reported that under optimum condition in a batch study PHB production was 6.37 g/L and in fed batch fermentation; highest 11.32 g/L PHB was produced when dairy waste was fed at different time intervals of 0th, 12th, and 24th hour, respectively, out of 36 h fermentation duration. Koller (2015) used hydrolyzed whey permeate as a source of carbon for PHA production with extremely halophile archaeon *Haloferax mediterranei*. The researcher also used waste generated, e.g. cell debris and spent fermentation broth after PHA recovery/downstream processing for subsequent PHA production. The author reported that costly nitrogen and phosphorous source yeast extract 29% in the fermentation medium can be replaced by cell debris from previous fermentation process. Salty supernatant was also used as a medium for recycling experiment after adjusting the final medium composition. In the original fermentation final PHA concentration achieved was 7.2 g/L and in the recycling experiment it was 2.28 g/L. The final PHA in biomass obtained in the original fermentation and recycling experiment was 66 and 70%, respectively. Pais et al. (2016) evaluated production of poly(3-hydroxybutyrate-*co*-3-hydroxyvalerate), P(3HB-*co*-3 HV) by *Haloferax mediterranei* in a highly saline medium where cheese whey was used as a substrate. In batch fermentation method active biomass concentration of 7.54 g/L, P(3HB-*co*-3 HV) content of 53% was achieved. Colombo et al. (2016) produced PHA from two fermented cheese whey medium (medium 1 with lactic, acetic, and butyric acids; medium 2 with acetic, propionic, butyric, lactic, and valeric acids) with mixed microbial culture. The cultures were enriched from activated sewage sludge and maximum PHA production of 60–70 g/kg cheese whey total solid. Most recently Ryder et al. (2020) generated bioplastic films from casein isolated from dairy wastewater using a dissolved air flotation method and combining this DAF-casein with κ-carrageenan, sodium carboxymethylcellulose, and gelatin.

8.4.4 Biosurfactants Production

Biosurfactants and bioemulsifiers are microbial amphiphilic compounds including neutral lipid, phospholipid, glycolipids, lipopolysaccharide, lipopeptides, polysaccharide–protein complexes, protein-like substances, and fatty acids which have surface and emulsifying activities. Biosurfactants can be produced by wide range of bacteria, fungi, and yeasts, have diverse nature chemically and different microorganisms can produce different types and amount of biosurfactant. Biosurfactant have application in bioremediation, biodegradation, oil recovery, in food and pharmaceuticals and others. Whey, fermented whey, permeate from membrane processing have been used as a cost-effective medium for biosurfactant production by bacteria, e.g. *Pseudomonas, Bacillus, Lactococcus, Enterococcus,* fungi like *Penicillium* spp.

Dubey et al. (2012) utilized curd whey for biosurfactant production by *Pseudomonas aeruginosa* PP2 and *Kocuria turfanesis* strain-J and stated that the biosurfactant produced have potentiality in bioremediation of pesticide contaminated soil. Rhamnolipid production was studied in whey (50% in water) and olive oil mill wastewater (30%) using *Ps. aeruginosa* and its recombinant (expressing *Vitreoscilla* haemoglobin gene *vgb*). Under optimum condition maximum rhamnolipid production achieved in whey was 9.6 and 13.3 g/L by parent strain and the recombinant, respectively (Colak and Kahraman 2013). Deproteinized cheese whey was used as a low-cost medium for biosurfactant production by a probiotic strain of *Lactobacillus* and maximum 0.8 mg/mL biosurfactant yield was achieved at lab scale fermenter. The biosurfactant was also identified to be glycolipid closely similar to xylolipids (Sharma et al. 2014). Patowary et al. (2016) studied rhamnolipid production by *Ps. aeruginosa* strain, SR17 in paneer whey and reported that yield of biosurfactant was 2.7 g/L, can be increased to 4.8 g/L by supplementing 2% glucose and mineral salts. The strain was efficient to use paneer whey and reduced the surface tension of the medium from 52 to 26.5 mN/m. The researchers also studied emulsification efficiency of the biosurfactant against *n*-hexadecane, olive oil, kerosene, diesel oil, engine oil and found more than 80% efficiency; in crude oil emulsification efficiency was 100%. A low-cost waste based medium containing 15% whey was developed and used for production of glycolipopeptide (biosurfactant) by *Lactococcus lactis* CECT-4434 and average production of 8.9 mg/L and surface tension reduction by about 18.1 mN/m was reported (Vera et al. 2018). Bioemulsifier production by *Enterococcus mundtii* Tw278 in a lactic whey based medium with highest yield of 270 mg/L was reported by (de Carvalho et al. 2019). Most recently, Truan et al. (2020) used milk whey along with barley for production of biosurfactant by *Penicillium sclerotiorum* Ucp 1361 and reported surface tension reduction from 72 to 27 mN/m. Permeate effluent from membrane separation of whey (to produce whey protein concentrate) was used as a medium for biosurfactant production by *B. methylotrophicus* and *B. pumilus* (Decesaro et al. 2020).

8.4.5 Bioenergy

Generation of biohydrogen and bio-methane gaining popularity as an alternate of sustainable and renewable energy source. Various agro industrial wastes including dairy waste water have been used for production of hydrogen, methane and to reduce organic load of the waste water.

Antonopoulou et al. (2008) fermented raw cheese whey in a continuous-type stirred tank bioreactor and produced H_2 2.49 L/L of cheese whey and then this mixed liquor from the first stage was subsequently digested anaerobically to generate CH_4 @ 17.9 L/L of cheese whey. H_2 production of 12.73 mM H_2/g COD was achieved from dairy waste water having initial COD 21.1 g level (Moreno-Dávila et al. 2011). Co-addition of hematite and nickel oxide nanoparticles improved the biohydrogen yield by 27% and production by 59% from dairy waste water (Gadhe et al. 2015). Kothari et al. (2017) used 75% dairy waste water for sequential production of H_2 and methane using *Enterobacter aerogenes* and methanogenic bacteria of cow dung and reported H_2 production of 0.562 L/L and 0.59 L CH_4/L. Da Silva et al. (2019) produced biohydrogen from natural fermentation of dairy wastewater using anaerobic fluidized-bed reactor, reported hydrogen yield of 2.56 mol H_2/mol carbohydrate and maximum hydrogen yield was achieved at organic loading rate of 28.7 ± 8.9 kg COD/m^3/d.

8.4.6 Biofuels

Majority of the biofuels currently used today are ethanol and biodiesel. Commercially available biofuels are mainly produced from corn starch, sugarcane, soybean, or oilseed rape the edible food crops, etc. In the dairy industry large amount of whey is generated from both cheese and paneer plant and only portion of this whey is used to produce products like whey powder, whey protein concentrate, and several dairy beverages. Research has been carried out to use whey, whey permeate as a substrate for ethanol production as it contains quite amount of lactose. Sludge from dairy effluent treatment plant is been used for generation of biodiesel (Balasubramanian et al. 2018). Dairy waste water, cheese whey has been used for generation of biomass for biodiesel production as well as for COD reduction. *Chlorella vulgaris* (Choi and Choi 2016), microalgae *Chlamydomonas polypyrenoideum* (Kothari et al. 2013), *Chlorella* sp. (Lu et al. 2015), and bacteria *Rhodococcus opacus* (Kumar et al. 2015) were grown in dairy waste water for biodiesel production.

Liu et al. (2016) developed a *L. lactis* strain CS4435 (MG1363 Δ3*ldh*, Δ*pta*, Δ*adhE,* pCS4268), which solely produce ethanol from lactose and used the strain to produce ethanol from residual whey permeate with corn steep liquor hydrolysate as a nitrogen source. In a fed batch fermentation the strain produced 41 g/L ethanol with yield of 70%. *K. marxianus* URM 7404 was used for ethanol production from cheese whey and whey permeate as a substrate under aerobic, microaerobic, and anaerobic condition. Under anaerobic condition, the ethanol production of 8.90 g/L in whey permeate after 18 h of fermentation and 8.08 g/L in cheese whey of 10 h fermentation was achieved, although ethanol yield was high in cheese whey (Murari et al. 2019). Ethanol production from cheese whey permeate was studied using *K. lactis* CBS2359 in shake cultures and maximum ethanol concentration obtained was 15.0 g/L, yield 0.47 g/g lactose, and productivity 0.31 g/L/h (Sampaio et al. 2019).

8.4.7 Polysaccharide Production

Wide array of microbes produce polysaccharides which are either excreted outside the cell (EPS; exopolysaccharide) or form a layer outside the cell as in case of capsular polysaccharide. Some of these polysaccharides such as dextran, xanthan, pullulan, gellan are commercially produced by microorganisms which have wide range of application in food (additive such as biothickener, gelling agent, texturizers, etc.), pharmaceuticals, and other industries. In fermented i.e., yoghurt and cheese, the in situ production of EPS by food grade LAB offers technological benefits like low TS and fat content in final products along with health benefits like cholesterol-lowering ability, antimicrobial, immune modulating properties and also as prebiotic. Whey is being used as a medium for the production of polysaccharides as a low-cost medium.

For cost-effective production of dextran by *Leuconostoc mesenteroides* BA08, a whey based medium supplemented with sucrose, yeast extract, and K_2HPO_4 was developed and under batch fermentation dextran production of 17.25 g/L was achieved (Lule et al. 2016). Kefiran was produced in sweet whey medium containing either glucose or sucrose from kefir grain and 5.2 g/L of its production was reported (Pais-Chanfrau et al. 2018). Hegazy et al. (2019) studied EPS production from *Alteribacillus bidgolensis* and *B. licheniformis* in salt whey from cheeses plant and milk permeate and reported its production of 52 g/L, 42 g/L in whey and 43 g/L and 36 g/L in milk permeate by both the strains, respectively. Malaka et al. (2020) recently reported EPS production of 69.60 ± 9.48 mg/L in milk whey by a strain of *Lb. delbrueckii* subsp. *bulgaricus* after fermentation of 16 h at 30 °C.

8.4.8 Production of Organic Acids, Enzymes, and Bioactive Peptides

Organic acids like lactic acid, acetic acid, propionic acids are produced by fermentation and whey has been used as low-cost substrate for production of organic acids. Lactic acid is industrially produced by fermentation with *Lb. delbrueckii* using whey as one of the substrate. Neutral protease treated cheese whey powder was used for production of lactic acid by strain of *Lb. bulgaricus* and under optimum condition in fed batch fermentation LA production was 113.18 g/L and productivity 2.36 g/L/h (Liu et al. 2018).

Pal and Nayak (2016) developed a sustainable technology to produce acetic acid and whey protein from cheese whey. Traditional fermenter was integrated with multistage fouling free membrane with cell, unused sugar and nutrient recycling and continuous product removal. The technology resulted higher yield of >98% and productivity of 96 g/L/h. The final osmosis stage concentrated acetic acid to 962 g/L. Bioproduction of acetic acid and propylene glycol was demonstrated by cultivation of *Lb. buchneri* using whey lactose and whey powder containing lactose and protein as an economical source of nutrient medium. At the end of the fermentation process, an equal concentration of acetic acid and propylene glycol (25–30 g/L) produced using cotton cheese cloth immobilized with *Lb. buchneri* (Veeravalli and Mathews 2018).

Indole-3-acetic acid, a plant growth-promoting substance was produced in a whey based medium by *Enterobacter* sp. DMKU-RP206 and under optimum condition 3963.0 mg Indole-3-acetic acid/L was produced (Srisuk et al. 2018).

Ngome et al. (2017) used skimmed whey as a low cost medium for production of propionic acid by mixed culture of *Lb. helveticus* and *Propionibacterium freudenreichii* and reported production of 3.78 g/L propionic acid.

Enzymes like cysteine protease (was produced in cheese whey supplemented with minerals by *Rhizopus oryzae*, Gul et al. 2012), food grade protease (was produced from dairy effluent by *Bacillus* spp., Madhu 2016), lipase, laccase (was produced by *Trametes hirsuta* in a ghee residue based medium, Khanam et al. 2013) were produced from dairy wastes.

Whey is one of the important sources of bioactive peptides with diverse physiological functions like antihypertensive, antimicrobial, antioxidative, antidiabetic, immunomodulatory, antiproliferative, opioid, and other biological activities (Mann et al. 2019).

8.5 Conclusion

Waste generated during all the steps in the dairy processing, starting with milk reception, processing, packaging, transportation, storage, distribution, and marketing, impacts the environment. Due to highly diversified nature of dairy industry, various product processing, handling, and packaging operations create wastes of different quality and quantity, which, if not treated, could lead to increased disposal and severe pollution problems. In general, wastes from the dairy processing industry contain high concentrations of organic material such as proteins, carbohydrates, and lipids, high concentrations of suspended solids, chlorides, or high biological oxygen demand (BOD) and chemical oxygen demand (COD). Milk powder and butter plants tend to have a strongly alkaline wastewater while the production of lactic acid in the wastewater from cheese, casein, and whey plants makes the wastewater from these plants acidic in nature. Generation of waste water load can be reduced by some advanced CIP systems which include recirculation system to regenerate spent cleaning solutions; using advanced technologies like microfiltration, ultrafiltration, and nanofiltration and proper water management practices. Application of biotechnology to convert the organic matters of waste to useful substances simultaneously reducing the organic load is a promising option. There are number of technology available now to produce substances like organic acids, enzymes, SCP, energy, fuels, biofertilizer, biosurfactant, etc., from dairy waste, however their commercial application is very limited as large-scale production is not economical due to factors like lower yield, higher initial cost, problem in purification, etc.

Although the dairy industry is not commonly associated with severe environmental problems, it must continually consider its environmental impact, particularly as dairy pollutants which are mainly of organic origin and works should be done to develop cost-effective processes for conversion of dairy waste to industrially useful compounds.

References

Al-Shammari SB, Bou-Hamad S, Al-Saffar A, Salman M, Al-Sairafi A (2015) Treatment of dairy processing wastewater using integrated submerged membrane microfiltration system. J Environ Anal Toxicol 5:278

Ana L, Torres-Sánchez, López-Cervera SJ, de la Rosa C, Maldonado-Vega M, Maldonado-Santoyo M, Peralta-Hernández JM (2014) Electrocoagulation process coupled with advance oxidation techniques to treatment of dairy industry waste water. Int J Electrochem Sci 9:6103–6112

Antonopoulou G, Stamatelatou K, Venetsaneas N, Kornaros M, Lyberatos G (2008) Biohydrogen and methane production from cheese whey in a two-stage anaerobic process. Ind Eng Chem Res 47(15):5227–5233

Arbeli Z, Brenner A, Abeliovich A (2006) Treatment of high strength dairy waste water in an anaerobic deep reservoir: analysis of the Methanogenic fermentation pathway and the rate limiting step. Water Res 40:3653–3659

Babu M, Raj SP, Nirmala CB, Deccaraman M, Sagadevan E (2014) Production of single cell protein using *Kluyveromyces marxianus* isolated from paneer whey. Int J Biomed Adv Res 5 (5):255–258

Balasubramanian R, Sircar A, Sivakumar P, Anbarasu K (2018) Production of biodiesel from dairy wastewater sludge: a laboratory and pilot scale study. Egypt J Pet 27(4):939–943

Borja R, Banks CJ (1995) Response of an anaerobic fluidized bed reactor treating ice-cream wastewater to organic, hydraulic, temperature and pH shocks. J Biotechnol 39(3):251–259

Bosco F, Chiampo F (2010) Production of polyhydroxyalcanoates (PHAs) using milk whey and dairy wastewater activated sludge: production of bioplastics using dairy residues. J Biosci Bioeng 109(4):418–421

Briao VB, Granhen Tavares CR (2007) Effluent generation by the dairy industry: preventive attitude and opportunities. Braz J Chem Eng 24(4):487–497

Britz TJ, Schalkwyk C, Hung YT (2006) Treatment of dairy processing wastewater. In: Wang LK, Hung YT, Lo HH, Yapijakis C (eds) Waste treatment in the food processing industry. Taylor and Francis Group, New York, pp 1–27

Carvalho F, Prazeres AR, Rivas J (2013) Cheese whey wastewater: characterization and treatment. Sci Total Environ 445:385–396

Chaudhuri SR (2018) Dairy effluent conversion into biofertilizer using tailor-made microbial consortium: the waste to wealth approach. J Food Microbiol Saf Hyg 3. https://doi.org/10.4172/2476-2059-C4-017

Choi HJ, Choi HJ (2016) Dairy wastewater treatment using microalgae for potential biodiesel application. Environ Eng Res 21(4):393–400

Colak AK, Kahraman H (2013) The use of raw cheese whey and olive oil mill wastewater for rhamnolipid production by recombinant *Pseudomonas aeruginosa*. Environ Exp Biol 11 (3):125–130

Colombo B, Sciarria TP, Reis M, Scaglia B, Adani F (2016) Polyhydroxyalkanoates (PHAs) production from fermented cheese whey by using a mixed microbial culture. Bioresour Technol 218:692–699

Crothers G (2007) Water use in the dairy processing industry. Retrieved from http://www.abare.gov.au/interactive/Outlook08/files/day_1/Crothers_food.pdf

da Silva AN, Macêdo WV, Sakamoto IK, Pereyra DDLAD, Mendes CO, Maintinguer SI, Caffaro Filho RA, Damianovic MHZ, Varesche MBA, de Amorim ELC (2019) Biohydrogen production from dairy industry wastewater in an anaerobic fluidized-bed reactor. Biomass Bioenergy 120:257–264

de Carvalho KG, Gómez JE, Vallejo M, Marguet ER, Peroti NI, Donato M, Itri R, Colin VL (2019) Production and properties of a bioemulsifier obtained from a lactic acid bacterium. Ecotoxicol Environ Saf 183:109553

Decesaro A, Machado TS, Cappellaro ÂC, Rempel A, Margarites AC, Reinehr CO, Eberlin MN, Zampieri D, Thomé A, Colla LM (2020) Biosurfactants production using permeate from whey ultrafiltration and bioproduct recovery by membrane separation process. J Surfactant Deterg 23 (3):539–551

Donkin J (1997) Bulking in aerobic biological systems treating dairy processing waste waters. Int J Dairy Technol 50:67–72

Dubey S, Joshi YP (2015) Characterization and treatment of ice cream industry wastewater using UASB reactor. Int J New Technol Sci Eng 2(5):69–76

Dubey KV, Charde PN, Meshram SU, Shendre LP, Dubey VS, Juwarkar AA (2012) Surface-active potential of biosurfactants produced in curd whey by *Pseudomonas aeruginosa* strain-PP2 and *Kocuria turfanesis* strain-J at extreme environmental conditions. Bioresour Technol 126:368–374

Durham RJ, Hourigan JA (2007) Waste management and co-product recovery in dairy processing. In: Waldron K (ed) Handbook of waste management and co-product recovery in food processing, vol 1. Woodhead/CRC Press, Cambridge/Boca Raton, pp 332–387

Farizoglu B, Keskinler B, Yildiz E, Nuhoglu A (2007) Simultaneous removal of C, N, P from cheese whey by jet loop membrane bioreactor (JLMBR). J Hazard Mater 146(1–2):399–407

Gadhe A, Sonawane SS, Varma MN (2015) Enhancement effect of hematite and nickel nanoparticles on biohydrogen production from dairy wastewater. Int J Hydrog Energy 40 (13):4502–4511

Gul A, Baig S, Naz M, Nadeem M (2012) Efficient utilization of dairy industry waste for hyper production and characterization of a novel cysteine protease. Pak J Zool 44:713–721

Halder N, Gogoi M, Sharmin J, Gupta M, Banerjee S, Biswas T, Agarwala BK, Gantayet LM, Sudarshan M, Mukherjee I, Roy A (2020) Microbial consortium–based conversion of dairy effluent into biofertilizer. J Hazard Toxic Radioact Waste 24(1):04019039

Hale N, Bertsch R, Barnett J, Duddleston WL (2003) Sources of wastage in the dairy industry. In: Guide for dairy managers on wastage prevention in dairy plants, IDF Bull, 382:7–30.

Hawkes FR, Donnelly T, Anderson GK (1995) Comparative performance of anaerobic digesters operating on ice-cream wastewater. Water Res 29(2):525–533

Hegazy A, El-Nawawy M, Ali A, El-Samragy Y (2019) Isolation and identification of halophilic bacteria producing exopolysaccharides from whey and milk permeate. Arab Univ J Agric Sci 27 (2):1491–1501

Hwang S, Hansen CL (1998) Characterization of and bioproduction of short-chain organic acids from mixed dairy-processing wastewater. Trans ASAE 41(3):795–802

Janczukowicz W, Zielinski MD, Bowski M (2008) Biodegradability evaluation of dairy effluents originated in selected sections of dairy production. Bioresour Technol 99(10):4199–4205

Kadu PA, Landge RB, Rao YRM (2013) Treatment of dairy wastewater using rotating biological contactors 3. Eur J Exp Biol 3(4):257–260

Kannahi M, Sangeetha A (2014) Physico chemical and bacteriological characterization of cheese processing effluent and their effect on *Vigna mungo* growth. Int J Pharm Sci Rev Res 29 (2):179–182

Kebbouche-Gana S, Gana ML (2014) Algerian yeast strains: isolation, identification and production of single cell protein from whey with strain candida kefyr. Int J Biosci Biochem Bioinforma 4 (3):160

Khanam R, Prasuna RG, Akbar S (2013) Evaluation of total phenolic content in ghee residue: contribution to higher laccase production. Microbiol J 3:12–20

Klemes J, Smith R, Kuk Kim J (2008) Handbook of energy and water management in food processing. Woodhead, Cambridge, pp 3–43

Koller M (2015) Recycling of waste streams of the biotechnological poly (hydroxyalkanoate) production by *Haloferax mediterranei* on whey. Int J Polym Sci 2015:370164

Kothari R, Prasad R, Kumar V, Singh DP (2013) Production of biodiesel from microalgae *Chlamydomonas polypyrenoideum* grown on dairy industry wastewater. Bioresour Technol 144:499–503

Kothari R, Kumar V, Pathak VV, Tyagi VV (2017) Sequential hydrogen and methane production with simultaneous treatment of dairy industry wastewater: bioenergy profit approach. Int J Hydrog Energy 42(8):4870–4879

Koyuncu I, Turan M, Topacik D, Ates A (2000) Application of low pressure nanofiltration membranes for the recovery and reuse of dairy industry effluents. Water Sci Technol 41 (1):213–221

Kumar S, Gupta N, Pakshirajan K (2015) Simultaneous lipid production and dairy wastewater treatment using *Rhodococcus opacus* in a batch bioreactor for potential biodiesel application. J Environ Chem Eng 3(3):1630–1636

Lee H, Song M, Yu Y, Hwang S (2003) Production of *Ganoderma lucidum* mycelium using cheese whey as an alternative substrate: response surface analysis and biokinetics. Biochem Eng J 5 (2):93–99

Liu YY, Haynes RJ (2011) Origin, nature, and treatment of effluents from dairy and meat processing factories and the effects of their irrigation on the quality of agricultural soils. Crit Rev Environ Sci Technol 41(17):1531–1599

Liu J, Dantoft SH, Würtz A, Jensen PR, Solem C (2016) A novel cell factory for efficient production of ethanol from dairy waste. Biotechnol Biofuels 9(1):33

Liu P, Zheng Z, Xu Q, Qian Z, Liu J, Ouyang J (2018) Valorization of dairy waste for enhanced D-lactic acid production at low cost. Process Biochem 71:18–22

Lu W, Wang Z, Wang X, Yuan Z (2015) Cultivation of Chlorella sp. using raw dairy wastewater for nutrient removal and biodiesel production: characteristics comparison of indoor bench-scale and outdoor pilot-scale cultures. Bioresour Technol 192:382–388

Lule VK, Singh R, Pophaly SD, Tomar SK (2016) Production and structural characterisation of dextran from an indigenous strain of *Leuconostoc mesenteroides* BA 08 in whey. Int J Dairy Technol 69(4):520–531

Madhu PC (2016) Utilization of dairy effluent for food grade protease production using Bacillus sp. Am J Biosci Bioeng 4(6):90–95

Malaka R, Maruddin F, Dwyana Z, Vargas MV (2020) Assessment of exopolysaccharide production by *Lactobacillus delbrueckii* subsp. *bulgaricus* ropy strain in different substrate media. Food Sci Nutr 8(3):1657–1664

Mann B, Athira S, Sharma R, Kumar R, Sarkar P (2019) Bioactive peptides from whey proteins. In: Whey proteins. Academic Press, Samford, pp 519–547

Moreno-Dávila IMM, Ríos-González LJ, Garza-García Y, Rodríguez-de la Garza JA, Rodríguez-Martínez J (2011) Biohydrogen production from diary processing wastewater by anaerobic biofilm reactors. Afr J Biotechnol 10(27):5320–5326

Murari CS, da Silva DCMN, Schuina GL, Mosinahti EF, Del Bianchi VL (2019) Bioethanol production from dairy industrial coproducts. BioEnerg Res 12(1):112–122

Myint KT, Otsuka M, Okubo A, Mitsuhashi R, Oguro A, Maeda H, Shigeno T, Sato K, Nakajima-Kambe T (2020) Isolation and identification of flower yeasts for the development of mixed culture to produce single-cell protein from waste milk. Bioresour Technol Rep 10:100401

Nayeem M, Chauhan K, Khan S, Rattu G, Dhaka RK, Sidduqui H (2017) Optimization of low-cost substrate for the production of single cell protein using *Kluyveromyces marxianus*. Pharm Innov J 6:22–25

Ngome MT, Alves JGLF, Piccoli RH, de Carmo DE, Pinto SA, Bernal OLM (2017) Inoculum concentration and inoculation time for propionic acid production from whey using mixed culture of *Lactobacillus helveticus* and *Propionibacterium freudenreichii* PS-1. Acta Sci Technol 39:543–550

Pais J, Serafim LS, Freitas F, Reis MA (2016) Conversion of cheese whey into poly (3-hydroxybutyrate-co-3-hydroxyvalerate) by *Haloferax mediterranei*. New Biotechnol 33 (1):224–230
Pais-Chanfrau JM, Acosta LDC, Cóndor PMA, Pérez JN, Guerrero MJC (2018) Small-scale process for the production of kefiran through culture optimization by use of central composite design from whey and kefir granules. In: Current topics in biochemical engineering. IntechOpen, London. https://doi.org/10.5772/intechopen.82257
Pal P, Nayak J (2016) Development and analysis of a sustainable technology in manufacturing acetic acid and whey protein from waste cheese whey. J Clean Prod 112:59–70
Pandian SR, Deepak V, Kalishwaralal K, Rameshkumar N, Jeyaraj M, Gurunathan S (2010) Optimization and fed-batch production of PHB utilizing dairy waste and sea water as nutrient sources by *Bacillus megaterium* SRKP-3. Bioresour Technol 101(2):705–711
Passeggi M, Lopez I, Borzacconi L (2009) Integrated anaerobic treatment of dairy industrial wastewater and sludge. Water Sci Technol 59:501–506
Patel S, Modi DA, Rathod NP, Chavda GR, Parmar DK (2013) Physico chemical analysis of effluent from Havmor ice cream industry. Int J Adv Biosci 1(1):07–09
Patowary R, Patowary K, Kalita MC, Deka S (2016) Utilization of paneer whey waste for cost-effective production of rhamnolipid biosurfactant. Appl Biochem Biotechnol 180(3):383–399
Prazeres AR, Carvalho F, Rivas FJ (2012) Cheese whey management: a review. J Environ Manag 110:48–68
Qasim W, Mane AV (2013) Characterization and treatment of selected food industrial effluents by coagulation and adsorption techniques. Water Resour Ind 4L:1–12
Rad SJ, Lewis MJ (2014) Water utilisation, energy utilisation and waste water management in the dairy industry: a review. Int J Dairy Technol 67(1):1–20
Rivas J, Prazeres AR, Carvalho F (2011) Aerobic biodegradation of precoagulated cheese whey wastewater. J Agric Food Chem 59(6):2511–2517
Ryder K, Ali MA, Billakanti J, Carne A (2020) Evaluation of dairy co-product containing composite solutions for the formation of bioplastic films. J Polym Environ 28(2):725–736
Sampaio FC, de Faria JT, da Silva MF, de Souza Oliveira RP, Converti A (2019) Cheese whey permeate fermentation by *Kluyveromyces lactis*: a combined approach to wastewater treatment and bioethanol production. Environ Technol 13:1–9
Scharnagl N, Bunse U, Peinemann K (2000) Recycling of washing waters from bottle cleaning machines using membranes. Desalination 131:55–63
Schwarzenbeck N, Borges JM, Wilderer PA (2005) Treatment of dairy effluents in an aerobic granular sludge sequencing batch reactor. Appl Microbiol Biotechnol 66:711–718
Sharma D, Saharan BS, Chauhan N, Bansal A, Procha S (2014) Production and structural characterization of *Lactobacillus helveticus* derived biosurfactant. Sci World J 2014:493548
Singh NB, Singh R, Imam MM (2014) Waste water management in dairy industry: pollution abatement and preventive attitudes. Int J Sci Environ Technol 3(2):672–683
Şişman T, Gur O, Dogan N, Ozdal M, Algur OF, Ergon T (2013) Single-cell protein as an alternative food for zebrafish, *Danio rerio*: a toxicological assessment. Toxicol Ind Health 29 (9):792–799
Sparling GP, Schipper LA, Russell JM (2001) Changes in soil properties after application of dairy factory effluent to New Zealand volcanic ash and pumice soils. Aust J Soil Res 39:505–518
Srisuk N, Sakpuntoon V, Nutaratat P (2018) Production of indole-3-acetic acid by Enterobacter sp. DMKU-RP206 using sweet whey as a low-cost feed stock. J Microbiol Biotechnol 28 (9):1511–1516
Strydom JP, Britz TJ, Mostert JF (1997) Two-phase anaerobic digestion of three different dairy effluents using a hybrid bioreactor. Water SA 23:151–156
Tamime AY, Robinson RK (eds) (1999) Yoghurt science and technology. Woodhead, Cambridge
Tetra Pak (2003) Dairy processing handbook. Tetra Pak Processing Systems, Lund
Tetrapak (1995) Dairy effluents. In: Dairy processing handbook. Tetrapak Printers, London, pp 415–424

Torrijos M, Vuitton V, Moletta R (2001) The SBR process: an efficient and economic solution for the treatment of wastewater at small cheese making dairies in the Jura Mountains. Water Sci Technol 43:373–380

Truan L, Marques N, Souza A, Rubio-Ribeaux D, Cine A, Andrade R, Silva T, Okada K, Takaki G (2020) Sustainable biotransformation of barley and milk whey for biosurfactant production by *Penicillium Sclerotiorum* Ucp 1361. Chem Eng Trans 79:259–264

Veeravalli SS, Mathews AP (2018) Exploitation of acid-tolerant microbial species for the utilization of low-cost whey in the production of acetic acid and propylene glycol. Appl Microbiol Biotechnol 102(18):8023–8033

Vera ECS, de Azevedo PODS, Domínguez JM, de Souza Oliveira RP (2018) Optimization of biosurfactant and bacteriocin-like inhibitory substance (BLIS) production by *Lactococcus lactis* CECT-4434 from agroindustrial waste. Biochem Eng J 133:168–178

Victoria Environmental Protection Authority (1997) The dairy processing industry. Environmental Protection Authority, State Government Environmental guidelines for of Victoria, Melbourne

Wendorff WL (2001) Treatment of dairy wastes. In: Marth EH, Steele JL (eds) Applied dairy microbiology, 2nd edn. Marcel Dekker, New York, pp 681–704

Yadav JSS, Bezawada J, Ajila CM, Yan S, Tyagi RD, Surampalli R (2014) Mixed culture of K*luyveromyces marxianus* and *Candida krusei* for single-cell protein production and organic load removal from whey. Bioresour Technol 164:119–127

Yadav JSS, Yan S, Ajila CM, Bezawada J, Tyagi RD, Surampalli RY (2016) Food-grade single-cell protein production, characterization and ultrafiltration recovery of residual fermented whey proteins from whey. Food Bioprod Process 99:156–165

Yadavalli R, Heggers GRVN (2013) Two stage treatment of dairy effluent using immobilized Chlorella pyrenoidosa. J Environ Health Sci Eng 11(1):36

Potential Value Addition from Cereal and Pulse Processed By-Products: A Review

9

Renu Yadav, Neelam Yadav, Pinki Saini, Devinder Kaur, and Rajendra Kumar

Abstract

India is a leading producer of cereals and its processing generates large amount of by-products and waste each year. Cereal and pulse processing produces by-products that can be maintained and protected or used as low-prize materials for extraction of useful components for nutraceuticals, dietary supplements, food coating and packaging materials, and feeds for sustainable availability and consumption. These by-products are also magnificent source of the most valuable constituents like antioxidants, bioactive compound, dietary fibers, β-glucans, enzymes, and many more. This chapter is well defining on the functional and bioactive constituents of major cereals and pulse by-products and how technologies can transform them to value-added products. Valorization of these by-products has immense potential to Indian economy.

Keywords

Cereal · Pulses · Bran · Husk · By-products · Waste management

R. Yadav (✉)
Amity Institute of Organic Agriculture, Amity University, Noida, Uttar Pradesh, India
e-mail: ryadav3@amity.edu

N. Yadav · P. Saini · D. Kaur
Centre of Food Technology, University of Allahabad, Prayagraj, India

R. Kumar
Division of Genetics, IARI, PUSA, New Delhi, India

M. Thakur et al. (eds.), *Sustainable Food Waste Management*,
https://doi.org/10.1007/978-981-15-8967-6_9

9.1 Introduction

Cereal and pulses are most staple foods in Indian diets. High demand and production of these crops also produce processing waste which requires judicious management. Wastage occurs at harvesting and processing stage as well as during the consumption. It has been well defined that cereal by-products depict a beneficial source of health-promoting dietary fiber, antioxidants, and biochemical. These could be recovered and supplemented to the customers as the ingredient of high-value food or nutraceuticals. The cereal and pulses by-products like husk/hulls, germs, bran, middling, broken kernels, etc. could reduce the environment and economic burden and significantly increase the sustainability of our diets and nutrition. This addresses a step forward to waste reduction in the food chain, but also the production of different value-added food products. The foremost objective of these by-products should be proper utilization along with the permanent supply of food with the nutritional demand in the most sustainable way along with keeping population health, social justice, and other principles of sustainability.

9.2 Cereal: By-Product Waste Utilization

The word "cereals" refers to crops of the *Poaceae* family like wheat, maize, rice, oat, rye, barley, millet, sorghum, and triticale (a hybrid of rye and wheat). Cereals and cereal waste comprise a significant fragment of the routine food of humans and animals. Besides their main end products, several by-products are impending from the superficial layers, characterized by higher micronutrient and husk substances. Cereals have a large volume of solid wastes resultant from processing. They are presently also used as animal fodder or are returned to the harvested field for land application. However, they may characterize a source of mixture with distinctive nutritional, physico-chemical, and functional properties, also give a high value for sustainable human diet (Hemery et al. 2007). Several value-added foods are produced by the incorporation of cereals by-products such as bread, biscuits, cookies, breakfast cereals, and pasta (Verma and Mogra 2013). This overview concerns mainly the potential utilization of by-products from agriculture and agro-industry. By-products are the excesses and wastes from the increasing and processing of raw agronomic products. By-products of cereals processing are steep, germ, bran, husk, spent grains, middling, and the endospermic tissue aleurone layer. These grain segments are a rich source of bio-functional particles, fiber, vitamins, lignans, minerals, phytoestrogens, and phenolic composite. The grain by-products are rich in numerous compounds like ferulic acid, β-glucans, arabinoxylans, sterols, polyphenols, oryzanol, and vitamins can hamper the presence of recent society diseases like diabetes, cardiac diseases, cancer, etc. (Salazar-López et al. 2019). Therefore, these by-products have different possible uses in nutrition and nonfood industries (Papageorgiou and Skendi 2018). Bran, germ, broken grains, weed seeds, and husk/hulls are major by-products of cereal processing (Table 9.1). Some of them are related to numerous grains, whereas some are used completely to describe a by-product of one grain. Bran includes the rough external shell of the kernel consisting of aleurone and pericarp with minor quantities of flour and higher amount of fiber and protein, and it is related to the corn, rice, and wheat bran (Murugan et al. 2013).

9.2.1 Rice Bran

Rice bran is one of the important by-products of paddy that has been exploited most in comparison to the bran of other cereals due to its high commercial value and sustainability. Paddy processing results in production of husk, endosperm, germ, and bran. Approximately 7–8.5% of the grain is bran. Rice bran constitutes various functional components having high nutritive value and enormous health properties as well. It is also rich in lipids, protein, and fiber (Hammond 1994). Rice bran contains 11–18% fat, 11–17% protein, 10–14% dietary fiber, 9% ash, and 45–60% nitrogen-free extract. Bran is rich source of micronutrients like Mg, K, Fe, Mn, B vitamins, and considered as a good source of choline and inositol (Hoffpauer et al. 2005). It is also an excellent source of vitamin E and minerals like aluminum, calcium, chlorine, iron, phosphorus, potassium, sodium, and zinc (Weicheng et al. 1996). Proximate composition of rice bran conducted by diverse studies is given in Table 9.2.

Bran produced by traditional mills has poor quality in terms of palatability therefore it is restricted to the use of cattle feed only. Rice bran has laxative effects (Saunder 1990), offers benefits like help in lowering of blood cholesterol (Hegsted and Kousik 1994; Kahlon et al. 1994), and decreases the rate of incidence of arthrosclerosis and diverticular disease. On the basis of overall composition, nutritional profile, and functional characterization, rice bran is being added in different food products that are more beneficial for those individuals who show allergenicity to other cereal grains containing properties of allergens (Sidhu et al. 2007). The problem seen in using rice bran is that it has property to deteriorate rapidly, thus it is very necessary to find a method that helps in preventing this rapid deterioration of rice bran and also to ensure the good quality material for further processing. In order to extend the shelf life, stabilization of the rice bran must be done by minimizing its FFA (free fatty acids) content, immediately after milling. The stabilization can be done by inactivating lipase and peroxidase enzymes with the help of different heat treatment. Depending on the type of heat treatment, the lipase enzyme may lead to either reversible inhibition or an irreversible denaturation.

One of the major drawbacks, limiting the nutritional quality of bran is the presence of anti-nutritional factors such as phytin, trypsin inhibitors, oxalates, tannins, polyphenols, hemagglutinins, and lectins. These undesirable anti-nutritional constituents restrict the direct utilization of bran in the diet. Rice bran has higher phytin content than that of other cereals like wheat bran, soy bran, and oat hulls. It also constitutes about 69% of the dietary fiber content of brown rice. During milling, higher quantity of tannins leads to its concentration in the bran resulting pigmented rice (Eggum et al. 1981). It is necessary to eliminate these undesirable attributes for

Table 9.1 Composition of cereal by-products

Compositions (%)	Brewers' spent grain	Corn husk	Corn cob	Corn dry distilleries	Rice husk	Wheat germ	Wheat husk	Rye husk	Psyllium husk
Hemicellulose	28	44.5	35	*	29	*	36	26	*
Cellulose	17	38.2	32	*	29	*	18	16	*
Lignin	28	6.6	20	*	24	*	16	13	*
Fiber	3	*	*	40	*	2.53	*	*	3.83
Protein	*	1.9	*	30	*	24.24	6	10	2.08
Fat	*	*	*	12	*	7.99	5	7	*
Ash	*	2.8	*	*	*	3.78	*	*	3.85

*Data not available

Source: Aliyu and Bala (2011), Ch et al. (2012), Lim et al. (2012), Liu (2011), Qaisrani et al. (2014) and Sidhu et al. (2007)

Table 9.2 Proximate composition of rice bran per 100 g (as is basis)

S. No.	References	Crude fat (g)	Crude fiber (g)	Crude protein (g)	Total ash (g)	Dry matter (g)	Energy (kcal)
1.	Singh et al. (2013)	16.8 ± 1.6	6.3 ± 0.5	10.53 ± 0.55	10.76 ± 1.05	85.8 ± 0.78	363.3 ± 6.39
2.	Malik and Chughtai (1979)	14.65	3.85	11.45	10.80	92.60	–
3.	Choo and Sadiq (1982)	9.50	12.00	11.00	9.80	–	297.00
4.	Rao and Reddy (1986)	18.10	7.60	12.00	17.40	–	–
5.	Ghazi (1992)	10.71	15.71	12.97	17.15	92.75	374.06
6.	NRC (1994)	11.00	3.58	12.20	–	–	309.00
7.	Nadeem (1998)	14.07	11.86	14.97	10.75	91.38	301.60
8.	Leeson and Summers (2001)	15.00	2.40	11.00	–	–	275.00
9.	Ambreen et al. (2006)	13.51	15.15	13.00	10.60	92.05	223.70

Values present are in Mean ± SD

achieving effective utilization of brans with respect to human nutrition. All the methods that have been executed in the past were focused on eliminating one or the other toxic factors only (Rehman and Mahmood 1996). On the basis of extensive literature survey, it was stated that all the anti-nutritional factors are protein in nature except phytates. Therefore, it was assumed that these toxic compounds which are protein in nature can be eliminated or reduced their toxicity by denaturing or modifying their structures. The processing treatments that can be applied for the removal of undesirable factors and stabilization of bran are extrusion cooking, wet, dry, microwave heating, and chemical treatment. Dry heating (toasting) apart from helping with flavor development improves micronutrients like various minerals and protein content of food and eliminates heat labile anti-nutrients.

Brans can be used as an ingredient in the formulations of products, resulting in various desirable improvements like texture modification and enhanced food stability during production and storage. The properties of hydration of dietary fiber which determine its appropriate level of utilization in products are very important because they play a role in maintaining and retaining a desirable texture of the products (Thebaudin et al. 1997). Stabilized rice bran is being successfully incorporated into several products like bread, muffins, and cookies at level up to 20% (Carroll 1990). Baked products provide many potential possibilities for the utilization of rice bran in foods. Utilization of rice bran is recommended in preparing breads, muffins, pancakes, cookies, cakes, pies, extruded snacks, and breakfast cereals (Sharma et al. 2004).

9.2.2 Rice Bran Oil (RBO)

Full fat rice bran contains 16.2–18.5% of oil and is used for various purposes after extracted. At present India produces 0.9 MT of rice bran oil and has potential to produce 1.6 MT (Panoth et al. 2019). Rice bran oil is an excellent source of antioxidants consisting of oryzanol, tocopherol, and tocotrienols. These antioxidants are responsible for its hypocholesteromic property. Bran is defatted to remove its oil content and can be utilized as defatted rice bran (DRB). Crude rice bran contains glycerides (80%), phospholipids, glycolipids, free fatty acids (FFAs), and many impurities, namely wax, resinous, and unsaponifiable materials. Crude oil is processed by degumming, dewaxing, deacidification, bleaching, and deodorization. Extraction of oil from bran is done by solvent extraction method using food grade hexane (Panoth et al. 2019).

RBO is considered as best quality cooking oil as it is hypoallergenic, has very high smoking point making it suitable for deep frying and its antioxidant provides 2.5 times oxidative stability and shelf life in comparison to other oils (Nayik et al. 2015). It also absorbs less oil during frying as it has less polymers (less greasy). It has balanced fatty acid profile with appreciable amount of monounsaturated (MUFA) and polyunsaturated fatty acid (PUFA).

9.2.3 Edible Coating from Bran Wax

Bran wax is a secondary by-product of rice bran. It is obtained during dewaxing process of oil refining from crude rice bran oil. Crude wax is also refined to remove resinous matter, oil, free fatty acids. Crude wax contains 20–70% oil, up to 20% free fatty acids, 25–65% wax, and 5–12% resinous materials. Refining of crude wax is done by defatting, bleaching, filtration, and vacuum drying. Refined wax is hard and has brownish yellow color. It has properties similar to carnauba wax used for coating fruits and vegetable. Wax coatings are thin layers of edible materials applied on fruits and vegetable to improve its shelf life and to prevent dehydration. Coating being edible can be eaten along with fruits and vegetables. Rice bran wax coating is being widely used to extend shelf life of fruits and vegetables (Shi et al. 2017; Zhang et al. 2017).

Rice bran wax is long chain saturated alcohol and C20–C29 fatty acids containing long chain aliphatic like palmitic acid. Policosanol is also a long aliphatic chain found in bran wax containing different components octacosanol, triacontanol, dotriacontanol found to suppress formation of low-density lipoprotein (LDL), prevents clustering of platelets, has anti-tumor activity, and acts as antioxidant in preventing oxidation of PUFA induced by high temperature and UVA (Acosta-Estrada et al. 2019).

9.2.4 Wheat Bran

Milling of wheat leads to a production of white flour and a side product called wheat bran and it can be utilized in fortification to ensure sustainable nutritional supply. The amount of wheat bran is dependent on the rate of extraction during milling, i.e., the total quantity of kernel recovered in the wheat flour. Wheat bran lies on the outer layers of the kernel and it constitutes insoluble AX, cellulose, starch, protein, β-glucan, and lignin (Hemery et al. 2007). Its effect is very well known as it increases fecal bulk and reduces intestinal transit time. Health claims related to the effects of bran have accepted by ESFA and it has provided that either a food is rich in fiber or wheat bran of about 10 g is consumed regularly, respectively. Wheat bran is majorly utilized to enhance the fiber content of processed foods as it is well known good source of dietary fibers. Various dietary fiber sources are being used in bread making, including bran, whole meal flour, pulse hulls, wheat fiber, maize, and oats. Availability of products like steamed bread fortified with wheat bran or whole meal flour is becoming common in supermarkets in Asia (Sibakov et al. 2013).

Wheat bran is being used in many products like banana chocolate, nut and spice cakes as a source of fiber (Rajchel et al. 1976). Highly acceptable products can be prepared by using fermented coarse wheat bran such as bread supplemented with up to 10% fermented wheat bran. Incorporation of different levels of wheat bran (0–20%) into residual flour was used to develop mineral enriched brown flour

(Butt et al. 2004). High-fiber *phulkas* or *chappatis* can be prepared without affecting its sensory properties by incorporating 5% wheat bran.

9.2.5 Barley Bran

Barley contains a high level of soluble dietary fiber, making it a desirable food component for health benefits. Barley grain and especially fractions containing high levels of β-glucan have been shown to have a hypocholesterolemic effect in chicks, rats, and humans. Studies suggest that barley bran may be effective in protecting against the risk of cancer as well.

Barley bran fractions are more suitable in biscuits with excellent sensory properties and good baking characteristics. Biscuits made with 100% barley bran contained an average 2.2% β-glucan and labeled as high-fiber products (Nagel-Held et al. 1997). Oat bran was added in wheat bread and patties and also used as fat replacer in processed meats.

Studies undertaken regarding nutritional benefits of dietary fiber revealed that the traditional cereal-based and other food products can conveniently provide 8–10 g of water-soluble fiber in the American diet towards a goal of doubling the total fiber intake to 20–35 g/day (Ranhotra et al. 1990). Among the barley products, those made with white flour were high in total fiber but contained a portion of the total fiber as soluble fiber. Marlett (1991) analyzed three barley varieties that had each been processed into a ready-to-eat (RTE) cereal product and one unprocessed (control) barley analyzed for soluble and insoluble dietary fiber content and composition. The total fiber differed between the unprocessed and processed barley, 15.7 versus 12.2–12.4% (dry wt.). The results suggested that processed barley into an RTE cereal product increased the analytical solubility of dietary fiber.

9.2.6 Oat Bran

Association of Adolescent and Child Care in India (AACCI) defined oat bran as it is a product produced by grinding clean groats or rolled oats and then separating oat flour by the process of sieving, bolting, and/or other suitable methods into fractions such that the oat bran fraction is equal to or less than 50% of the initial material and having total β-glucan content of approximately 5.5% on dry-weight basis, and at least 16.0% total dietary fiber content on dry-weight basis, and this way one-third of the total dietary fiber is soluble fiber (Robert and Walter 1995).

Oat bran contains high soluble fiber and shows effects in lowering blood cholesterol levels. Wheat, corn, and rice bran have high insoluble fiber, which helps in preventing constipation. Bran can be sprinkled into many products listing from hot cereal and pancakes to muffins and cookies. Various popular products as high-fiber cereals and bars are also enriched with bran (Palmer 2008). The chief fraction of alimentary fiber present in oats having a great importance for the human health is β-glucan. B-glucan is a polymer found in less quantity in several tissues of many cereals (Brennan and Cleary 2005). Consumption of oat bran as a part of diet exerts beneficial physiological activity possibly due to the presence of its β-glucan content. β-glucan has an important role in increasing absorption of water and moisture of bread (Salehifar and Shahedi 2007). Oat bran contains 10% β-glucan on a dry basis. It shows many beneficial effects on human health such as the reducing cholesterol levels (Wood et al. 1989).

9.3 By-Products of Milling

9.3.1 Husk/Hulls

Rice contains 20% husk and generally used as fuel for the parboiling process of paddy. Moreover, this is used as biomass for production of energy by using thermo-chemical (combustion, pyrolysis, and gasification) and biochemical (biomethanation of biomass) processes (Papageorgiou and Skendi 2018). Johar et al. (2012) have extracted cellulose fibers and nanocrystals from rice husks and applied in the production of synthesis of antimicrobial and medical resources, biosensing, enzyme immobilization, green catalysis, production of drug delivery system in therapeutic drugs. Rye and psyllium husk are used as a source of vitamins (tocopherols, tocotrienols, folates, etc.), dietary fiber (arabinoxylans, β-glucans, etc.), and different phytochemicals (phytosterols, phenolic compounds, etc.), these have encouraging health effects. Psyllium showed ability to lower significantly total cholesterol, low-density cholesterol, constipation, and diarrhea in humans. Many value-added products have done the incorporation of cereal husk like bread, cakes, cookies, breakfast cereals, and pasta (Verma and Mogra 2013).

Composition of cereals by-product is given in Table 9.1, and shows that they are having hemicelluloses, cellulose, lignin, and starch in abundant and fraction of protein and fat content (Bledzki et al. 2010). Corn husks are the major solid wastes that are obtained processing from sweet corn. Traditionally they are used for animal feed and/or land application. But according to study the corn husks could supply as the latent substrate for manufacture of citric acid from *Aspergillus niger* (Hang and Woodams 2000) and also could be used as the probable resource of pigment anthocyanins (Li et al. 2008). Oyeleke and Jibrin (2009) reported that guinea corn husk and millet husk could be a medium for the production of ethanol by using fermenting organisms *Z. mobilis* and *A. niger*. The rice-husk flour has also been used for thermoplastic polymer composite, polypropylene as the matrix, as the reinforcing filler. The husk from wheat contains tricin; those can selectively demolish two cancer cell lines (colon cancer and cirrhosis in a fibrotic liver) in vitro showing zero side effects for normal cell lines (Seki et al. 2012; Moheb et al. 2013).

9.3.2 Middlings and Broken Kernels

Middlings and broken kernels are by-products of milling process. The broken kernel is portion that remains after grinding and sifting, rich in proteins, dietary fibers, phytochemicals, and vitamins and used as animal feed after mixing with bran. Middlings of barley have increased fiber, i.e., β-glucan content. Blandino et al. (2013) have formulated wheat bread with the incorporation of pearled wheat and reported improved natural antioxidant compound, e.g., phenolics, alkylresorcinols, dietary fibers, and β-glucans content in final product. Middlings and broken kernels of milling of wheat, maize, barley, rye, sorghum, and oats are traditionally used in protein supplement for animal feeding. Middling and hominy feed are used for the production of ethanol due to its high starch content and also used as low-protein feed ingredient. According to Huang et al. (2014) by-products of wheat make nearly 25–30% of the total wheat. Generally, they are utilized as feed for animal, as they offer energy, protein, and some minerals. The difference among the nutritional value of wheat by-products was due to variations in the variety of the wheat being processed, growing conditions, and variation in the technique of processing (Huang et al. 2014).

9.3.3 Cereal Germs

Cereal germs composed of fat, protein, crude fiber, dietary fiber vitamins, and minerals are well known for their contributions for value addition to foods and feeds. Germ contains 2–3% of the total weight of grain kernel. Corn germs are concentrated source of phytosterols. Corn germ oil is the richest source of Vitamin E (tocopherols and tocotrienol), and it is approximately 28 times higher in corn germ oil as compare to corn fiber oil. Corn oil has comparatively low phenolic content, and the majority of phenolics of corn oil are trans-cinnamic acid (0.9 mg/kg), vanillin (2.8 mg/kg), and ferulic acid (0.5 mg/kg). The germ is also affluent source of functional phytochemicals especially flavonoids, sterols, octacosanols, and glutathione and unsaturated fatty acids (USFA), chiefly oleic, linoleic, and α-linoleic acids. Cereal germ is used as substrate for manufacturing of several enzymes like lipase, acid phosphatase, protease, etc. Wheat germ contains high amount of vitamin E and lipids, so it is used for oil production, while defatted wheat germ is a superior source protein and amino acid high amounts of albumins, globulins, glycine, aspartic acid, glutamic acid, arginine, leucine, alanine, proline, and lysine, and lower concentration of isoleucine, valine, methionine, and arginine (Brandolini and Hidalgo 2012). This was observed that wheat germ oil is a tremendous source of natural antioxidants, bioactive and antimicrobial components. Therefore, they could be used for production of functional, nutraceuticals, and other food products due to its exceptional biological activities and health beneficial properties. According to Arshad et al. (2007), wheat germ could be utilizing into production of value-added products and used to fight against malnutrition. In another study wheat germ oil was supplemented to rats, it was found that it reduces lipid peroxidation which might be a helpful technique for reducing cardiovascular diseases. Huang et al. (2010) used defatted wheat germ for the production of glutathione and other wheat germ protein. Rice germ is rich in phytic acids, vitamin E and B vitamins, and aminobutyric acid (GABA). Millet germs are rich in albumin and globulin type proteins. Cereal germs are rich in protein, fat, ash and some of them have appreciable amount of fiber (corn about 18% protein, and 35% fat); rice (protein 24%, fat 26%, and 5% fiber); rye (protein 41%, fat 11% and 3% fiber); sorghum (protein 19%, and fat 28%); wheat (about 27% protein and 7.2% fat with 3% fiber) so they could be used as an ingredient in protein rich value-added food products. Rice-husk ash has been incorporated successfully for development of biodegradable films (Julio Harda et al. 2017).

9.3.4 By-Product of Malting

The by-products of malting are malt sprouts that consist of sprouts, roots, and malt hulls and known as a source of protein (Bajpai and Patil 1997; Chi and Zhao 2003). Dried distiller's grains characterize the major by-product from the production of distilled alcohol (Table 9.1). After fermentation of cereals, protein, fat, vitamins, non-starch polysaccharides, and minerals remain in the dried distiller's grains that make an excellent source for the formulation of value-added bread and bakery produce. Another application of malt by-products is as animal feed. Sorghum dry distiller's grain also reduced cholesterol and non-high-density lipoprotein plasma cholesterol. Brewers spent grains are regularly used as cattle feed. Because brewer's grains provide protein, fiber, and energy, so they can be used as functional components in various dietary supplements. Spent grains also have plentiful phenolics such as p-coumaric acid, ferulic acid, syringic acid, p-hydroxybenzoic acid, and vanillic acid. Spent grains enclose several functional groups such as carboxyl, hydroxyl, and amine (used for biosorbents). The chief component of desiccated hops is fiber (xylose, mannose, galactose, pectin, uronic acid, rhamnose, arabinose, and glucose), bitter acids, protein, ash, fats, salts, tannins, and polyphenols. Spent hopes are used to produce spices, carbohydrates, and organic acids (Oosterveld et al. 2002). Radosavljević et al. (2019) concluded that malting by-products could be used for the production of lactic acid and also for animal feed. Spent grains can be used as substrate for microorganisms cultivation and further for the production of enzymes like alpha-amylase by *Bacillus licheniformis, Bacillus subtilis*, xylanase by *Streptomyces avermitilis* and *Aspergillus awamori*, feruloyl esterase by *Aspergillus oryzae*, and *Streptomyces avermitilis*, and cellulase by *Trichoderma reesei* (Mussatto 2009). As the spent grain hydrolysates are rich in sugars and other nutrients so it is used as fermentation medium for the production of various components like xylitol by *Candida guilliermondii*, ethanol by *Saccharomyces cerevisiae;* arabitol, xylitol, ethanol, and glycerol by *Debaryomyces hansenii*, and lactic acid by *L. delbrueckii, L. pentosus, or L. rhamnosus* (Cruz et al. 2007). Other uses of malt by-products are adsorbent for removing volatile organic compound emissions, as a source of biogas, or removal of organic material from waste matter. Moreover, these bioactive components could be utilized as food ingredients and nutraceuticals in the food industry.

9.4 Pulses: By-Product Waste Utilization

In India, the pulse milling industry is ranked as the third biggest cereal processing industry (Ramakrishnaiah et al. 2004). The conversion of pulses into *dal* is mostly done by dividing/splitting the whole seed. Out of the total pulse and legume production throughout the country, about 75% are converted into *dal*. Processing of pulse is considered as a small scale industry with various *dal* mills located all over the country. The dry pulse milling process is outlined in Fig. 9.1. First of all, the pulses are thoroughly cleaned by removing all the mud and stones. After the removal of the outer layer, the pulses are split in half and sometimes polished to improve the finish. Some of the pulses like chickpea, mung, and urad undergo milling to produce flour. The pulses contain about 11–14% of the husk, 2–5% of germ, and the remaining portion makes up the endosperm of the seed. The rate of extraction of pulse processing is found to be about 70–88% of the raw material. In India, the annual generation of by-products by the dhal mills is about 2.5 million tons which contain nearly 40% of beneficial cotyledon material (Ramakrishnaiah et al. 2004). The major by-products generated by the milling of pulses are 6–13% brokens, 7–12% powder and germ mixture, and 4–14% husk.

In order to improve the product's quality and make it acceptable for the population, legume seeds are not only dehulled but the splitting of its cotyledons is also done. Dehulling helps in removing tannins which improves the digestibility of proteins (Bressani and Elfas 1980). In whole pulses, hulls play an important role by contributing as an insoluble dietary fiber (DF). Hulls have high DF content, extending from 75% in chickpeas to 89% in peas (Dalgetty and Baik 2006). These by-products act as a potential resource of various biologically active compounds that can be utilized in food industries as they have the favorable technological aspect and/or nutritional content. The utilization of pulses by-products in bakery and other value-added food materials can improve its nutritional value (Tiwari et al. 2011). With an increase in pulse processing, its by-product disposal also poses an emerging

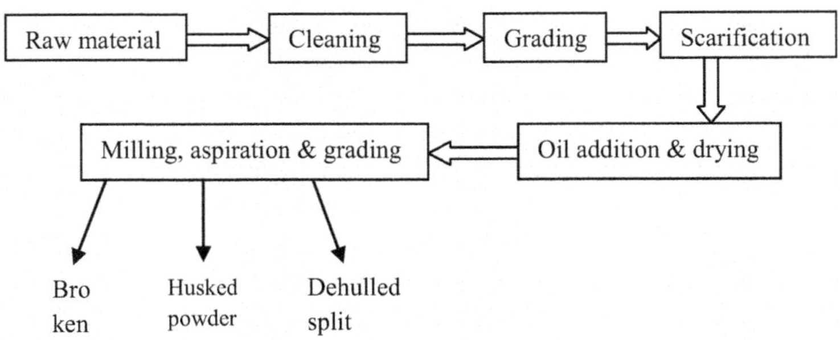

Fig. 9.1 Dry milling process of pulses

problem as the materials obtained from plants are prone to spoilage caused by microorganisms, thereby, restricting its utilization. Moreover, the cost of drying, storing, and shipping of by-products are also some of its limiting factors, from the economical viewpoint (Schieber et al. 2001). Hence, these agro-industrial by-products are mostly used as a feed for animals or utilized in the manufacturing of fertilizers. To exploit the rich potential of these waste materials efficiently, inexpensive and environmental-friendly utilization methods are required (Lowe and Buckmaster 1995). The by-products of pulses have abundant dietary fiber, polyphenols, minerals with high antioxidative potential, which makes them suitable for manufacturing food products (Luzardo-Ocampo et al. 2019). By identifying the appropriate technologies for secure management, categorization, and assessment of the functional as well as phytochemical properties of pulse, its market value can be increased manifolds.

9.5 Value-Added Food Products

Nowadays, agricultural waste products are regarded as a favorable provenience of functional components (Carle et al. 2001). Various possibilities are being searched to utilize valuable components from the by-products of food. In this concern, a wide range of researches have been done on the utilization of agro-industrial waste of pulses with an attempt to include its by-product in a variety of food products. For instance, Tiwari et al. (2011) used the dehulled flour of *Cajanus cajan* L., commonly called as pigeon pea (PPDF) and pigeon pea by-product flour (PPBF) in place of wheat flour to formulate biscuits. The protein content of the newly formed biscuits was increased considerably, also, a significant effect on its physical attributes and sensory characteristics was observed. The potential utilization of milling by-products of pigeon pea in the formulation of biscuits was explored by this study. It demonstrated the fruitful supplementation of pulse materials in baked products which was found to increase the nutritional value of the formulated product, along with providing an appropriate use of by-products. Fibers of the pea hull are being utilized in bakery products to a greater extent. The commercially available pea hull fibers consist of total dietary fiber and soluble fiber in a range of about 80–88% and 2–10%, respectively, which is been incorporated in various food products and is considered as the most studied among all the other pulse fibers. In the year 1988, the earliest and most extensive research was executed on wheat bread (high-fiber content) supplemented with 5–20% of pea hulls to check the quality parameters of loaf and dough. The researchers found that an increment in the pea hull fiber content in the composite flour results in an increased water holding capacity. The color of the crust also turned lighter and the specific volume of loaf decreased (Sosulski and Wu 1988). The assimilation of 15% (w/w) of pea hulls results in attaining composite flour having high total dietary fiber (TDF) content (15% approx.). Hence, these by-products have the potential to be utilized as an ingredient for the process of fortification in order to enhance the functionality of different food products. Agro-waste obtained by the processing of cereals and legumes had been incorporated for developing nutritionally rich and cheap deep-fried eatables (Tiwari et al. 2009). The study demonstrated that the addition of some legume flours (i.e. red gram, black gram, and green gram) with rice flour helps in improving the nutritive value of the formulated products. Another study showed that the incorporation of pea hull (3%) to wheat flour could slightly help in increasing the water holding capacity along with the resistance of dough, and also decrease the dough extensibility (Wang et al. 2002). The volume of the loaf was decreased, however, the bread crumbs were soft and other quality attributes were also found to be acceptable. Saini et al. (2017) carried out a research to assess the effect of incorporating chickpea husk on physical, sensory, and nutritional characteristics, along with the antioxidant nature of the formulated biscuits. Incorporation of 20% chickpea husk produced fiber-rich biscuits having moderately required overall acceptability. The addition of chickpea husk results in a significant increase (from 2.15% to 10.48%) in the total dietary fiber content of the newly formed biscuits. The 2,2-diphenyl-1-picryl-hydrazyl-hydrate (DPPH) activity jumped from 40.76 to 87.44%, whereas an increment (from 23.48 to 48.11 mg ascorbic acid per 100 g) in ferric reducing antioxidant power (FRAP) was also observed. The obtained results demonstrated the formulation of nutritionally rich (high protein and high-fiber content) biscuits that are considerably acceptable in terms of sensory characteristics and have high antioxidant potential. The use of such composite flours can be significantly used in various other bakery products apart from biscuits.

Dalgetty and Baik (2006) isolated indissoluble and dissoluble legume food cotyledon fiber fractions and compared with pulse hull fibers for wheat bread enrichment. An examination of the baking and mixing properties of the dough was done along with the bread staling properties during storage. It was found that the bread incorporated with fibers of the pulse hull had more moisture content than the control. They concluded that incorporating 5% hull (insoluble fibers) or 3% soluble fiber in the regular wheat flour can significantly increase the overall fiber content and can improve the moisture content of the bread. Pea cotyledon fiber was studied as a texture-modifier in beef patties having low-fat content (10% and 14% fat) to enhance organoleptic quality (Anderson and Berry 2000). Pea fibers were not found to be influential in retaining fat but helped in improving the cooking yield and tenderness of beef patties. Bose and Shams-Ud-Din (2010) also studied the incorporation of chickpea husk to formulate crackers by substituting wheat flour at the levels of 0–20% and evaluated its baking properties. The study demonstrated that the highest spread ratio of crackers was achieved by adding about 3% of the processed husk. The study illustrated that an increment in the levels of processed husk results in increasing the fat, ash, moisture, and total fiber content of the crackers, while the total carbohydrate and protein content decreased.

9.6 Extraction of Bioactive Components

One of the most important and common operations in the food processing industry is extraction which is mostly used for liberating essential components suspended in the complex food matrix, like vegetables, fruits, cereals, pulses, and their waste products (Barzana et al. 2002). Many compounds are obtained by the process of extraction which can be utilized in the form of food additives or providing some health benefits for human beings (Osada et al. 2001). The seed coat of pulse contains variety of polyphenolic compounds such as flavonols, flavanols, flavone glycosides, proanthocyanidins (polymeric and oligomeric) which contribute to its antioxidant potential. Among all the other pulses, broad beans possess the excessive amount of unbound flavanols. The average concentration of total flavanols was found to be about 154.5 mg/100 g (fresh weight) of broad beans. Most of the antioxidants (40–80%) in other pulses are not free and can only be released with the help of alkaline hydrolysis which is particularly detected through FRAP (ferric reducing antioxidant power).

The fractions of horse gram and chickpea are obtained by dehulling it mechanically and the evaluation of its phytochemical content was carried out to demonstrate its potential application (Sreerama et al. 2010). The hulls were found to be rich not only in phenolics but also in trypsin content and showed inhibiting actions of alpha-amylase. The by-products of dehulled chickpeas were collected from ten processing plants in Iran which contain about 0.01% total tannins, 0.14% total phenolics, 3.1% ether extracts, and 5.1% soluble sugars (Maheri-Sis et al. 2007). The husk of black gram is a waste product of the milling industries which makes up to 9% of the total yield. This husk is used to extract an enzyme peroxidase which is highly thermostable and has abundant biomedical, analytical, and industrial applications (Ajila and Prasada Rao 2009). The extraction, purification, and chemical characterization of xylogalacturonans obtained from pea hulls was done by Le Goff et al. (2001) at low (acidic) pH. Their research indicated that pea hulls contain about 925 mg/g of sugar, comprising 659 mg/g of cellulosic glucose and 90 mg/g of uronic acid. Muralikrishnaa and Tharanathana (1994) extracted and characterized pectic polysaccharides obtained from the field bean husks (*Dolichos lablab*), pea (*Pisum sativum*), and cowpea (*Vigna sinensis*) by using both hydrochloric acid (pH 2.0) and 0.5% EDTA (Ethylenediaminetetraacetic acid) separately. The yield obtained from HCl was 1.43% and that obtained from EDTA was 5.37%. Therefore, they concluded that the extraction of pectic polysaccharide by EDTA gave more yield and was more viscous when compared with acid extraction. Lai et al. (2010) extracted the water-soluble polysaccharides mannose (MP1) along with rhamnose and galactose (MP2) by ultrasonic techniques from the hull of mung bean and studied their antioxidative activity. They evaluated the scavenging activity of free radicals, reduction power, and inhibitory activities of 1,2,3-phentriol by self-oxidation and found that MP1 demonstrated high potential for antioxidant activity when compared with BHA. The seed coat of lentils is made up of tannins and contains gallocatechin and catechin units, with proanthocyanidin (65–75%) being the most abundant polymer fraction when compared with monomeric and oligomeric fractions (Dueñas et al. 2003). Mateos-Aparicio et al. (2010) extracted the polysaccharides from the cell wall of broad bean pod and pea pod (by-products of legume) by employing sequential alkaline isolation technique using different extraction solutions. The isolated polysaccharides of the cell wall gave three extracts rich in pectin. Among them, two extracts contain pectin-hemicellulose complex and another one consists of a cellulose-rich deposit.

9.7 Solid-State Fermentation

Solid-state fermentation is an environment-friendly procedure which is found helpful in solving the disposal problems of solid waste. The yields of the products are generally lower in submerged fermentation when compared with solid-state fermentation, as reviewed by many authors. The conditions of solid-state fermentation produce biocatalysts by using substrates obtained from plants which are proved to be beneficial in increasing productivity, cost-effective manpower, time and medium constituents along with some environmental benefits like producing minimum effluent, less waste generation, etc. (Gervais et al. 1996).

Application of *Rhizopus oligosporus* in mixed solid-state fermentation on dehulled foxtail millet (*Setaria italica*) and chickpea (*Cicer arietinum*) was found to be effective in reducing the anti-nutritional content like phytic acid and soy antigenic protein present in the corn-soybean mixed feed (Shi et al. 2017). In maize flour, the digestibility of in vitro protein was also increased and reached to 83.6% from 78.5% by the process of solid-state fermentation (SSF) (Reyes-Moreno et al. 2004). In chick pea, an increase in the B vitamins (thiamine, niacin, and vitamin B6) was observed (Sultana et al. 2011). Fermentation causes compositional changes in cereals, which affects the digestibility, bioactivity, and antioxidant potentiality (Heiniö et al. 2003a). The application of solid-state fermentation was also found beneficial for increasing the total phenolic content, antioxidant activity, and antihypertensive potential of cereals and legumes (common beans & hyacinth bean) (Guzmán-Uriarte et al. 2013; Sadh et al. 2017).

The viability of waste husk was investigated by utilizing the husk of some pulses, namely, green gram, Bengal gram, black gram, and red gram as a substrate for producing laccase with the help of oyster mushroom (*Pleurotus ostreatus* 1804) by employing solid-state fermentation process. FTIR and CHNS analysis were done to characterize the pulse husk (solid substrate) before and after the process of fermentation to evaluate the components such as total carbohydrate, cellulose, and lignin. At the optimum culture conditions, the yield of laccase becomes twice which was enhanced by the inducer. It was proved that the green gram is the most excellent source for the production of laccase, having a maximum yield of about 2200 U when compared with all the other sources researched (Prasad et al. 2011; Chandrasekar et al. 2016). An enzyme tannase is significantly utilized in the food processing industries for manufacturing beer, instant tea, fruit juices, grape wine, and coffee-flavored soft drinks. It accelerates the breakdown of such tannins that can undergo hydrolysis such as ethyl gallate, methyl gallate, and tannic acid (Pandey et al. 2000).

Paranthaman et al. (2009) used husk of the red gram as a substrate for producing tannase by optimizing the incubation period and temperature in the solid-state fermentation process. The maximum tannase was produced by *Aspergillus niger* (43 U/g/min at 35 °C), having an incubation period of 96 h with a specific activity of 45 U/g/min in immobilized form.

9.8 Conclusion

The application of by-products in food and feed has been enhanced due to the recognition of its potentiality. The chapter has demonstrated the possible utilization of the by-products obtained from the processing of cereals and pulses. Further researches are needed to evaluate the interaction of the phytochemicals (extracted from the by-products) with different food components and their stability. Applications of by-products will not only reduce the food insecurities by adding wastes as a novel food source but also helps in improving various health issues that have a direct effect on the health systems and economic growth. Further studies are needed to explore the effective utilization of under-utilized pulse/bean and cereal processing by-products as food ingredients as well as for value addition. Exploration of the nutraceutical potential of bioactive compounds is an emerging area where the use of cereals and pulse processing by-products can lead to the development of new commercial applications. But numerous developmental procedures are still required concerning food, feed, and waste regulations to make the best possible utilization of former food products and valorizing the by-products of cereals from the food web without affecting the safety of food and feeds. Formulation of some novel methodologies is also necessary for the extraction of bioactive compounds from food waste, more willing than its dumping, to ensure the sustainability of nutrition and diet.

References

Acosta-Estrada BA, Gutierrez-Uribe JA, Sema-Saldivar SO (2019) Minor constituents and photochemicals of kernel in corn. AACC International Press, pp 369–403

Ajila CM, Prasada Rao UJS (2009) Purification and characterization of black gram (*Vigna mungo*) husk peroxidase. J Mol Catal B Enzym 60:36–44

Aliyu S, Bala M (2011) Brewer's spent grain: a review of its potentials and applications. Afr J Biotechnol 10(3):324–331

Ambreen N, Hanif NQ, Khatoon S (2006) Chemical composition of rice polishing from different sources. Romer Labs, Pakistan Rawalpindi, Pakistan. Pak Vet J 26(4):190–192

Anderson ET, Berry BW (2000) Sensory, shear, and cooking properties of lower-fat beef patties made with inner pea fiber. J Food Sci 65:805–810

Arshad MU, Anjum FM, Zahoor T (2007) Nutritional assessment of cookies supplemented with defatted wheat germ. Food Chem 102(1):123–128

Bajpai B, Patil S (1997) Induction of tannin under acyl hydrolase (EC 3.1.1.20) activity in some members of fungi imperfecti. Enzyme Microb Technol 20:612–614

Barzana E, Rubio D, Santamar RI, Garcia-Correa O, Garca F, Ridaur-Sanz VE (2002) Enzyme-mediated solvent extraction of carotenoids from marigold flower (*Tagetes erecta*). J Agric Food Chem 50:4491–4496

Blandino M, Sovrani V, Marinaccio F, Reyneri A, Rolle L, Giacosa S, Arlorio M (2013) Nutritional and technological quality of bread enriched with an intermediated pearled wheat fraction. Food Chem 141(3):2549–2557

Bledzki AK, Mamun AA, Volk J (2010) Barley husk and coconut shell reinforced polypropylene composites: the effect of fibre physical, chemical and surface properties. Compos Sci Technol 70(5):840–846

Bose D, Shams-Ud-Din M (2010) The effect of chickpea (*Cicer arietinim*) husk on the properties of cracker biscuits. J Bangladesh Agric Univ 8(1):147–152

Brandolini A, Hidalgo A (2012) Wheat germ: not only a by-product. Int J Food Sci Nutr 63 (S1):71–74

Brennan CS, Cleary LJ (2005) The potential use of cereal beta-glucans as functional food ingredients. J Cereal Sci 42:1–13

Bressani R, Elfas LG (1980) Nutritional values of legume crops for humans and animals. In: Summerfield RJ, Bunting AH (eds) Advances in legume science. FAO, London, pp 57–66

Butt MS, Qamar MI, Anjum FM, Abdul A, Randhawa MA (2004) Development of mineral enriched brown flour by utilizing wheat milling by-products. Nutr Food Sci 34:161–165

Carle R, Keller P, Schieber A (2001) Method for obtaining useful materials from the byproducts of fruit and vegetable processing. Patent application. WO 01/78859 A1

Carroll LE (1990) Functional properties and applications of rice bran in bakery products. Food Technol 44:74–76

Ch VS, Reddy R, Nagalakshmi D, Rao J (2012) Evaluation of sweet sorghum (*Sorghum bicolour* (L.) *moench*) bagasse by chemical, in sacco and in vivo techniques in graded murrah buffalo bulls. J Vet Adv 2(8):418–423

Chandrasekar V, Ganapathy S, Karthikeyan S (2016) Enhancing alpha amylase activity of finger millet (*Eluesine coracana*) for improving baking property through solid state fermentation. Adv Life Sci 5(10):4069–4076

Chi Z, Zhao S (2003) Optimization of medium and cultivation conditions for pullulan production by a new pullulan-producing yeast. Enzyme Microb Technol 33:206–221

Choo BS, Sadiq MM (1982) Indigenous feed stuffs and poultry feeds. Poultry Prod Res, Sindh, Karachi

Cruz JM, Moldes AB, Bustos G (2007) Integral utilisation of barley husk for the production of food additives. J Sci Food Agr 87:1000–1008

Dalgetty DD, Baik BK (2006) Fortification of bread with hulls and cotyledon fibers isolated from peas, lentils, and chickpeas. Cereal Chem 83:269–274

Dueñas M, Sun B, Hernández T, Estrella I, Spranger MI (2003) Proanthocyanidin composition in the seed coat of lentils (*Lens culinaris L*). J Agric Food Chem 51:7999–8004

Eggum BO, Alabata EP, Juliano BO (1981) Protein utilization of pigmented and nonpigmented brown and milled rice by rats. Plant Foods Hum Nutr 31:175–179

Gervais P, Marechal P, Molin P (1996) Water relations of solid state fermentation. J Sci Ind Res 55:343–357

Ghazi AR (1992) Measurement of true ME (TME) of indigenous feeding stuffs commonly used poultry rations. M.Sc. Thesis, Univ. Agri. Faisalabad, Pakistan

Guzmán-Uriarte ML, Sánchez-Magaña LM, Angulo-Meza GY, Cuevas-Rodríguez EO, Gutiérrez-Dorado R (2013) Solid state bioconversion for producing common bean (*Phaseolus vulgaris* L.) functional flour with high antioxidant activity and antihypertensive potential. Food Nutr Sci 4:480

Hammond N (1994) Functional and nutritional characteristics of rice bran extracts. Cereal Foods World 39:752–754

Hang YD, Woodams EE (2000) Corn husks: a potential substrate for production of citric acid by *Aspergillus niger*. LWT-Food Sci Technol 33(7):520–521

Harda J, Amorim CA, Braga PL, Machando LDB, Oliveira RR, Neto AC, Macedo JRN, Silvia LGA, Rosa DS (2017) Characterization of biodegradable mulch black films incorporated with organics fertilizers and rice husk ash. TMS 2016-146rd Annual meeting & exhibition, February 26–March 2017-San Diego, CA, USA
Hegsted M, Kousik CS (1994) Rice bran and rice bran oil may lower heart disease risk by decreasing cholesterol synthesis in the body. La Agric 37:16–17
Heiniö RL, Katina K, Wilhelmson A, Myllymäki O, Rajamäki T (2003a) Relationship between sensory perception and flavour-active volatile compounds of germinated, sourdough fermented and native rye following the extrusion process. LWT-Food Sci Technol 36:533–545
Le Goff A, Renard CMGC, Bonnin E, Thibault JF (2001) Extraction, purification and chemical characterisation of xylogalacturonans from pea hulls. Food Chem 45:325–334
Heiniö RL, Katina K, Wilhelmson A, Myllymäki O, Rajamäki T (2003b) Relationship between sensory perception and flavour-active volatile compounds of germinated, sourdough fermented and native rye following the extrusion process. LWT-Food Sci Technol 36:533–545
Hemery Y, Rouau X, Lullien-Pellerin V, Barron C, Abecassis J (2007) Dry processes to develop wheat fractions and products with enhanced nutritional quality. J Cereal Sci 46(3):327–347
Hoffpauer DW, Light Heart LCC, Crowley LA (2005) New application for whole rice bran. Cereal Foods World 50:173–174
Huang JN, Song GH, Sun Q, Zhan CB, Wei H (2010) Extraction of glutathione and wheat germ protein from defatted wheat germ. Food Sci 22
Huang Q, Shi CX, Su YB, Liu ZY, Li DF, Liu L, Huang CF, Piao XS, Lai CH (2014) Prediction of the digestible and metabolizable energy content of wheat milling by-products for growing pigs from chemical composition. Anim Feed Sci Technol 196:107–116
Iriondo-DeHond M, Miguel E, Del Castillo MD (2018) Food by-products as sustainable ingredients for innovative and healthy dairy foods. Nutrients 10(10):1358
Johar N, Ahmad I, Dufresne A (2012) Extraction, preparation and characterization of cellulose fibres and nanocrystals from rice husk. Ind Crop Prod 37(1):93–99
Kahlon TS, Chow FI, Sayre RN (1994) Cholesterol-lowering properties of rice bran. Cereal Foods World 39:99–103
Lai F, Wen Q, Li L, Wu H, Li X (2010) Antioxidant activities of watersoluble polysaccharide extracted from mung bean (*Vigna radiata L.*) hull with ultrasonic assisted treatment. Carbohydr Polym 81:323–329
Leeson S, Summers JD (2001) Nutrition of the chicken. University Books, Guelph
Li CY, Kim HW, Won SR, Min HK, Park KJ, Park JY, Rhee HI (2008) Corn husk as a potential source of anthocyanins. J Agric Food Chem 56(23):11413–11416
Lim JS, Manan ZA, Alwi SRW, Hashim H (2012) A review on utilisation of biomass from rice industry as a source of renewable energy. Renew Sust Energ Rev 16(5):3084–3094
Liu K (2011) Chemical composition of distillers grains, a review. J Agr Food Chem 59 (5):1508–1526
Lowe ED, Buckmaster DR (1995) Dewatering makes big difference in compost strategies. Biocycle 36:78–82
Luzardo-Ocampo I, Cuellar-Nuñez ML, Oomah BD, Loarca-Piña G (2019) Pulse by-products. Food wastes and by-products. pp 59–92. https://doi.org/10.1002/9781119534167.ch3
Maheri-Sis N, Chamani M, Sadeghi AA, Mirza-Aghazadeh A, Safaei AA (2007) Nutritional evaluation of chickpea wastes for ruminants using in vitro gas production technique. J Anim Vet Adv 6:1453–1457
Marlett JA (1991) Dietary fiber content and effect of processing on two barley varieties. Cereal Foods World 36:576–578
Malik MY, Chughtai MID (1979) Chemical composition and nutritive value of indigenous feedstuffs. Pakistan Assoc Adv Sci:11–45
Mateos-Aparicio I, Redondo-Cuenca A, Villanueva-Suárez MJ, Zapata-Revilla MA, Tenorio-Sanz MD (2010) Pea pod, broad bean pod and okara, potential sources of functional compounds. LWT-Food Sci Technol 43:1467–1470

Moheb A, Grondin M, Ibrahim RK, Roy R, Sarhan F (2013) Winter wheat hull (husk) is a valuable source for tricin, a potential selective cytotoxic agent. Food Chem 138(2–3):931–937
Muralikrishnaa G, Tharanathana RN (1994) Characterization of pectic polysaccharides from pulse husks. Food Chem 50:87–89
Murugan K, Chandrasekaran SV, Karthikeyan P, Al-Sohaibani S (2013) Current state of the art of food processing by products. CRC Press, Boca Raton, p 35
Mussatto SI (2009) Biotechnological potential of brewing industry by-products. In: Biotechnology for agro-industrial residues utilization. Springer, Dordrecht, pp 313–326
Nadeem MA (1998) Evaluation of poultry feed stuffs for TME and available amino acids under tropical conditions. PhD Thesis, Univ. Agri. Faisalabad, Pakistan
Nagel-Held B, Welling B, Seibel W, Brack G, Schildbach R (1997) Production of barley fractions with high nutritive value and their utilization in baked products. 111. Production of semi-hard biscuits with milled barley products. Getreide-Mehl-und-Brot 51:180–182
Nayik GA, Majid I, Gull A, Muzaffar K (2015) Rice bran oil, the future edible oil of India: a mini review. J Rice Res 3:4. https://doi.org/10.4172/2375-4338.1000151
NRC (1994) Nutrient requirements of poultry, 9th edn. National Academy Press, Washington, DC
Oosterveld A, Voragen AGJ, Schols HA (2002) Characterization of hop pectins shows the presence of an arabinogalactan-protein. Carbohydr Polym 49:407–413
Osada K, Hoshina S, Nakamura S, Sugano M (2001) Cholesterol oxidation in meat products and its regulation by supplementation of sodium nitrite and apple polyphenol before processing. J Agric Food Chem 48:3823–3829
Oyeleke SB, Jibrin NM (2009) Production of bioethanol from guinea cornhusk and millet husk. Afr J Microbiol Res *3*(4):147–152
Palmer S (2008) The top fiber-rich foods list today's dietitian. 10(7):28
Pandey A, Soccol CR, Nigam P, Brand D, Mohan R, Roussos S (2000) Biotechnological potential of coffee pulp and coffee husk for bioprocesses. Biochem Eng J 6:153–162
Papageorgiou M, Skendi A (2018) Introduction to cereal processing and by-products. In: Sustainable recovery and reutilization of cereal processing by-products. Woodhead Publishing, pp 1–25
Paranthaman R, Vidyalakshmi R, Alagusundaram K (2009) Production on tannin acyl hydrolase from pulse milling by-products using solid state fermentation. Acad J Plant Sci 2(3):124–127
Prasad KK, Chaganti SR, Mohan SV, Sarma PN (2011) Solid state fermentation of laccase from new pulse husks: process optimization and bioprocess study. Int J Innov Biol Chem Sci 2:22–34
Panoth A, Lavanya D, Naik MG, Venkata CN (2019) Waste to wealth: potential of rice bran wax as edible coating. Indian Food Ind Mag 38(1):38–43
Qaisrani TB, Butt MS, Hussain S, Ibrahim M (2014) Characterization and utilization of psyllium husk for the preparation of dietetic cookies. Int J Mod Agric 3(3):81–91
Radosavljević M, Pejin J, Pribić M, Kocić-Tanackov S, Romanić R, Mladenović D, Mojović L (2019) Utilization of brewing and malting by-products as carrier and raw materials in l-(+)-lactic acid production and feed application. Appl Microbiol Biotechnol 103(7):3001–3013
Rajchel CL, Zabik ME, Everson E (1976) Wheat bran and middlings-a source of dietary fibre in banana, chocolate, nut and spice cakes. Baker's Digest 49:27
Ramakrishnaiah N, Pratape VM, Sashikala VB, Narasimha HV (2004) Value addition to by-products from dhal milling industry in India. J Food Sci Technol 41:492–496
Ranhotra GS, Gelroth JA, Astroth K (1990) Total and soluble fiber in selected bakery and other cereal products. Cereal Chem 67:499–501
Rao PV, Reddy MJ (1986) Evaluation of chemical and nutrient composition in raw, de-oiled and parboiled rice polishing and maize. Indian J Poult Sci 21(1):72–74
Rehman MYA, Mahmood S (1996) Anti-nutritional factors, metabolizable energy and chemical composition of rice bran. MARDI Res J 24:135–145
Robert LA, Walter JW (1995) Compositional changes in trypsin inhibitors, phytic acids, saponins and isoflavones related to soyabean processing. J Nutr 125:581S–588S
Reyes-Moreno C, Cuevas-Rodríguez EO, Milán-Carrillo J, Cárdenas-Valenzuela OG, Barrón-Hoyos J (2004) Solid state fermentation process for producing chickpea (*Cicer arietinum* L)

tempeh flour. Physicochemical and nutritional characteristics of the product. J Sci Food Agric 84:271–278

Sadh PK, Saharan P, Duhan S, Duhan JS (2017) Bio-enrichment of phenolics and antioxidant activity of combination of *Oryza sativa* and *Lablab purpureus* fermented with GRAS filamentous fungi. Resource-Efficient Technologies

Saini P, Kaur D, Yadav N, Kumar R, Sanghmitra (2017) Effect of incorporation of chick pea husk on quality characteristics of biscuits. Corporation of chick pea husk on quality characteristics of biscuits. Int J Food Nutr Sci 6(2):2320–7876

Salazar-López NJ, Ovando-Martínez M, Domínguez-Avila JA (2019) Cereal/grain by-products. Food wastes and by-products, pp 1–34. https://doi.org/10.1002/9781119534167.ch1

Salehifar M, Shahedi M (2007) Effects of oat flour on dough rheology, texture and organoleptic properties of Taftoon bread. J Agric Sci Technol 9:227–234

Saunder RM (1990) The properties of rice bran as a food stuff. Cereal Foods World 35:632–636

Schieber A, Stintzing FC, Carle R (2001) By-products of plant food processing as a source of functional compounds – recent developments. Trends Food Sci Technol 12:401–413

Seki N, Toh U, Kawaguchi K, Ninomiya M, Koketsu M, Watanabe K, Kusukawa J (2012) Tricin inhibits proliferation of human hepatic stellate cells in vitro by blocking tyrosine phosphorylation of PDGF receptor and its signaling pathways. J Cell Biochem 113(7):2346–2355

Sharma HR, Chauhan GS, Agarwal K (2004) Physico-chemical characteristics of rice bran processed by dry heating and extrusion cooking. Int J Food Prop 7:603–614

Shi C, Zhang Y, Lu Z, Wang Y (2017) Solid-state fermentation of corn-soybean meal mixed feed with *Bacillus subtilis* and *Enterococcus faecium* for degrading anti-nutritional factors and enhancing nutritional value. J Anim Sci Biotechnol 8:50

Sibakov J, Myllymäki O, Suortti T, Kaukovirta-Noorja A, Lehtinen P, ,Poutanen K, (2013) Comparison of acid and enzymatic hydrolyses of oat bran β-glucan at low water content. Food Res Int 52(1): 99-108

Sidhu JS, Kabir Y, Huffman FG (2007) Functional foods from cereal grains. Int J Food Prop 10 (2):231–244

Singh P, Yadav N, Mishra PK, Sheikh S (2013) Utilization of rice bran for the development of value added Indian Sweet. Int J Agric Food Sci 3(2):76–79

Sosulski FW, Wu KK (1988) High-fiber breads containing field pea hulls, wheat, corn, and wild oat brans. Cereal Chem 65:186–191

Sreerama YN, Neelam DA Sashikala VB, Pratape VM (2010) Distribution of nutrients and antinutrients in milled fractions of chickpea and horse gram: seed coat phenolics and their distinct modes of enzyme inhibition. J Agric Food Chem 58:4322–4330

Sultana N, Azam M, Amin M, Shams B, Satter M, Masum S (2011) Vitamin B and essential minerals contents of mixed solid state fermented millet and Bengal gram by *Rhizopus oligosporus*. Bangladesh J Sci Ind Res 46(1):1–8

Thebaudin JY, Lefebvre AC, Harrington M, Bourgeois CM (1997) Dietary fibers: nutritional and technological interest. Trends Food Sci Technol 8:41–48

Tiwari BK, Tiwari U, Brennan CS, Jagan Mohon R, Surabi A, Alagusundaram K (2011) Utilisation of pigeon pea (*Cajanus cajan* L) byproducts in preparation of biscuits. LWT-Food Sci Technol. https://doi.org/10.1016/j.lwt.2011.01.018

Tiwari U, Gunasekaran M, Jaganmohan R, Alagusundaram K, Tiwari BK (2009) Quality characteristic and shelf life studies of deep-fried snack prepared from rice brokens and legumes byproduct. Food Bioprocess Technol. https://doi.org/10.1007/s11947-009-0219-6

Verma A, Mogra R (2013) Psyllium (*Plantago ovata*) husk: a wonder food for good health. Int J Sci Res 4(90):1581–1585

Wang J, Rosell CM, Benedito de Barber C (2002) Effect of the addition of different fibers on wheat dough performance and bread quality. Food Chem 79:221–226

Weicheng H, John HW, Tai-Sun SJ, Samuel G (1996) Comparison of isopropanol and hexane for extraction of vitamin E and oryzanols from stabilized rice bran. JAOCA 73(2):1653–1655

Wood PJ, Anderson JW, Braaten JT, Cave NA, Scott RW, Vachon C (1989) Physiological effects of β-D-glucan rich fractions from oats. Cereal Foods World 34:87882
Zhang L, Chen F, Zhang P, Lai S, Yan H (2017) Influence of rice bran wax coating on the physicochemical properties and pectin nanostructure of cherry tomatoes. Food Bioprocess Technol 10(2):349–357

Waste from Oil-Seed Industry: A Sustainable Approach

10

Suka Thangaraju, Manoj Kumar Pulivarthi, and Venkatachalapathy Natarajan

Abstract

Oilseeds are one of the major agricultural commodities. Oil plays an important role in maintaining the human health and growth. For the production of good quality of oil, the oilseeds have to be cleaned before refining processes and there should be effective refining steps including cleaning of seeds, mechanical expression, solvent extraction, degumming, neutralization, bleaching, and deodorization. These processes result in good quality of oil, along with a huge quantity of by-products. Disposal of these by-products leads to environmental pollution and causes many other problems. Instead, there are many possible ways to utilize the by-products like husk, hull, cake, pomace or meal, gum, soap stock, spent bleach earth/clay, and distillate. The main aim of this chapter is to know the possible sustainable approach to use the by-products from each step.

Keywords

Colorants · Oilseeds cake · Pectin · Lipids · Antioxidants

S. Thangaraju · V. Natarajan (✉)
Department of Food Engineering, Indian Institute of Food Processing Technology, Thanjavur, Tamil Nadu, India
e-mail: venkat@iifpt.edu.in

M. K. Pulivarthi
Department of Grain Science & Industry, Kansas State University, Manhattan, KS, USA

M. Thakur et al. (eds.), *Sustainable Food Waste Management*,
https://doi.org/10.1007/978-981-15-8967-6_10

10.1 Introduction

Oilseed crops are one of the major agricultural commodities grown worldwide. Global oilseed production for the year 2019–20 is forecast at 577 million metric tons. Presently rapeseed, canola, sunflower, peanut, soybean are the major commercial oilseeds. In India total production of oilseeds for the crop year 2019–20 is about 36,364 thousand metric tons which are contributed mostly by cottonseed, peanut, rapeseed, soybean, and sunflower (USDA 2020).

Any oilseeds are composed of lipids, carbohydrates, fiber, vitamins, and protein. Oils include triacylglycerol with some free fatty acids (FFA), mono and diacylglycerides along with phytosterols, tocopherols, tocotrienols (Dunford 2004). Even though oils and fats have an unhealthy connotation with the public, they play an important role in maintaining health and human growth. There are defined processing procedures followed by industries. In general, oilseeds should be cleaned to remove dirt, leaves, stem, stalk, and any other foreign materials and then seeds should be separated from their outer shell, husk, or hull. Size reduction and flaking of oilseeds are some of the important preliminary steps which aid in easy removal of oil using the solvent extraction method. Oilseeds are also cooked to denature proteins, inactivate enzymes, and release the oil from cells. There are several methods of extracting oil from oilseeds in which mechanical expression and solvent extraction are the most common methods. Meal or cake is one of the major by-products from mechanical expression and solvent extraction. The mechanical expressed oil is the crude oil which contains both desirable and undesirable compounds. Tri-acylglycerides, tocopherols, and phytosterols are some of the desirable components. Undesirable components include free fatty acids, phospholipids, lipid oxidation products which are removed during oil refining steps to reduce their adverse reaction on oil quality. There is a sequence of steps involved in the refining of crude oil.

Degumming, neutralization, bleaching, deodorization, and winterization are the processes commonly used for the edible oil refining process. Phospholipids also referred to as gum, settle down in the bottom during storage and harms quality (color and flavor) of the oil. Therefore, the gum is removed from crude oil by degumming process and produces degummed oil and gum or lecithin as a by-product. According to standards, any edible refined oil should have FFA value less than 0.1% which is done by deacidification or neutralization process which gives neutral oil and soap stock as a by-product. In the beginning, bleaching was done to remove colored components from oil. But nowadays bleaching is done to remove any undesirable compounds like aldehydes, ketones, peroxides, trace metals, and some contaminants like pesticides, aromatic hydrocarbons. Deodorization is the next step in the refining process which aims to produce bland and stable oil by removing FFA, peroxides, aldehydes, and ketones from bleached oil using steam distillation. This step removes volatile and odoriferous compounds and deodorizer distillate (DD) as a by-product. The final step in the oil refining process is winterization in which refined bleached oil is winterized to remove waxes (by-product), hence reduces the solidification of product in refrigerated storage. As explained here, oilseeds processing and oil refining steps produce various by-products which can be further processed to manufacture value-added products. Figure 10.1 explains oilseeds processing and oil refining steps and their associated by-product in each step.

10.2 Utilization of By-products from Oilseed Processing

The most important oilseed crops that are grown primarily to produce edible oil and protein-rich residues are rapeseed, sunflower, flax, peanut, and sesame. Apart from these oilseeds, crops like soybean, palm fruit, and cotton also contribute greatly to the oil production even though they are not primarily grown for oil extraction. Generally, soybean is known to be cultivated primarily for its protein content, whereas the cotton is cultivated for fiber production. Oil is a by-product in case of both soybean and cotton crops. Furthermore, linseeds are utilized for the production of edible oil as well as the industrial oil, while the peanuts are consumed by humans directly apart from using them for oil extraction (Day 2004; Mullen et al. 2015).

Any wastes from a food processing plant can be generally classified as solid and liquid wastes. As some of these wastes generated from the oilseed industry are hazardous to our environment, there are certain treatments given to the waste to protect the ecosystem from harmful substances. Hence it is the responsibility of the food processor to implement best disposal practices and effectively utilize the waste generated by value addition. It is always wise and sustainable approach to convert wastes into valuable by-products (Galanakis 2015).

The wastes from oilseed processing plants have been significantly increasing from the past few decades. As the oilseeds are majorly processed by either mechanical expression or solvent extraction, wastes like oilseed meal, seeds, oil sludge, peels, etc., are formed during the extraction process. Table 10.1 enlists different types of most common oilseed wastes generated during different processing steps of oil extraction.

10.2.1 Oilseed Cake/Meal

The oilseed cake/meal is a by-product obtained in any oil extraction industries while recovering crude oil from the oilseeds through mechanical expression or solvent extraction. Usually, the meal obtained through solvent extraction is not considered human friendly, due to the chemical treatments it undergoes. Hence the safety of the recovered oilseed meals through this process is still doubtful and complicated (Galanakis and Schieber 2014; Mullen et al. 2015).

The nutritional quality of defatted oil cake is very rich since only the oil from the whole seed is removed through the extraction process. Therefore, all other nutritional qualities from the oilseed are mostly retained in the cake. The oilseed cake is rich in proteins, fiber content, antioxidants, colorants, and other health beneficial components. Usually, the majority of the industries are utilizing oilseed cake as an animal feed. However, there is tremendous scope for it to be utilized in

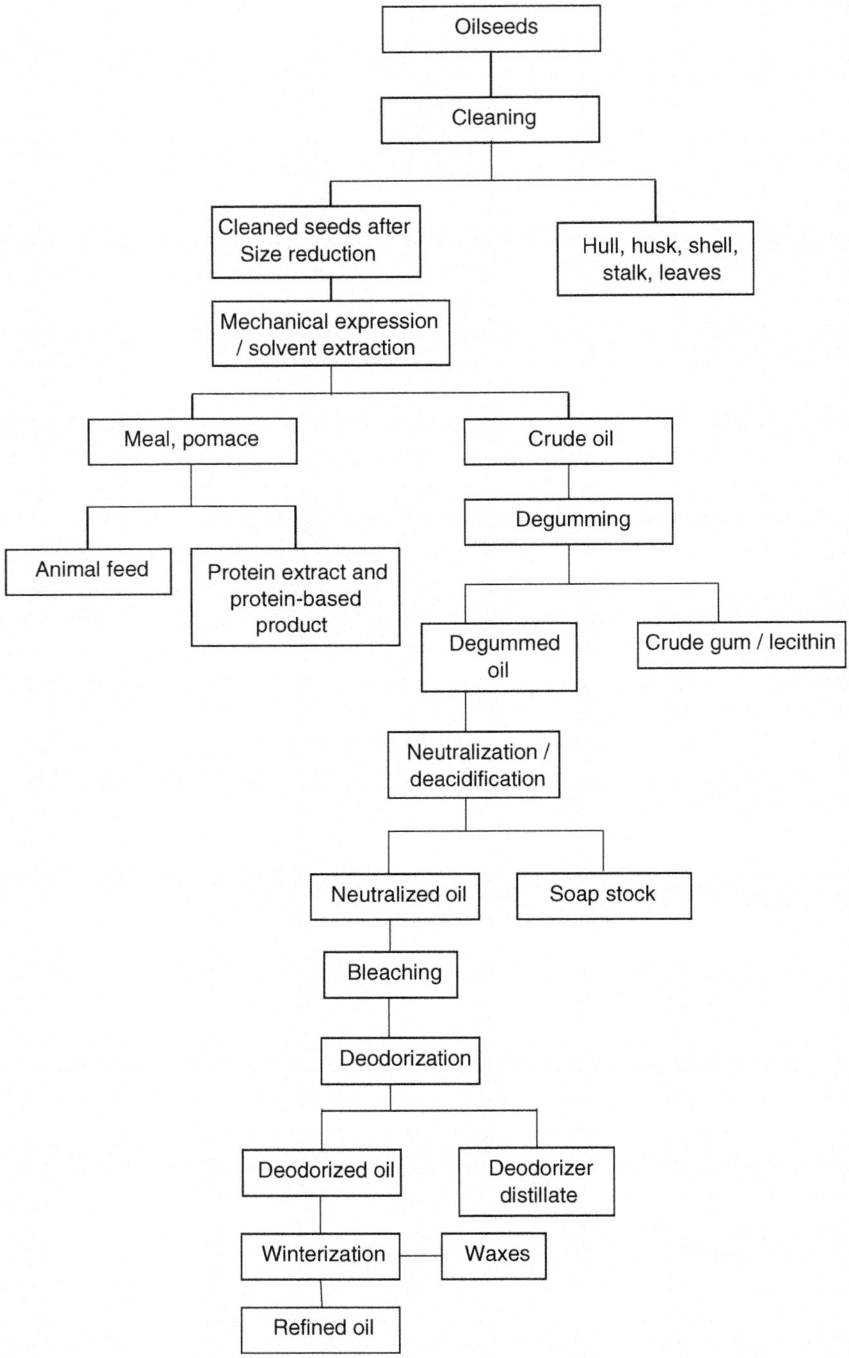

Fig. 10.1 Steps involved in processing and refining of oilseeds

Table 10.1 Waste generated at different processing steps of oil extraction (Source: Mullen et al. 2015)

S. No	Processing step	Waste obtained
1	Cleaning	Stems, pods, leaves, broken kernels, dirt, small stones, extraneous seeds, stalks, coir
2	Dehulling	Hull/husk
3	Cooking	Wastewater
4	Oil extraction	Oilseed cake

recovering more beneficial and value-added products. Therefore, these oilseed cakes can be further subjected to various treatments to make them more suitable and safer for human consumption. For example, oilseed cake can be utilized to enhance the nutritional profile of gluten-free foods (Radočaj et al. 2014).

10.2.1.1 Proteins

The nutritional quality of proteins is generally determined by amino acid profile, it must be noted that the proteins extracted from oilseed cake lack in sulfur-containing amino acids. Usually, the protein quality of any oilseed cake is measured by comparing with soybean's seed cake, which has a very good amino acid profile suitable for children according to WHO (Tan et al. 2011; Rodrigues et al. 2012). Even though the oilseed cake is nutrient-rich, the presence of some anti-nutritional factors (like trypsin inhibitors, tannins, etc.) limits the usage of the products extracted for human consumption. For example, the presence of a yellow pigment named gossypol in cottonseed oil cake restricts usage of nutritious oilseed cake as both human food and animal feed (Liadakis et al. 1993). These factors interfere with the digestibility of proteins, sensory properties and even restrict the absorption of micro and macronutrients. A study conducted by Pojić et al. (2014) suggests that a pre-treatment can be given to the oilseed cake to extract the valuable substances separately by eliminating the anti-nutrient components.

The application of proteins extracted from oilseed cakes is increasing lately in food, pharmaceutical, and cosmetic industries (Moure et al. 2006; Rodrigues et al. 2012). So, the utilization of valuable substances like protein for human consumption from oilseed meal is completely dependent on efficiency in processes developed to eliminate anti-nutrients. The extracted protein can be classified into three categories based on the amount of protein present (Oreopoulou and Tzia 2007). They are:

1. Flours (up to 50% protein)
2. Concentrates (65–70% protein)
3. Isolates (90–95% protein)

10.2.1.2 Antioxidants

The oilseed cake is also a potential source for extraction of antioxidants (Oreopoulou 2003). For example, the presence of lignans in sesame cake makes it an excellent

Table 10.2 Type of solvents used to extract different antioxidants

S. No	Solvent	Antioxidants
1	Hexane, petroleum ether (non-polar)	Tocopherols and phenolic terpenes
2	Ethyl ether and ethyl acetate	Flavonoid aglycons, phenols of low-molecular weight, and phenolic acids
3	Ethanol or ethanol-water mixtures (higher polarity)	Flavonoid glycosides and phenols of higher molecular weight

source for antioxidant activity (Suja et al. 2005). The solvent extraction process can be adapted to isolate the antioxidants with an appropriate solvent. The type of solvent used plays a major role in obtaining different antioxidants with great yield and activity. Some of the most efficient solvents used in the extraction of different antioxidants are given in Table 10.2.

10.2.1.3 Non-food Applications

In addition to the food applications, the protein extracts are also being used in the development of biodegradable films for food packaging. The soy protein from soy seed cake is found to be a very suitable one for producing good quality smooth films with greater flexibility and transparency (Gontard and Guilbert 1994). The protein isolates also have the potential to be used as adhesives (Mullen et al. 2015). To enhance the gluing strength water resistance, Huang and Sun (2000) have isolated soy protein by modifying with urea and guanidine hydrochloride solution.

10.2.2 Hull

10.2.2.1 Heating Source

Hull is another important waste by-product obtained from oil extraction industries. In general, it is more often utilized as a fuel for the source of heat. This heat is utilized mainly for the distillation process to separate the solvent from extracted crude oil (Seiler and Gulya 2015). Before using them directly, the hull is modified slightly by adding some binders and glue-like substances to make small solid blocks (briquettes) which are suitable for boilers. For example, the hull portion in a sunflower seed is comprised to be 20–30%. It is being used as a boiler feed in a few processing industries by making slight modifications to the boiler (Seiler and Gulya 2015).

10.2.2.2 Colorants

Even colorants can be derived from the hull of oilseeds rich in anthocyanins. The hull of a purple seeded sunflower can be used to extract a red dye which is approved as a food colorant by the FDA (Seiler and Gulya 2015). The hull of these sunflower seeds consists of anthocyanin content of 6–16 g/kg of the hull which is almost equivalent to that of well-known sources like grape skin and beet pulp. As the trend now is moving towards usage of natural colorants instead of synthetic ones, it is a sustainable approach to extract the colorants from oilseed for better applications. However, the stability of these natural colorants is low compared to the synthetic ones. Hence the applications of these dyes are narrow and still more developments must come into existence for widening the applications.

10.2.2.3 Pectin

A well-known food thickener used in making jellies is pectin. It can also be extracted from the oilseed head. The amount of pectin present in a sunflower seed head after the removal of seed is 150–250 g/kg (Seiler and Gulya 2015). Unlike the high methoxyl pectin extracted from apples and citrus fruits, the sunflower pectin is low methoxyl. This is favorable in making low sugar jellies that are gaining attention among the diet food markets nowadays.

10.3 Wastewater

It is important to implement different processing methods to recover valuable products at different processing steps of the oil processing industry. The absence of such measures will result in loss of all the valuable nutrients like carbohydrates, lecithin, phenols, vitamins, sterols, etc., in the wastewater generated during the processing. The right way of disposal of these wastewaters is most essential for protecting the environment from harmful substances. The wastewater obtained from the oil processing plant is very toxic and polluted because of high levels of BOD, COD, dissolved solids, oil and fat residues, organic nitrogen, and ash residues. At the same time, the wastewater is also highly nutritious as discussed before with all the nutrients. Hence, important co-products like sugars, sugar alcohol, polymer building blocks, polymerins, phenols, sterols, biogas, isoflavones, biodiesel, etc. can be recovered from wastewaters of an oil processing plant. After recovering the co-products, there are also various promising methods developed for treating the wastewaters to decontaminate and detoxify for the safe disposal of treated water into the environment. Some of these treatments include anaerobic digestion, biological removal of phenolic compounds with help of fungi, a combination of aerobic and anaerobic treatment, a combination of biological and chemical oxidation treatment, chemical oxidation, composting, utilization of biosurfactants produced from microbes, and genetic engineering (Arienzo and Violante 2007).

10.4 Extraneous Material

Apart from the above-mentioned well-known by-products, there are also some extraneous materials like plant stems, pods, leaves, damaged/spoiled kernels, dirt, small stones, extraneous seeds, stalks, coir, etc. obtained during the oil extraction process. Most of this extraneous material is obtained at the initial stages of processing steps like cleaning and screening. Large amounts of dust are also generated during these unit operations. Even while collecting the seeds by threshing from the plant for oil extraction, large quantities of plant material attached to the seeds are wasted. Nearly 1 tonne of empty fruit bunches is left behind for the manufacturing of 1 tonne of crude palm oil. This plant waste is rich in lignin, cellulose, and hemicellulose contents. Most commonly, the solid wastes obtained are either utilized as boiler feed to generate steam or converted into compost. It is also seen rarely to be landfilled in some small-scale oilseed processing industries. Even charcoal can be produced by subjecting this plant material to pyrolysis (Lua and Guo 1999).

On an average around 0.7–0.8 tonnes of solid waste is generated per tonne of raw material. This is mostly converted into either valuable by-products or converted into fuel as discussed in the above paragraph.

10.5 Wastes from the Oil Refining Process

10.5.1 Gum

Gum is a by-product which is obtained after the degumming process in the refining of crude oil. It is also termed as phospholipids or lecithin. It is a mixture of different phospholipids like phosphatidic acid, phosphatidylcholine, phosphatidylinositol, phosphatidylserine, phosphatidylethanolamine. The existence of phospholipids in oil can lead to discoloration, produces off-flavors and add up to the loss of neutral lipids while neutralizing. Phospholipids are good emulsifiers, which creates a bond between oil molecules and leads to an increase in viscosity and loss of oil during a further refining process. Viscosity is another important determining factor in the overall quality and stability of the oil. Hence, the separation of phospholipids from crude oil is an important step in the refining process for the production of good quality oil. Gum can be removed from the oil by water, acid (citric acid, acetic anhydride, phosphoric acid, oxalic acid), and ultrafiltration processes which cannot assure the required reduction of phosphorous levels after refining and these methods of degumming are not sufficient for oils which are having higher levels of non-hydratable phospholipids. Enzyme-assisted degumming is a quite distinct process because it involves the removal of both hydratable and non-hydratable phospholipids by hydrolyzing it to the corresponding lysophospholipids (Zufarov et al. 2008; Lamas et al. 2016).

Soybean is one of the major sources of lecithin, followed by rapeseed, sunflower, cottonseed, corn and now rice bran lecithin is an emerging source of lecithin in commercial production. The gum received after the degumming process is crude lecithin which further has to be purified to remove carbohydrates, proteins, and other contaminants. The purified gum/lecithin has various applications which include food uses (bakery and confectionary, emulsifiers, release agents, instant powders, encapsulation, etc.), animal feed, food supplements and pharmaceuticals, and other applications in cosmetics, paint industries as a lubricant, releasing agent, emulsifier, adhesives, and absorbents (Clarke 2007).

Lecithin is mainly useful in terms of rheological aspects. It has its application in the bakery as an emulsifying agent to homogenize fat/oil and water, as a wetting agent to contribute immediate wetting to dry powders and it helps in decrease in mixing time, as an antioxidant to stabilize fats. Doughnuts prepared by adding lecithin as an additive show better improvement in make-up and are stable and elastic and also it forms a thin skin which is less tender. The main advantage of using lecithin as an additive is that it has no consequences on adsorption, fermentation, baking time. Margarine is one of the most important ingredients in bakery and confectionary industry. Lecithin helps in formulation on margarine/low-fat spread by acting as an anti-spattering agent and generally, margarine has to be of good quality without spattering of fat. Bread quality is mostly determined based on volume and freshness and lecithin especially hydrolyzed lecithin helps in enhancing both the properties of bread by physical linkage of wheat gluten with phospholipids by forming H-bridges like lipoprotein. This results in a better lubrication of protein by which elasticity and volume of dough are increased. Lecithin also has its application on pretzels by acting as a lubricating agent. It helps to reduce the stickiness of pretzel dough. It also helps in increasing the gloss and less out of spec on the surface of pretzel after production. In case of cookies, lecithin helps in dispersion of the shortening during preparation. Addition of 0.5% of lecithin reduces 30% fat for the production of low-fat cookies. Lecithin here helps in reducing the stickiness of the dough (Van Nieuwenhuyzen and Tomás 2008).

Chocolate is a dry emulsion with sugar and cocoa particles present in the continuous fat phase of 30–34% cocoa butter, here sugar is hydrophilic and cocoa particle is lipophilic. After mixing and refining, the ingredients are conched to a paste-like mass. After conching, 0.3–0.5% of lecithin is added to the mixture and it helps in viscosity reduction, flow improvement during tempering, molding, or enrobing. During chocolate processing, β-fat crystals are produced while tempering and further in the molding process which is distributed in liquid fat. Lecithin helps in lubrication of sugar, cocoa particles, and fat crystals which makes the flow of chocolate mass more liquid (Schantz and Rohm 2005).

Lecithin has been used in instant powders like infant foods as a wetting and dispersing agent. Phospholipids present in lecithin reduce the surface tension in the fat–water interface and interlinkage between lecithin and proteins is created. The hydrophilic nature of lecithin helps in dispersion of powders in milk or water. Lecithin is also used for encapsulation of high-value agents, for example: Liposome encapsulation of flavours, anti-oxidants, drugs for pharmaceutical purposes. It helps in the controlled release of flavours, drugs, etc. (Viñado et al. 2019; Chen et al. 2020). Also, lecithin has various applications in animal, fish, and poultry feed. Lecithin alone or in combination with other stabilizers has been used as stabilization of milk replacers for calves, piglets, and other newborn animals. Phospholipids are important nutrients for both larvae and shrimps. Phospholipids are good in absorbing dietary fat and fat-soluble vitamins and also help in intermediary metabolism and fatty acid metabolism (Van Nieuwenhuyzen and Tomás 2008).

Lecithin has other industrial applications like in cosmetics, paint, ink, dye, etc. Incorporation of lecithin in cosmetics improves the surfactant properties of the product, which is the most valued property in the cosmetics industry. The extracted pigments and particulates coated with lecithin have a smooth texture, increase adhesion to skin and also improve color stability. Lecithin can act as a surfactant, so it has been used in magnetic recording equipment. It helps in the dispersion of magnetic particles on the pigment surface. It enhances both physical and magnetic properties. Lecithin is also used in paints, coatings productions for their ability to disperse pigment uniformly. Lecithin binds to pigment surface and allows wetting of pigment by the equipment by which it has been dispersed (Williams 1945; List 2015).

10.5.2 Lipids/Soap Stock

The next step in refining process after degumming is called neutralization or deacidification in which degummed oil is continuously mixed with dilute sodium hydroxide to remove free fatty acids (FFA), phosphatides, and other unsaponifiable materials. FFA is the most valued components and the composition varies according to the source of oil, condition of oilseeds while processing, oil removal methods, type of solvent used, duration of extraction, and the condition during the refining process. The removal of large quantity of lipids during refining process, leads to the production of good quality refined oil. Addition of sulfuric acid to lipids/soap stock stabilizes the lipid and also reduces its weight which helps in shipment. This process is called acidulation. Vegetable oils are refined with caustic potash (KOH), acidulation with sulfuric acid, and followed by neutralization with ammonia.

Lipids have its major application in animal feed. They are added to the meal to increase its weight and fat content in meal. Lipids not only have higher caloric value but also provide essential fatty acids. Raw and acidulated lipids are mixed in different ratios with animal tallow for production of soap with different characteristics. Palm oil and coconut oil are important fatty acid sources for soap production. Lipids have also been used in the growth of microorganisms. Soap stock has been used in the production of lipase by *Aspergillus niger* by solid state fermentation method (Ribeiro dos Santos et al. 2014). It is also used as a stabilizer for granular shoulders to avoid edge rutting. Soap stock is used as a carbon source for the production of carotenoid producing yeast (*Rhodotorula rubra*). Application of soap stock has increased carotenoids yield up to 5.36 folds compared to the production using glucose (Alipour et al. 2017). Soap stock has been used in the production of green epoxides from fatty acids. Soap stock having high unsaturated free fatty acids had been used for enzymatic production of natural epoxides in microchannel bioreactor. The yield of epoxidation enhanced 2.8 folds (Mashhadi et al. 2018).

10.5.3 Spent Bleaching Earth

Bleaching is a process in which the color pigments present in the oil are removed by passing the oil through beaching clay/earth. Spent bleaching earth (SBE) after the refining process has up to 20–40% w/w absorbed oil. So, spent clay is a good source of recoverable oil. Spent clay has a disadvantage. It is susceptible to immediate combustion when in direct contact with air especially for the clay with more unsaturated oils. It is also an environmental concern, as a fire hazard and threat to the groundwater when it is discarded as landfill. Recovery of oil from spent clay can be done by steam treatment, aqueous extraction, solvent extraction, and pressure extraction methods.

SBE had being used in the production of poultry feed with the inclusion of up to 7.5%. This had no deleterious effect on feed efficiency or growth rate. The metabolizable energy of spent clay is about 2870 kcal/kg and it varies on the amount of oil present in it. Spent clay and its disposable problem are a big environmental concern for all refinery industries. Spent clay has its application in the production of lubricating grease. It is used as a thickener and waste cooking oil as a base oil for production of grease. SBE has very fine particles which can absorb oil and hold the oil within it. The ability to hold and release oil when needed is the most important property for any thickener (Abdulbari et al. 2011). The more recent approach for spent clay is that it can be converted into biofertilizers. And the fertilizer characteristics in relation with physical, chemical, and biological effects on crops and surrounding environment (nutrients, water pollutants, climate, etc.) have been studied and it shows better effects on agricultural yield, soil fertility, and crop productivity. SBE with other potential nutrients can lead to sustainable agriculture in future (Loh et al. 2017).

10.5.4 Distillates from Deodorizer

Deodorization is another physical refining step in which off-flavors/unwanted flavors are stripped off by steam distillation method. The by-product formed is distillate which is like sludge appearance. Deodorizer distillate is a mixture of tocopherols, sterols, fatty glycerides, hydrocarbons, and some amount of fatty acids. The applications of distillate depend on the composition of the mixture. The tocopherols and sterols present in the sludge can be used as a natural source of vitamin E and drug manufacturing, respectively (Sherazi et al. 2016).

Deodorizer distillate is a very cheap source of several health beneficial components like tocopherols, sterols, free fatty acids, etc. These valuable components are being used in different foods, pharmaceutical, and cosmetics industries (Sherazi et al. 2016). Squalene is one of the high value-added products which can be purified from distillate sludge using centrifugal partition chromatography method. The purity of isolated squalene was found to be 95.5%, recovery was 76.3%, and the productivity was calculated as 234 mg/h/L (Xynos et al. 2016). Biodiesel had been produced by Naz et al. (2014) using fatty acids from corn oil distillate. Tin alginate as a catalyst and methyl esterification method was used. With 4% of catalyst, 1:12 of oleic acid to methanol ratio, 98.7% fatty acid, methyl ester was recovered in two hours of reaction time.

10.6 Sustainable Approach

It is important to efficiently utilize the oilseeds since a lot of natural resources and energy goes into the production process. Even though our component of interest is mainly the oil or fat content, the goal has to utilize the remaining parts of the seed to the fullest by keeping the waste generated as low as possible. As discussed in the above chapter, sometimes the waste material is just burned to produce heat. Although it is the easiest way of handling the waste, it is an uneconomical and major loss of valuable alternative products that can be obtained otherwise. Even to date, the waste management strategies or techniques used in most of the oilseed industries are traditional leading to under-utilization of nutritious waste. The goal of any food processor should be to adopt a holistic approach for sustainable waste management. This can be achieved by adopting new innovative technologies like genetic engineering, computer-controlled processing, production of valuable products by biotechnology, utilization of enzyme technology for easy separating of different components of waste, etc. Apart from these new technologies, it is also important to minimize the waste even before recovering the by-products, implementing best practices in place to keep the waste generated separately at each point of processing step and also by developing best methodologies for getting high recovery rates. Thus, by gradually implementing these changes into our oilseed processing industries, we will succeed in achieving a sustainable, environment friendly and safe world.

References

Abdulbari HA et al (2011) Lubricating grease from spent bleaching earth and waste cooking oil: tribology properties. Int J Phys Sci 6(20):4695–4699. https://doi.org/10.5897/IJPS11.561

Agriculture prices, National Agricultural Statistics Services, Agriculture Statistics Board, United States Department of Agriculture. https://www.nass.usda.gov/Publications/Todays_Reports/reports/agpr0220.pdf

Alipour S et al (2017) β-carotene production from soap stock by loofa-immobilized Rhodotorula rubra in an airlift photobioreactor. Process Biochem 54:9–19. https://doi.org/10.1016/j.procbio.2016.12.013

Arienzo M, Violante P (2007) Improving waste management and co-product recovery in vegetable oil processing. In: Handbook of waste management and co-product recovery in food processing. Woodhead Publishing, pp 534–570

Chen GJ et al (2020) Effects of soybean lecithin supplementation on growth performance, serum metabolites, ruminal fermentation and microbial flora of beef steers. Livest Sci:104121. https://doi.org/10.1016/j.livsci.2020.104121

Clarke Z (2007) Lecithin. xPharm: The Comprehensive Pharmacology Reference (1):1–3. https://doi.org/10.1016/B978-008055232-3.62016-1

Day L (2004) In: Wrigley C, Corke H, Walker CE (eds) Lipid chemistry. Encyclopedia of grain science. Elsevier Academic Press, Oxford

Dunford NT (2004) Effects of oil and oilseed processing techniques on bioactive compounds. Nutritionally enhanced oil and oilseed processing. AOCS Press, Champaign, pp 25–37

Galanakis CM (2015) The universal recovery strategy. In: Food waste recovery. Academic Press, pp 59–81

Galanakis CM, Schieber A (2014) Editorial. Special issue on recovery and utilization of valuable compounds from food processing by-products. Food Res Int 65:299–230

Gontard N, Guilbert S (1994) Bio-packaging: technology and properties of edible and/or biodegradable material of agricultural origin. In: Food packaging and preservation. Springer, Boston, MA, pp 159–181

Huang W, Sun X (2000) Adhesive properties of soy proteins modified by urea and guanidine hydrochloride. J Am Oil Chem Soc 77(1):101–104

Lamas DL, Constenla DT, Raab D (2016) Effect of degumming process on physicochemical properties of sunflower oil. Biocatal Agric Biotechnol. https://doi.org/10.1016/j.bcab.2016.03.007

Liadakis GN, Floridis A, Tzia C, Oreopoulou V (1993) Protein isolates with reduced gossypol content from screw-pressed cottonseed meal. J Agric Food Chem 41(6):918–922

List GR (2015) Soybean lecithin: food, industrial uses, and other applications. In: Polar lipids. Elsevier, Amsterdam, The Netherlands, pp 1–33

Loh SK, Cheong KY, Salimon J (2017) Surface-active physicochemical characteristics of spent bleaching earth on soil-plant interaction and water-nutrient uptake: a review. Appl Clay Sci 140:59–65. https://doi.org/10.1016/j.clay.2017.01.024

Lua AC, Guo J (1999) Chars pyrolyzed from oil palm wastes for activated carbon preparation. J Environ Eng 125(1):72–76

Mashhadi F, Habibi A, Varmira K (2018) Enzymatic production of green epoxides from fatty acids present in soap stock in a microchannel bioreactor. Ind Crops Prod 113:324–334. https://doi.org/10.1016/j.indcrop.2018.01.052

Moure A, Sineiro J, Domínguez H, Parajó JC (2006) Functionality of oilseed protein products: a review. Food Res Int 39(9):945–963

Mullen AM, Álvarez C, Pojić M, Hadnadev TD, Papageorgiou M (2015) Classification and target compounds. In: Food waste recovery. Academic Press, pp 25–57

Naz S et al (2014) A green approach for the production of biodiesel from fatty acids of corn deodorizer distillate. RSC Adv 4(89):48419–48425. https://doi.org/10.1039/c4ra08108k

Oreopoulou V (2003) Extraction of natural antioxidants. In: Tzia C, Liadakis G (eds) Extraction optimization in food engineering. Marcel Dekker Inc., New York, pp 317–334

Oreopoulou V, Tzia C (2007) Utilization of plant by-products for the recovery of proteins, dietary fibers, antioxidants, and colorants. In: Utilization of by-products and treatment of waste in the food industry. Springer, Boston, pp 209–232

Pojić M, Mišan A, Sakač M, Dapčević Hadnađev T, Šarić B, Milovanović I, Hadnađev M (2014) Characterization of byproducts originating from hemp oil processing. J Agric Food Chem 62 (51):12436–12442

Radočaj O, Dimić E, Tsao R (2014) Effects of hemp (Cannabis sativa L.) seed oil press-cake and decaffeinated green tea leaves (Camellia sinensis) on functional characteristics of gluten-free crackers. J Food Sci 79(3):C318–C325

Ribeiro dos Santos R et al (2014) Characterization of different oil soap stocks and their application in the lipase production by *Aspergillus niger* under solid-state fermentation. J Food Nutr Res 2 (9):561–566. https://doi.org/10.12691/jfnr-2-9-6

Rodrigues IM, Coelho JF, Carvalho MGV (2012) Isolation and valorisation of vegetable proteins from oilseed plants: methods, limitations and potential. J Food Eng 109(3):337–346

Schantz B, Rohm H (2005) Influence of lecithin-PGPR blends on the rheological properties of chocolate. LWT-Food Sci Technol 38(1):41–45. https://doi.org/10.1016/j.lwt.2004.03.014

Seiler GJ, Gulya TJ (2015) Sunflower: overview. In: Wrigley C, Corke H, Seetharaman K, Faubion J (eds) Encyclopedia of food and grains, vol 1, 2nd edn. Elsevier, Waltham, pp 247–253

Sherazi STH, Mahesar SA, Sirajuddin (2016) Vegetable oil deodorizer distillate: a rich source of the natural bioactive components. J Oleo Sci 65(12):957–966. https://doi.org/10.5650/jos.ess16125

Suja KP, Jayalekshmy A, Arumughan C (2005) Antioxidant activity of sesame cake extract. Food Chem 91(2):213–219

Tan SH, Mailer RJ, Blanchard CL, Agboola SO (2011) Canola proteins for human consumption: extraction, profile, and functional properties. J Food Sci 76(1):R16–R28
Van Nieuwenhuyzen W, Tomás MC (2008) Update on vegetable lecithin and phospholipid technologies. Eur J Lipid Sci Tech 110(5):472–486. https://doi.org/10.1002/ejlt.200800041
Viñado A et al (2019) Erratum: crude soybean lecithin as alternative energy source for broiler chicken diets (Poultry Science (2019) 98(11) (5601–5612), (S0032579119457659), (10.3382/ps/pez318))'. Poul Sci J 98(12):7172. https://doi.org/10.3382/ps/pez553
Williams KA (1945) Industrial oil and fat products. Analyst. https://doi.org/10.1038/157822a0
Xynos N et al (2016) A single-step isolation of squalene from olive oil deodorizer distillates by using centrifugal partition chromatography. Sep Sci Technol (Philadelphia) 51(5):830–835. https://doi.org/10.1080/01496395.2015.1119843
Zufarov O, Schmidt S, Sekretár S (2008) Degumming of rapeseed and sunflower oils. Acta Chim Slov 1(1):321–328

Wealth from Meat Industry By-Products and Waste: A Review

11

Rukhsaar Sayeed and Pratibha Tiwari

Abstract

Slaughterhouses and meat processing industries produce large amounts of waste water with high fat, grease and protein content. Waste from animal by-products can be a significant source of physical and biological animal contamination. Waste material is also a potential source of food for pests, which may give rise to further microbiological contamination. These contaminations can pose a great impact on the health of consumers and hence must be prevented and reduced. By slaughtering and processing of animals for meat, large qualities of waste and by-products are generated, which need to be adequately processed and utilized. Animal by-products and wastes are a good source of renewable energy as well and its production is economically feasible. This chapter deals with the management of waste and utilization of by-products of meat, poultry and fish processing industries.

Keywords

Slaughterhouses · By-products · Animal waste · Meat processing industries

R. Sayeed (✉)
Amity Institute of Food Technology, Amity University, Noida, Uttar Pradesh, India

P. Tiwari
Maharishi Markandeshwar University, Mullana, Ambala, Haryana, India

M. Thakur et al. (eds.), *Sustainable Food Waste Management*,
https://doi.org/10.1007/978-981-15-8967-6_11

11.1 Introduction

Food industries across the globe are under immense pressure to efficiently minimize and manage the wastes generated by them. The wastes with regards to the meat industry may be defined as "carcasses or animal parts that are not intended for direct human consumption" (Commission of the European Communities 1990). Consumption of meat by the world population is related to diet, living standards and production of livestock. The livestock industry including fish and poultry plays a vital role in economic growth of a country. It contributes to Agricultural Gross Domestic Product (AGDP) and serves as means of income for the common population. The Global production of meat including broiler meat for the year 2017 was around 263 million tonnes and is projected to increase by 16% in 2025 (OECD FAO 2016).

India is the 6th largest meat producing country in the world and produces around 6.90% of the total meat, owing to its large livestock population (Birthal 2008). According to the data released by the Ministry of Agriculture, India presently possesses 190.9 million cattle, 135.2 million goats, 108.7 million buffaloes, 10.3 million pigs, 65.07 million sheep and 729.2 million chickens (19th Livestock Census). The statistics suggest that the largest meat producer species is poultry, followed by bovines (cattle and buffaloes), goats, sheep and pigs (Table 11.1). The major importers of Indian meat are Vietnam, Malaysia, Thailand, Australia, UAE, Saudi Arabia and Egypt. Although meat is considered as an important component of human diet with very high biological value, its consumption has always been the subject of social and cultural prejudice in India.

Because of the increasing global demand for animal products, Indian meat sector has an opportunity to increase its exports for meat and meat products. Among the Indian agricultural produce, buffalo meat is an important component of export and is valued at USD 29 million annually, which makes India the world's biggest exporter of buffalo meat (FICCI). India exports fresh and frozen meat to 54 countries. Overall, there has been significant contribution of livestock products to GDP of the country and accounts for more than 40% of total agriculture sector and more than

Table 11.1 Production and domestic supply of meat

Specie	Production (1000 MT)	Import (1000 MT)	Export (1000 MT)	Total domestic supply	Total per capital supply (kg/year)
Poultry	2358	0	6	2352	1.88
Mutton and goat	747	0	21	725	0.58
Pig meat	354	1	0	354	0.28
Other	180	0	1	171	0.14
Meat offal	529	0	54	475	0.38
Total meat	6215	1	1589	4619	3.69

12% of GDP (Jayathilakan et al. 2012). This chapter discusses the waste management and by-product utilization in meat, poultry and fish processing industries.

11.2 Meat Industry and Waste Generated

Although the contribution of meat industries towards the economy of the country is enormous, the environmental risks and hazards posed by slaughtering and processing of meat cannot be undermined. They have the potential to generate large amounts of solid waste and waste waters with high organic matter content. Furthermore, inefficient utilization of by-products may lead to increased environmental hazards and cataclysmic health problems. The first step in meat processing involves slaughtering of animals in abattoirs (slaughterhouses), followed by secondary operation like deboning, grinding and processing for final consumption. Wastes generated in slaughterhouses consist of those parts and portions of a slaughtered animal that cannot be sold as meat or used in meat products and are unfit for consumption by humans. Both solid and liquid wastes are generated. Solid wastes include bones, skin, tendons, viscera, hoofs, horns, nails, etc. and liquid wastes include blood, waste water generated as a result of washing carcasses, cleaning and sanitizing equipment and facilities, etc. Table 11.2 lists the specific quantities of waste generated with respect to type of animal (Russ and Pittroff 2004). The waste characteristics depend largely upon the type of animal being slaughtered. So, the first step for determining the type of treatment(s) and evaluating costs is to measure the organic load of the effluent. Organic load can be measured and expressed in terms of BOD (Biological oxygen demand), COD (Chemical oxygen demand), SS (Suspended solids), TDS (Total dissolved solids), FOG (fats, oils and grease).

In India there are presently 1176 slaughterhouses, 80 registered abattoirs-cum-meat processing plants and 28 registered meat processing plants approved by APEDA. There are 10 fully integrated plants that are eco-friendly and have the capacity of producing 50,000–1,20,000 tons of meat per annum. The slaughterhouses have been classified into 3 categories: small, medium and large, based on the number of animals slaughtered. The classification basis is mentioned below:

(i) *Small*: Less than 50 large animals, i.e. bovines per day, or less than 300 small animals, i.e. goat and sheep per day (any day in a week).

Table 11.2 Specific waste index for abattoirs

Animal specie	Specific waste index[a]
Calf	0.87
Cow	0.56
Pig	0.2
Sheep	0.1

a: Mass of accumulated waste divided by the mass of saleable product (Russ and Pittroff 2004)

(ii) *Medium*: 50–200 large animals, i.e. bovines per day, or 300–1000 small animal, i.e. goat and sheep per day (any day in a week)
(iii) *Large*: More than 200 large animals, i.e. bovines per day, or more than 1000 small animal, i.e. goat and sheep per day (any day in a week).

11.3 Waste Water Characteristics from Meat and Poultry Processing Industry

Water is a pre-requisite for proper functioning of slaughterhouses and animal and fowl processing industries. All the operations carried out in abattoirs and processing industries are largely depended upon the availability of water. Water is used in slaughterhouses for washing carcasses, hide and hair removal, evisceration, defeathering, scalding, cutting, rendering, cleaning and sanitizing of equipment and facilities. Some waste water is also produced by associated facilities like cafeteria, toilets, lairages, refrigeration and chilling equipment and from packaging and storage areas. This waste water (effluent) generated by meat industries has much higher BOD, nitrogen, phosphorus and grease concentration than domestic sewage. The standards for effluent discharge for meat and sea food industries have been provided by CBCP and are given in Table 11.3.

Based on the CPCB advice, the Central Government has revised the above standards and notified the revised standards under the Environment (Protection) Rules 1986 on 28.10.2016, which are is given in Table 11.3.

11.4 Waste Water Treatment

The quantities of waste water generated vary with the number and type of animals slaughtered. The waste waters may contain pathogens including *Salmonella* sp. *Shigella* sp., amoebic cysts and parasite eggs, pesticide residues, high levels of chloride (up to 77,000 mg/l). In India, most of the slaughterhouses do not implement

Table 11.3 Standards for effluent discharge from meat industries, abattoirs and sea food industries

S. No.	Industry	Effluent parameter	Standard (mg/l, except for pH)
1.	Abattoirs or meat processing units or both	pH	6.5–8.5
		Biological oxygen demand (BOD) [3 days at 27 °C]	30
		Chemical oxygen demand (COD)	250
		Suspended solids	50
		Oil and grease	10
2.	Sea food industry	Biological oxygen demand (BOD) [3 days at 27 °C]	30
		Suspended solids	50
		Oil and grease	10

proper effluent treatment methods, owing to which extremely harmful effluent is discharged into land and water bodies and thus can lead to "*eutrophication*". In such cases, the guidelines for effective management and treatment of effluent laid down by the Ministry of Urban Development in the *Manual of Sewage treatment* shall be followed.

The most effective method for treating such waste waters is to develop a design that combines various treatments including:

(i) *Pre-treatment* or *preliminary* treatment of water which removes solids (both large and small), fats and grease and is capable of reducing up to 35% BOD, 90% fats and 65% solids. This can be done by using screens, with pore sizes of 10 mm and 4 mm for larger and smaller solids, respectively. Remaining suspended solids, fats, oils and grease can be separated from the water by using a method known as *dissolved air flotation*. It works by producing micro-air-bubbles that get attached to the suspended matter, forming a scum that rises to the surface and can be removed easily. This method can be aided by the addition of chemicals like ferrous sulphate, ferric chloride, ferric sulphate, aluminium sulphate (alum), calcium carbonate (lime), sodium carbonate (soda ash), polyelectrolytes, etc. Proteins and remaining organic matter can be removed by using *physio-chemical* treatment which involves the use of cationic (Fe^{3+} and Al^{3+} salts) and anionic coagulants (sodium alginate, sodium hexametaphosphate and lignosulfonate). The change in pH causes precipitation and agglomeration of proteins into larger flocs. This is followed by physical removal of flocs using DAF or sedimentation.

(ii) *Biological or secondary treatment* of waste water can be performed by using biological treatment systems which involve providing and maintaining controlled and favorable environment for mixed culture of micro-organisms. These micro-organisms utilize the organic matter present in the waste water to reproduce and synthesize new cells. Secondary treatment can be *anaerobic* or *aerobic*. In anaerobic processes, the BOD is reduced by the bacteria in absence of oxygen and is thus performed in enclosed system. This system is of great advantage if the BOD is higher than 2000 mg/l. In this process some of the carbon in the waste water is transformed into a revalorizable mixture of carbon dioxide and methane or biogas. In aerobic treatment, the bacteria utilize air to breakdown the complex organic matter and to remove the residual BOD and suspended solids. The main treatments include *activated sludge, lagoons, trickling filters, evaporation* and *irrigation*. Both the systems are robust and efficient but anaerobic treatment is economical and cost efficient than aerobic treatment due to its lower energy consumption and production of biogas.

(iii) *Tertiary treatment* involves processes that render the effluent free of contaminants or at least in permissible levels. The effluent after tertiary treatment is discharged into the environment.

11.5 Novel Methods for Waste Water Treatment

(A) *Radio-frequency diathermy/electro-coagulation/short wave electrolysis:*
Electro-coagulation (EC) system is a novel and affordable technique used for treatment of waste water and industrial processed water. This treatment technology removes TSS (total soluble solids), heavy metals, bacteria, emulsified oils, total petroleum hydrocarbon and other contaminants from waste water. There are 3 steps involved:
   (i) *Dissolution* of water into H^+ and OH^-.
   (ii) *Coagulation* of suspended solids and emulsified oils and dissolved metals.
   (iii) *Floc* removal from water in downstream solid separation and filtration process steps.

   A fully automated electro-coagulation system has no requirement of coagulants and filters. It can be integrated into existing treatment processes for rapid and more efficient effluent treatment.

(B) *Membrane separation:*
Membrane separation process is used to separate fine particles and dissolved substances from waste waters. The separation of particles depends on their molecular size. There are 4 membrane separation processes used in waste water treatment from meat plants: Ultrafiltration (UF), microfiltration (MF), nanofiltration (NF) and reverse osmosis (RO).

   Ultrafiltration is used to separate oil, fats and grease; microfiltration for recovery of slurries and nanofiltration and reverse osmosis for purification, disinfection and desalination.

11.6 Solid Waste Management

As discussed earlier, the solid wastes generated in the slaughtering and processing of meat is a threat to the environment. The amount of solid waste typically generated in slaughterhouses in India has been given in Table 11.4.

The solid waste can be classified into two categories and for the efficient treatment these two steams should always be segregated. The classification is mentioned below:

(i) *Type I/Vegetable matter:* this type consists of agricultural residues, dung, stomach and intestine content, rumen content, etc.
(ii) *Type II/Animal matter:* this type consists of meat trimmings, offal, bones, tissues, etc.

Table 11.4 Quantities of waste generated by slaughterhouses per day

Type of slaughterhouse	Daily waste generated
Small	0.5–1 tonne/day
Medium	2–6 tonnes/day
Large	6–7 tonnes/day

Table 11.5 Typical characteristic of solid waste

Parameter	Value (%)
Total solids	15–40
Total volatile solids	70–92
Total nitrogen	2.5–4.2
Phosphorous	0.2–0.4
Potassium	0.3–0.4
Organic carbon	22–28
Moisture content	60–85

Table 11.5 reports the characteristics of solid waste generated from slaughterhouses. The usual practice is to dispose of this solid waste in landfills along with other municipal wastes. Such a malpractice can lead to environmental contamination due to emission of methane gas and carbon dioxide on decomposition.

It is necessary that the solid waste be managed and dealt with properly by adopting the techiques mentioned below:

(A) *Composting:*

Composting provides a better alternative over land filling for the disposal of meat industry waste. Composting is the process of biological decomposition of organic matter into relatively stable form under aerated conditions, brought about by the action of micro-organisms such as bacteria, fungi and protozoa. It reduces waste volumes, kills the pathogens present in the waste, converts nitrogen from ammonia to stable organic forms and improves quality of the waste. Almost all wastes generated by slaughterhouses can be composted. Typically, the compost pit is 4–5 feet deep. The first 6 inches underneath should be layered with coarse materials like banana stems, grass, straw, twigs, etc. to provide proper ventilation. Then alternate layering of Type I and Type II waste should be done. In case the waste consists of large organs like liver and kidneys, then it is advisable to grind them to small pieces of 2–3 inches in order to increase the surface area. There is no requirement for water in the initial stage because the wastes have high moisture content of around 70% which is adequate for the start of microbial activity. It is advisable to use an inoculum of microbes (mesophiles and thermophiles) specially selected for the purpose of decomposition. This practice aids in speeding up the process and improves humus quality. To ensure optimum conditions for the bacteria during composting, the solid waste should be regularly aerated by mixing and/or inverting the material. The total time required for the completion of the process is usually 45 days depending on factors such as type and quantity of waste, aeration, temperature and inoculation level. The entire process is divided into two stages:

(i) *Stabilization:* Stabilization further consists of three phases. The first phase involves vigorous multiplication of mesophilic, and breakdown of easily oxidized carbon compounds to relatively simpler compounds of lower molecular structure. This results in a rapid rise in temperature. The second stage is the thermophilic phase in which at high temperature organic compounds are attacked by the thermophilic bacteria that are not easily degraded. Third stage is marked by a gradual decrease in temperature to ambient and leads to mesophilic phase, in which the growth of actinomycetes and fungi is observed. The biological degradation of remaining organic compounds occurs in this stage.

(ii) *Maturation:* Mineralization of organic matter at a very slow rate occurs in this stage. This process continues until the volatile matter content is reduced below 50% and carbon: nitrogen (C: N) ratio of 10: 1 is achieved.

The final decomposed, nutrient-rich product is referred to as *humus*, which can be used as a soil amendment to improve soil quality, fertilize gardens and improve growing conditions for plants.

(B) *Incineration:*

Incineration is the process of waste treatment process that involves combustion of organic matter present in the waste. This is a high temperature treatment also known as "thermal treatment". The process of incineration converts the inorganic constituents of waste into *ash*. This ash may take the form of particulates carried by the *flue gas*. The flue gas needs to be cleaned of pollutants before releasing into the atmosphere and the non-combustible residue is disposed to landfills. Incineration can be used to treat the solid waste from meat industries, slaughterhouses and the wastes left after treatment of effluent. The waste is incinerated at high temperatures between 850 °C and 1100 °C in specially designed chambers. The process of ignition is initiated and sustained by supplying an auxiliary fuel. Proper aeration and temperature control are vital for effective combustion. Direct incineration of wet waste is unfeasible due to difficulties in ignition. This may result in unsteady and incomplete combustion of organic substances with the emission of large amounts of volatile pollutants. Therefore, it is recommended to pre-treat the waste by dewatering the solids and palletizing them (Chiang et al. 2012). There are concerns from experts about using incineration as a treatment for waste due to rising environmental issues and health problems faced by the workers. The emission of toxic heavy metals like manganese, vanadium, chromium, arsenic, lead, mercury, nickel and cadmium is a cause of serious concern.

11.7 By-Product Production and Utilization

India is bestowed with a large livestock resource and therefore has the capacity of generating a lot of offal/bone waste after slaughtering and processing the animals. The waste produced is estimated to be approximately 21 lakh tones/annum (Jayathilakan et al. 2012). Only one-third part of the slaughtered and processed animal can be referred to as meat. The rest comprises waste and by-products. Every part of the slaughtered animal, except dressed meat (fifth- quarter or offal) may be

Table 11.6 Approximate yield of by-products from small and large animals

By-product	Cattle/buffalo (% yield)	Sheep/goat (% yield)
Boneless meat	28.57	–
Dressed weight	–	40.00
Bone, head, feet/hoofs	22.85	6.40
Lungs and oesophagus	2.00	1.00
Hide/skin	7.57	9.20
Paunch content	16.00	22.00
Blood	3.14	3.00
Liver	1.42	1.20
Fat/fatty tissue	1.71	2.00
Heart	0.42	0.40
Spleen	0.28	0.20
Kidney	0.14	0.40
Stomach	0.71	–
Pancreas	–	0.16
Head meat and brain	0.28	–
Gut and tripe/other offals	4.00	8.00
Tongue	0.28	–
Casings	2.00	–
Urine, body fluids, bile, dung	3.50	–

defined as "animal by-product" (Ockerman and Hansen 2000). The approximate yield of by-products from animals is given in Table 11.6 (Chatterjee et al. 1991). These by-products should be appropriately utilized to minimize wastage and contribute to the economy of livestock industries. Further, animal by-products may be categorized into *edible* and *inedible*.

(i) *Edible by-products:* Edible by-products (EBPs) can be consumed as food and are fit for human intake. This category includes variety meats, such as poultry giblets, heart, liver, brain, intestines, kidney, meat trimmings from head, etc. Blood can also be included in this category, as it is used as an edible by-product in some parts of the world. Edible by-products are processed and stored under sanitary conditions. Before processing, EBPs are thoroughly examined and should be free from any infections like echinococcosis, fasciolosis, tuberculosis, etc. and structural abnormalities (Yibar et al. 2015; Gonulalan et al. 2004). Because of the possible presence of disease-causing harmful micro-organisms, the offal is washed and heated to high temperatures of more than 100 °C, which ensure microbial destruction.

(ii) *Inedible by-products:* Inedible by-products (IEBPs) are unsuitable for human consumption and are either discarded or re-processed as secondary by-products. IEBPs include skin, hides, bones, feathers, horns, hair, hooves, snout, ear, etc. Dead animals and rejected parts and organs are also included. These are commonly used to manufacture pet food, MBM, gelatin, etc. This category has been further divided into *elementary* and *secondary* by-products (Sharma and Sharma 2011). Primary or elementary by-products are directly harvested after the animal is slaughtered and includes skin, hide, hoofs, bones, blood, pancreas, intestines and horns. Secondary by-products are derived from elementary by-products and may include bone meal, fat, gelatin, etc. from bones, casings from intestines and others.

The EBPs are widely accepted by consumers due to their high nutritive value and affordability; however, their utility varies by country, traditions, religion and culture. Regulatory requirements also restrict the use of some by-products concerning matters of food safety and quality. The uses of both edible and inedible by-products are described below:

(A) *Utilization of bones:*

Bones account for 20–30% of the total live weight of animals, while the bone marrow may constitute 4–6% of the carcass weight (West and Shaw 1975). Bones are composed of organic and inorganic matter. The organic matter includes proteins. The inorganic matter includes calcium and phosphorus which are essential for the growth of plants. Thus, bones can be used as "fertilizers". The traditional practice for utilization of bones includes preparation of nutritious soups, stews and broth. The bones which come as a waste product from slaughterhouses and processing industries can be utilized much profitably by extracting gelatin from them. Recovery of all the meat adhering to the bone had been a challenge for meat processing industries, until recently a novel separation technique was developed. This technique is referred to as "mechanical deboning" and the tissue/meat recovered is known as "mechanically deboned", "mechanically separated" or "mechanically removed". This mechanically separated meat can be added to other products like patties, burgers, sausages, nuggets, etc. but it affects the quality and flavour of the product if used in higher percentages than required. The end product may become darker and mushy. Therefore, mechanically separated meat must be incorporated into products at levels prescribed by the meat industries, viz. 5–20% in ground beef and burgers; 10–40% in sausages. There are regulations concerning the use of mechanically deboned meat in products. These regulations vary across countries. An example is that mechanically separated meat cannot be incorporated in baby food, hamburger, meat pies and ground beef in the USA. In Denmark, if the level of usage exceeds 2%, it has to be mentioned on the label. Likewise, in Australia, the presence of mechanically deboned meat has to be mentioned on the label of products intended for export.

Another breakthrough use includes extraction of "Bone morphogenic protein" or "BMP". This protein can be used in aesthetic, dental and facial surgeries. BMP is reportedly safe to use and has no antigenic sensitivity.

Meat and bone meal (MBM) is widely used to provide nutrition and proteins to animals. It has replaced proteinaceous feeds because of high mineral, essential amino acid and cyano-cobalt-amine (Vitamin B_{12}) content. Apart from being used as food, bones which are solid hard materials can be made into buttons, handles, bowls, decorative items like hairpins, hair combs, pendants (BBC 2000). The energy released from burning bones can be used as "fuel".

(B) *Utilization of blood:*

Blood is a body fluid that is responsible for delivering oxygen and essential nutrients to the cells and transporting carbon dioxide and metabolic wastes away from the same cells. It is composed of plasma and blood cells (red blood cells, white blood cells and platelets). It also contains high level of protein (17%), hormones, glucose, carbon dioxide and mineral ions (particularly iron), which makes it an important edible by-product. Blood represents about 4–8% of the animal's live weight. Blood from the slaughter industries may be utilized as listed below:

(i) In food industries blood is used as an emulsifier, colour enhancer, binding agent, stabilizer, meat curing agent, as a nutritional component and clarifier (Silva and Silvestre 2003; Toldrá et al. 2012; Ofori and Hsieh 2014).
(ii) It is used to make pan cakes, blood curd, blood sausages, blood cake, black pudding (Wan et al. 2002; Ghost 2001; Davidson 2014).
(iii) It is used as a vitamin stabilizer and lysine supplement.
(iv) Blood plasma contains high level of albumin (60%) and therefore has the ability to form a gel (Silva and Silvestre 2003).
(v) Frozen blood plasma can be added to ham and hot dogs at suitable concentrations to enhance the colour of the product (Autio et al. 1985).
(vi) Spray dried plasma is an excellent foaming and leavening agent and can be used to replace eggs in baking industries (Hsieh and Ofori 2011).
(vii) Transglutaminase (TGase) extracted from bovine blood can be used as a binding agent to improve the binding ability of meat products at low temperatures.
(viii) In feed industry, blood meal can be used in pet food and feedstuff (Bah et al. 2016; Toldrá et al. 2012).
(ix) In medicinal and pharmaceutical industries blood is used as a medium for the growth of microbes (probiotics), nutrient for tissue culture, vaccine stabilizer, production of porphyrin derivative, biological assays, production and purification of thrombin.
(x) Blood albumin is used to replenish fluid and blood loss.
(xi) In serology laboratories it is used as a reagent for blood clotting.
(xii) Other applications: plywood adhesive, finishing agent, mordant and fertilizer.

The blood intended to be used in food products must be collected from disease free, healthy animals. The resulting product may be dark in colour, rendering it unacceptable.

(C) *Utilization of skin and hides:*

Hides and skin represent about 7% of the live weight of large animals and around 11% of the live weight of small animals (Jayathilakan et al. 2012). Hides and skin are valuable by-products obtained from the animals and are mainly converted to leather after reprocessing for the production of secondary by-products like shoes, bags, clothing, seat covers, among other things (DAFF 2012). India ranks first in the production and export of hides and skin and their products, with a share of 10–12% in the World's leather market. Miscellaneous uses include production of sausage casings and skin, glue, cosmetics and edible gelatin. Collagen extracted from hides/skin can help coagulate blood during surgery and prevent excessive blood loss. Pig skin, which is very similar to human skin, can be used for skin grafting in case of burns and ulcers.

(D) *Utilization of Feathers:*

Feathers are a distinctive feature of the avian species and account to about 7% of their live body weight (Lortscher et al. 1957). Feathers are a rich source of protein. They contain about 90% protein, 8% water and 1% fat. Feathers can be used as livestock feed, bedding, ornamental purposes, sporting equipment and as filler in chemical fertilizer. However, the protein complex needs to be broken down to protein hydrolysates before utilizing the feathers as an animal feed. This is done by hydrolysing the complex protein (keratin) structure to make it digestible. These protein hydrolysates are used as protein fortification agent for concentrated soups and beverages as well as valuable component of solid and liquid seasoning. This can also be used as a diet for patients for faster and better postoperative recovery because of high biological value and digestibility. These protein hydrolysates find use in various industries such as textile (for dyeing), paper (for coating), paint (as ingredient), match stick (for shaft), leather industry (as filter syntax), animal/poultry feed, detergents (as foaming and sequestering agent), cosmetics (face creams and lotions), microbiology (media ingredients), pharmaceuticals, tissue culture techniques, etc.

(E) *Utilization of glands and organs:*

(i) As food: Animal organs and glands offer a wide variety of flavours and textures, and often have a high nutritional value. They are highly prized as food in many parts of the world, particularly Southeast Asia. Those used as human foods include the brain, heart, kidneys, liver, lungs and spleen. They also include the tongue, the bovine pancreas and udder, the stomach and uterus of pigs, the rumen, reticulum, omasum and abomasum of sheep and cattle and the testes and thymus of sheep and pigs (Liu 2002). Various parts of the digestive and excretory tracts like oesophagus, large and small intestines, caecum, rectum and bladders are processed into natural casings that serve as containers or packaging materials for comminuted meat products like sausages, salami, etc. The intestines of sheep and calves are used for the manufacture of catgut, to make internal surgical sutures.

(ii) In Pharmaceutical Industries: Animal glands and organs are traditionally used as medicine in many countries, including China, India and Japan. The endocrine glands secrete hormones (i.e. enzymes that regulate the body's metabolism). These include the liver, lungs, pituitary, thyroid, pancreas, stomach, parathyroid, adrenal, kidney, corpus luteum, ovary and follicle. Brains, nervous systems and spinal cords are a source of cholesterol which is the raw material for the synthesis of vitamin D3. Cholesterol is also used as an emulsifier in cosmetics (Ejike and Emmanuel 2009). Bile consists of acids, pigments, proteins, cholesterol, etc., and can be obtained from the gall bladder. It is used for the treatment of indigestion, constipation and bile tract disorders. It is also used to increase the secretory activity of the liver. Bile from cattle or pigs can be purchased as a dry extract or in liquid form. Some ingredients of bile, such as prednisone and cortisone, can be extracted separately, and used as medicines.

The liver is the largest gland in animals. Liver extract is produced by mixing raw ground liver with slightly acidified hot water. The stock is concentrated into a paste in a vacuum at a low temperature, and is used as a raw material by the pharmaceutical industry. Liver extract can be obtained from pigs and cattle, and has been used for a long time as a source of vitamin B12, and as a nutritional supplement used to treat various types of anaemia. (Colmenero and Cassens 1987; Devatkal et al. 2004a, b). Heparin can be extracted from the liver, as well as the lungs and the lining of the small intestines. It is used as an anticoagulant to prolong the clotting time of blood. It is also used to thin the blood, to prevent blood clotting during surgery and in organ transplants.

Progesterone and oestrogen can be extracted from pig ovaries. It may be used to treat reproductive problems in women. The pancreas provides insulin, which regulates sugar metabolism and is used in the treatment of diabetes. Glucagon extracted from the cells of the pancreas is used to increase blood sugar, and to treat insulin overdoses or low blood sugar caused by alcoholism. Chymotrypsin and trypsin are used to improve healing after surgery or injury.

11.8 Fish Waste/By-Product Utilization

Processing of fish leads to enormous amounts of waste. It is estimated that fish processing waste after filleting accounts for approximately 75% of the total fish weight. About 30% of the total fish weight remains as waste in the form of skins and bones during preparation of fish fillets. This waste is an excellent raw material for the preparation of high value products including protein foods. The important products that can be developed from the wastes and rejections of the fish processing industry is summarized below.

(A) Fish meal: Fish meal is highly concentrated nutritious feed supplement consisting of high-quality protein, minerals, vitamins B group and other vitamins and other unknown growth factors. Fish meal is rich in essential amino acids. It is produced by cooking, pressing, drying and grinding the fish, by-catch fish, miscellaneous fish, filleting waste, waste from canneries and waste from various other processing operations.

(B) Fish oil: The main source of fish body oil in our country is oil sardine. The method of extraction followed is cooking the fish in iron vessels and pressing and separating the oil. Apart from sardine oil, fish body oil is also obtained from the fish meal plants operating in the country. During the peak season fish has oil content of 17%. By the wet rendering process the fish will yield, on average 12% oil having analytical characteristics similar to other fish oils. Fatty acid composition of oil revealed that they contain high amounts of polyunsaturated fatty acids (PUFA). At present the medicinal values of fish oils are well known.

(C) Fish ensilage: When fish is available and which cannot be used for direct consumption for several reasons, it is used for production of fish meal. This has got ready market as an animal feed. Fish silage can be defined as a product made from whole fish or parts of the fish to which no other material has been added other than an acid and in which liquefaction of the fish is brought about by enzymes already present in the fish (Raa and Gildberg 1982). Almost any species of fish can be used to make fish silage though cartilaginous species like shark and rays liquefy slowly. The ensilage can be used as a fish meal replace for the production of feeds.

(D) Fish scales: Scales in fresh water fish generally constitute 1–2% of the body weight and commercially are not of much importance. Though the scales look unhygienic and litter the market, they can be used for ornamental and other purposes. Activated charcoal can be obtained by burning fish scales.

(E) Fish Hydrolysate: This is also a liquefied fish product but it differs from silage. These are products produced by a process employing commercially available proteolytic enzymes for isolation of protein from fish waste. By selection of suitable enzymes and controlling the conditions the properties of the end product can be selected. Hydrolysates find application as milk replace and food flavouring. Enzymes like papain, ficin, trypsin, bromelain and pancreatin are used for hydrolysis. The fish protein hydrolysate has desirable functional properties with potential applications as emulsifiers and binder agents, and can be used in place of diary based and plant based protein hydrolysates as well as protein powders currently available in market place.

(F) Fish Collagen: Collagen extracted from fish skin, a polymer that is a by-product of food manufacture, has various industrial applications in cosmetology and medicine. Dermis fish collagen presents an interesting new source of collagen as it is a by-product of food fabrics and already has cosmetic uses.

11.9 Waste-to-Energy (WTE) or Energy-from-Waste (EFW): Meat Waste as Fuel Source

The availability of wet biomass as waste from industrial processes and the need to meet the environmental standards stand for the main stimuli towards investigating all options in order to dispose this waste. The thermal recycling of residues as secondary fuel is of increasing interest for power plant operators. Due to sanitary, environmental problems and operational costs related to the discharge, land disposal and re-use of wastes, the utilization of this "biofuel" (dried sludge) for steam generation has shown to be a viable alternative. This type of fuel has a high heating value, and it is a renewable energy source. Biodiesel fuel acquired from the oils and fats of meat and fish is a substitute for, or an additive to diesel fuel derived from petroleum. There is an extensive literature on biogas production from cattle manure, piggery waste waters and by-products of aquaculture (Arvanitoyannis and Kassaveti 2008).

11.10 An Insight into the Process of Rendering

Rendering is the process of converting the by-products of meat and poultry processing into more stable, useful and saleable product. The type II waste or animal waste that includes edible and inedible proteins and fats (inedible offal, meat trimmings, discarded meat, bones, offal, tissues, etc.) can be rendered. Feathers, hatchery by-products (dead embryos, infertile and stale eggs) and carcasses of condemned animals can also be processed in rendering system. Such waste mainly constitutes of fat, solids and water. Sheep, poultry, pork and beef are the main sources of plants. Rendering plants can be either *integrated* or *independent*. The integrated rendering plants work conjunctionally with slaughterhouses or fowl processing plants. The independent rendering plants outsource their raw materials from various slaughterhouses, restaurants, butcher shops, farms, supermarkets, etc. (USEPA 1995). The rendering process separates the fat, solids and water physically and simultaneously dries the material yielding a fat commodity (tallow, lard, grease, etc.) and a protein meal like MBM (meat and bone meal). The fat recovered during rendering is used for the manufacture of soap and grease. If the fat is rendered from healthy and eatable parts of animals, it can be used edible purposes. The bone meal is used as a feed for farm animals, pet food and as fertilizers.

The process of rendering can be wet or dry. The first step is size reduction that involves cutting, chopping and comminuting large animal parts like head, bones and large internal organs by shredders or other specialized machinery. Cleaning of intestines, stomach and other organs containing manure is also done prior to rendering. These two preliminary steps are common irrespective of the type of rendering. In *wet rendering*, water or steam is used to aid the process, whereas in *dry rendering* surplus moisture is removed from the waste before processing. In dry rendering suspended proteins and water-soluble extracts are not discarded. In this way the output of dry rendering process is approximately 20% higher than the wet rendering.

(i) *Rendering for edible products*:
Continuous edible rendering of healthy animal parts yields tallow and edible lard as the main by-product. After finely chopping the edible parts of the animal, the mixture is subjected to heat treatment with or without the addition of water or steam. The resulting thick slurry is then centrifuged to separate water and fat from solids. The fat is separated from water in the second stage of centrifugal separation. The temperature for rendering edible fat can be either low (below 120 °F/49 °C) or high (between 180 °F/82 °C and 210 °F/100 °C). High temperature rendering provides better separation of fats and proteins but the proteins obtained are of low quality and therefore cannot be sold as an edible product (Prokop 1985). The residue left behind after rendering fat is known as *greaves*. The offal can alternatively be processed and cooked to obtain a thick stew that is subsequently canned and sold as tinned cat and dog food.

(ii) *Rendering for inedible products*:
The materials that are not suitable for human consumption are processed under inedible rendering. Usually, the inedible material is processed by using the dry rendering method. The process of dry rendering is explained above. The ground material dehydrated and drained to remove fat followed by pressing to drive off residual fat and moisture. The remaining solid is ground into a meal.

11.11 Conclusion

The utilization of animal by-products is often ignored; however, these items contribute a significant value to the livestock and meat industries. Non-utilization of animal by-products in a proper way may create major aesthetic and catastrophic health problems. Value addition of animal by-products has two benefits. Firstly, the meat industry gets additional revenue by processing them to industrial, household and cosmetic products; livestock feed additives; pet foods; pharmaceutical and medical supplies, etc. that otherwise would have been unrealized. Secondly, the costs of disposing of these secondary items are avoided. Value addition can also sometimes act as a cushion to cover losses suffered in the trade. Furthermore, although the development of synthetic substitutes in the middle of the twentieth century decreased the value of many animal by-products, but their importance in the pet food industry and the medical/veterinary field are contributing to an increase in by-product values in recent years. Utilization of these by-products as fertilizer contributes a lot in organic farming and could reduce our dependence on synthetic fertilizers. Animal by-products and wastes are a good source of renewable energy as well its production is economically feasible. The utilization needs become significantly stronger due to competition. This is important because increasing profit and decreasing the cost is required in the future for the meat industry to remain viable. These contributions and efforts are also necessary for the meat industries to change in an innovative manner and to widen the opportunities to utilize by-products. However, the saying "the packer uses everything but the squeal" has always existed in the meat industry and will continue to influence the utilization of meat by-products.

References

Arvanitoyannis IS, Kassaveti A (2008) Fish industry waste: treatments, environmental impacts, current and potential uses. Int J Food Sci Technol 43(4):726–745

Autio K, Lyytikainen H, Malkki Y, Kanko S (1985) Penetration studies of blood globin gels. J Food Sci 50(3):615–617

Bah CS, Bekhit AEDA, Carne A, Mcconnell MA (2016) Composition and biological activities of slaughterhouse blood from red deer, sheep, pig and cattle. J Sci Food Agric 96:79–89

BBC environment correspondent Tim Hirsch (2000) "BSE carcasses burned for electricity", last modified June 3. http://news.bbc.co.uk/2/hi/uk_news/775324.stm

Birthal PS (2008) Linking smallholder livestock procedures to markets: issues and approaches. Indian J Agric Econ 63:19–37

Chatterjee AK, Dempster IF, Sharma N (1991) Review of meat industry in India. FAO/UNDP Project No. TCP/IND/8955

Chiang KY, Chien KL, Lu CH (2012) Characterization and comparison of biomass produced from various sources: suggestions for selection of pretreatment technologies in biomass-to-energy. Appl Energy 100:164–171

Colmenero FJ, Cassens RG (1987) Influence of an extract of liver on color and shelf stability of sliced bologna. Meat Sci 21(3):219–230

Commission of the European Communities (1990) Council Directive 90/667/EEC. Office Journal No. L 63, 27. 12: 51

Davidson A (2014) The Oxford companion to food, 3rd edn. Oxford University Press, Oxford

Department of Agriculture, Forestry and Fisheries (DAFF) (2012) A Profile of the South African Hides. Skins and Leather Market Value Chain

Devatkal S, Mendiratta SK, Kondaiah N (2004a) Quality characteristics of loaves from buffalo meat, liver and vegetables. Meat Sci 67(2):377–383

Devatkal S, Mendiratta SK, Kondaiah N, Sharma MC, Anjaneyulu ASR (2004b) Physicochemical, functional and microbiological quality of buffalo liver. Meat Sci 68(5):79–86

Ejike CECC, Emmanuel TN (2009) Cholesterol concentration in different parts of bovine meat sold in Nsukka, Nigeria: implications for cardiovascular disease risk. Afr J Biochem Res 3 (4):095–097

Ghost R (2001) Fractionating of biological macromolecules using carrier phase ultrafiltration. Biotechnol Bioeng 74:1–11

Gonulalan Z, Kose A, Yetim H (2004) Effects of liquid smoke on quality characteristics of Turkish standard smoked beef tongue. Meat Sci 66:165–170

Hsieh YHP, Ofori JA (2011) Blood-derived products for human consumption. Revel Sci 1:14–21

Jayathilakan K, Sultana K, Radhakrishna K, Bawa AS (2012) Utilization of by products and waste materials from meat, poultry and fish processing industries: a review. J Food Sci Tech Mys 49:278–293

Liu DC (2002) Better utilization of by-products from the meat industry. Extension Bulletins. Food and fertilizer Technology Center for the Asian and Pacific region (FFTC publication database)

Lortscher LL. Sachsel GF, Wilkelmy D Jr, Filbert RB Jr (1957) Processing poultry by products in poultry slaughter plants. US Department of Agriculture Marketing Research Department, No. 181. Washington, DC

Ockerman HW, Hansen CL (2000) Animal by-product processing and utilization. CRC Press, Boca Raton

OECD, FAO Agricultural outlook, 2016- 25

Ofori JA, Hsieh YHP (2014) Issues related to the use of blood in food and animal feed. Crit Rev Food Sci Nutr 54:687–697

Prokop W (1985) Rendering systems for processing animal by-product materials. Papers from the Symposium on Animal Fats presented at the 74th AOCS Annual Meeting held in Chicago, Illinois, May 8–12, 1983. J Am Oil Chem Soc 62(4):804–811

Raa J, Gildberg A (1982) Fish silage: a review. CRC Crit Rev Food Sci Nutr 18:283–419

Russ W, Pittroff RM (2004) Utilizing waste products from the food production and processing industries. Crit Rev Food Sci Nutr 44(2):57–62

Sharma BD, Sharma K (2011) Outlines of meat science and technology. Jaypee Brothers Medical Pub (p) Ltd, p 360

Silva VDM, Silvestre MPC (2003) Functional properties of bovine blood plasma intended for use as a functional ingredient in human food. LWT-Food Sci Technol 36(5):709–718

Toldrá F, Aristoy MC, Mora L, Reig M (2012) Innovations in value-addition of edible meat by-products. Meat Sci 92:290–296

US Environmental Protection Agency (1995) Emission factor documentation for AP-42 Section 9.5.3. Meat rendering plants final report. EPA Contract No. 68-D2-0159, Washington, DC

Wan Y, Ghost R, Cui Z (2002) High resolution plasma protein fractionation using ultrafiltration. Desalination 144:301–306

West GC, Shaw DL (1975) Fatty acid composition of Dall sheep bone marrow. Comp Biochem Physiol B Biochem Mol Biol 50(4):599–601

Yibar A, Selcuk O, Senlik B (2015) Major causes of organ/carcass condemnation and financial loss estimation in animals slaughtered at two abattoirs in Bursa Province, Turkey. Prev Vet Med 118:28–35

Post-Harvest Management of Climacteric Fruits in India: The Promising Road Map for Future

12

Komal Mathur and Parul Chugh

Abstract

India is quintessential name globally after China in natural produce of fruits and vegetables. Fifteen agronomic zones of India ensure the supply of these edibles round the year. On the contrary, the bounteous natural produce goes unutilized due to poor post-harvest management and infrastructure in India. Productivity of fruits and vegetables is much below and only half of that in developed countries like the USA. Moreover the share of India in export market is negligible. There is tremendous scope for doubling productivity of this produce in India, as in developed countries and emerge as global leader. Being an agro-based economy, it is the earnest need of India to adopt latest technologies and upgrade expeditiously, replacing the obsolete existing approaches to overcome tremendous post-harvest nutrition losses and earn revenues by capturing the global market. Here, causes of post-harvest losses in climacteric fruits and current approaches to overcome them in India are compared with developed countries like the USA. Future roadmap for promising economic and nutrition returns from fruits and vegetables is suggested here through agricultural reforms.

Keywords

Post-harvest management · Climacteric fruit · Ethylene · Respiration · Cell wall hydrolases

K. Mathur (✉) · P. Chugh
Amity Institute of Biotechnology, Amity University Uttar Pradesh, Noida, Uttar Pradesh, India
e-mail: kmathur@amity.edu

M. Thakur et al. (eds.), *Sustainable Food Waste Management*,
https://doi.org/10.1007/978-981-15-8967-6_12

12.1 Introduction

Both fruits and vegetables, unlike non-vegetarian food which is non respiring dead cells, are living biological systems consuming energy through their cellular respiratory tracts. Following harvest, fruits and vegetables continue their living processes (Slavin and Lloyd 2012). Ethylene is ripening hormone which shows its effects even at small amounts (Iqbal et al. 2017).

Depending on the ethylene production, fruits and vegetables can be divided as:

- **Low C_2H_4 producing**: citrus, leafy vegetables, root vegetables, potato.
- **Moderate C_2H_4 producing**: Banana, guava, mango.
- **High C_2H_4 producing**: Apples, avocado, papaya, kiwi.

Shelf life after harvest depends on the rate of consumption of their food reservoirs and their rate of water loss (Mishra and Guru 2017). Fruits may be classified as climacteric and non-climacteric. Ethylene phytohormone is the main trigger for climacteric fruits ripening. It is characterized with peak in rate of ethylene production and respiration, e.g., in fruits like tomato, apple, pear, and melon (Kumar et al. 2014). However, non-climacteric fruits like orange do not exhibit the respiratory or ethylene peaks during ripening (Paul et al. 2012).

Ripening is a controlled biochemical differentiation process leading to prominent ethylene production and other physical and biochemical changes caused by ethylene. Fruit ripening is the product of the respective plant's hormonal signals. Biosynthesized ethylene is the hormone responsible for bringing the signal (Liu et al. 2015). It is provided by all parts of the plants during the plant's life, and is controlled during its growth phases. Ethylene is also responsible for various reactions in plants, such as abscission and seed germination (Chang 2016). Ethylene is synthesized by a complex method of transforming amino acid methionine with the help of various enzymes (Fig. 12.1). Oxygen helps to transform ACC to ethylene (Vanderstraeten and Van Der Straeten 2017).

Mature unripe climacteric fruits exhibit climacteric burst in ethylene evolution and triggers cascade of enzymatic changes that cause change in texture, color, taste and turn mature unripe fruit into a mature ripe fruit. After changes in the physiochemical properties, maturation transforms the fruit to edible state, ready for consumption (Maduwanthi and Marapana 2019).

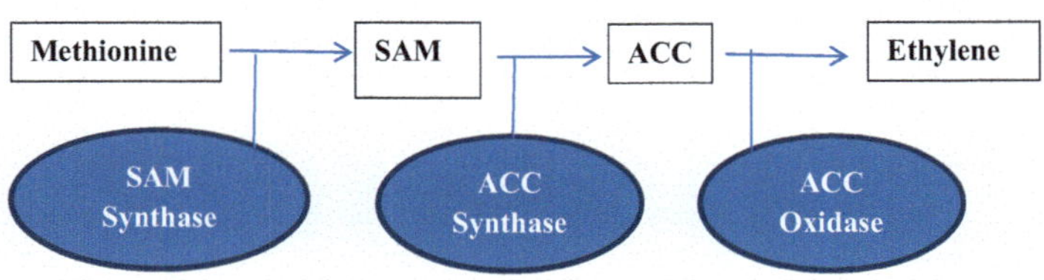

Fig. 12.1 Ethylene biosynthetic pathway

Ethylene induced enzymatic changes during climacteric fruit ripening are as follows:

Taste

The fruit's flavor improves at maturation. Because of the presence of acids the fruit is a little tart or sour at the initial level. Kinase enzymes make the acidic fruit neutral as the fruit ripens. The fruit is sweet as it matures because of the amylase enzyme, which transforms all the starch present in the fruit into simple sugars when it matures (Yang et al. 2018).

Color

When they mature, the fruit's color changes from green. The coloring pigments are released as hydrolase enzymes break the chlorophyll down. These new pigments include carotenoids and xanthophylls that give ripened fruit a color. The color has a tremendous impact on attracting animals to support the seed dispersal (Khoo et al. 2011).

Odor

Hydrolases are also necessary for transforming large organic molecules into highly volatile, smaller organic aromatic compounds that contribute to the distinctive aroma of the matured fruits and also help to attract animals, which later aid in seed dispersion (Baldassarre et al. 2015).

Hardness

Unripe fruits tend to be hard. This toughness is due to pectin present in the primary cell wall. Excessive softening during maturation is associated with increased expression of protein and carbohydrate degrading enzymes in the cell wall, such as pectinases, pectin methyl esterase, etc. (Paniagua et al. 2014).

Fruit ripening is genetically regulated resulting in well-coordinated biochemical changes. Processes. Physical and physiological changes become the reason of post-harvest losses from farm to fork (Prasanna et al. 2007). Despite being an agribusiness country, India is slow in implementing the new technologies. For India to implement the new technologies to reduce post-harvest losses and minimize huge nutritional and economic losses is the urgent need of the hour.

12.2 India Is the Global Fruit Basket

India is a basket of fruit & vegetables worldwide. India produces 99.97 million tons of fruits and 191.77 million tons of vegetables. This produce contributes to 12.6 and 14% of global produce, respectively. India is naturally endowed with 15 diverse geographical agronomic zones with distinct seasons, making it possible to grow wide range of vegetables and fruits (Fig. 12.2).

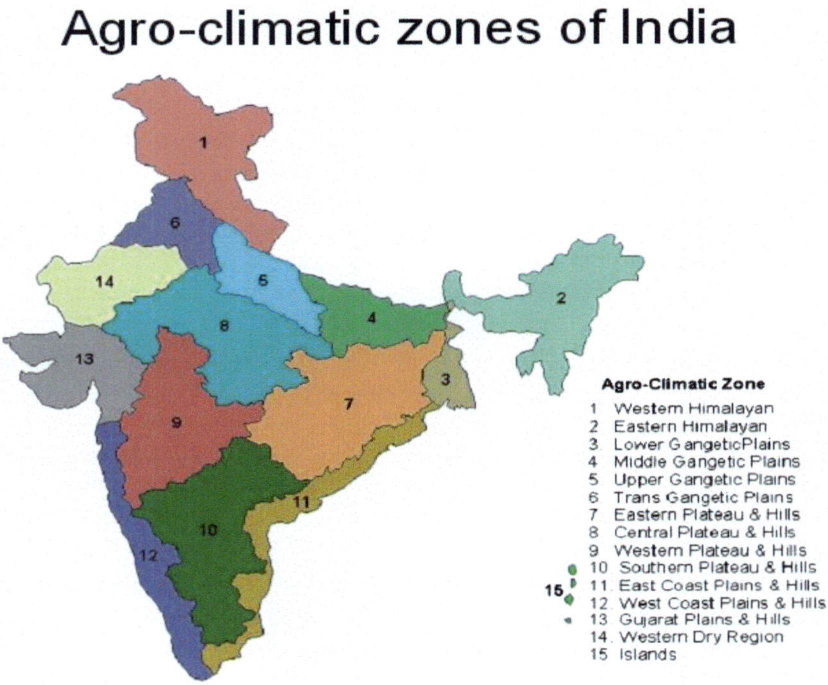

Fig. 12.2 Indian map depicting 15 diverse agronomic zones (Source: Maps of India)

India tops globally in banana, papaya, and mangoes production. Their global share is 25.7, 43.6 and 40.4%. Among vegetables ginger and okra are the top produce globally (Horticulture Statistics 2018).

12.3 Post-Harvest Losses in India

Though there is bounteous natural produce in India, conversely when it comes to managing the harvest, India is one of the biggest losers of the natural produce due to very poor post-harvest management leading to heavy nutritional and economic losses (Singh et al. 2014).

Inadequate infrastructure, inadequate management, market instability, or simply farmers' carelessness before they reach the consumer in the retail market are the major causes of post-harvest losses (Kiaya 2014). Different sites of post-harvest losses include: farmers field, packaging, transportation, and marketing.

Estimation of economic loss is essential to think of innovative strategies for sustainable food supply and overpowering the losses. McKinsey and Co. and CII have reported at least 50% of fruits and vegetables losses in the country amounting to Rs.23,000 crores annual economic losses (McKinsey and CII Report 2001).

12.4 Nature of Post-Harvest Losses

Rapid respiratory nature of living fruit tissues, high moisture content, tender texture become the main reasons of their high perishability (Gibbs and Steele 2018). If a high value nutritious produce is not handled properly, it can deteriorate fast. Post-harvest losses have been categorized as follows:

12.4.1 Metabolic

Physiological deterioration occurs spontaneously through enzymatic action. All fresh grown crops are living cells. The normal respiration cycle involves breaking down food reserves and aging these organs. Respiration is the most degrading cycle of harvested fruits that results in the breakdown of complex carbohydrates into CO_2 and H_2O; intake of food reservoirs leading to pulpiness of the fruits, if not managed (Gibbs and Steele 2018).

The ripening of climatic fruits is further induced by ethylene, which in effect initiates a cascade of enzymatic pathways leading to maturation, namely amylase activation, cell wall hydrolases, pigment changes, flavor, and volatile aromatic growth. Oxygen facilitates biosynthetic processes in both the respiratory and ethylene (Nascimento 2017). Heat produced during breathing contributes to enhanced enzymatic activity, and respiratory water also invites microbial growth on nutritious fruits and increased fruit deterioration levels (Fig. 12.3).

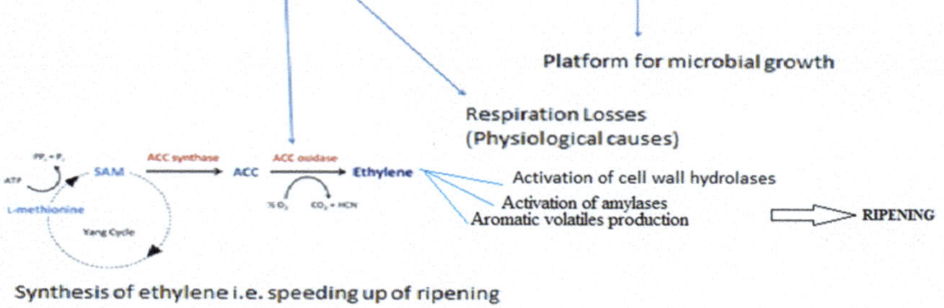

Fig. 12.3 Respiration and ethylene key physiological causes of climacteric fruits deterioration

12.4.2 Mechanical

Fruits and vegetables are highly hydrated and have delicate texture that makes them vulnerable to mechanical injury, inadequate packaging and storage, unsafe containers lead to bruising, cutting, cracking, which impact wounding and reduce acceptable quality for their customers (El-Ramady et al. 2015).

12.4.3 Developmental

Sprouting, rooting, seed germination are the developmental changes that occur in the living edibles that lead to deterioration in nutritional and market quality.

12.4.4 Parasitic Diseases

Owing to high moisture and nutrient content in fruits and vegetables, they become vulnerable to parasitic invasion supporting the growth of fungi, bacteria, insects, and other species. (Mahajan et al. 2014). For example: *Fusarium* spp. causes Fusarium wilt disease, *Plasmodiophora* and *Spongospora* species cause Club root disease in cabbage.

12.4.5 Environmental Factors

Physiological disorders can occur because of environmental factors such as temperature, relative humidity, oxygen content of the ambient gas, and mineral deficiency. Low atmospheric humidity encourages fruit shriveling through transpiration (Do nascimento nunes 2008). Likewise high temperature and oxygen raise the rate of respiration and ethylene production leading to increased consumption levels of food reserves and processes of maturation.

12.4.6 Lack of Market Demand

Due to poor supply chain management, planning, market information, inadequate carriage and storage amenities, produce may be rotting in processing areas if it is not transported by farmers to consumers who need it in remote locations (Mercier et al. 2017).

12.4.7 Others

It includes miscellaneous factors, namely information gaps between farmers, contractors, policy makers, inadequate infrastructure, poor entrepreneurial urge in rural areas, inadequate technical support, technology gaps, cold chain management market intelligence, and market information service (MIS).

Despite being the second largest producer in the world India incurs enormous economic and nutritional losses of fruits and vegetables. Nutritionally important vitamins and minerals are lost in transit from harvest to market (Sawicka 2019). These damaged goods often do not attract foreign buyers and less reward the country's export remuneration.

To improve the situation, awareness among producers, farm employees, administrators, and exporters of economic implications of losses is crucial. It is fundamental to know the handling and storage of live fruits. Critical infrastructure such as storage, handling, sorting, packaging, transportation, and marketing services needs to be provided by government. This could be implemented by both the private and public sectors.

12.5 Current Global Approaches Employed to Check Post-Harvest Losses

Sustainable farming is a central part of the sustainable development framework to achieve food security with minimum stress on biodiversity and natural habitats. Sustainable agriculture has taken a global standpoint on Agenda 21. It was signed at the 1992 World Summit in Rio de Janeiro. This definition necessitates a detailed understanding of the roles of agro-ecosystems (Tuomisto et al. 2017). Resolving the chronic issue of hunger is not just about creating new agricultural technologies and practices. Even poor farmers are unable to afford the costly technologies. They will have to consider different types of approaches that are focused on locally accessible and low cost. Sustainable escalation encompasses use of best possible technologies which may include best genotypes, agronomic observes, and post-harvest management to optimize yields and overcome losses (Sharpley et al. 2015). Some of the sustainable post-harvest technologies adopted are discussed below:

12.5.1 Physical Treatment

12.5.1.1 Pre-Cooling

Pre-cooling is done to reduce harvested product's field heat, which is detrimental to preserving fruit and vegetable quality and is done to prolong processes of ripening and senescence. Farm heat refers to the heat in the harvested fruits due to the absorption of on-site solar radiation. Prompt field level pre-cooling, until packaging, maintains the weight and extends tomato storage life. By using pre-cooling procedures, physiological weight loss (PLW) in storage can be decreased from 6 to 2.5% (Snowden 2008).

12.5.1.2 Methods of Pre-Cooling

Many successful methods are in commercial use across countries for rapid removal of field heat from produce. Method selection depends primarily on the product's perishability and refrigeration equipment.

12.5.1.3 Hydro-Cooling

Hydro-cooling is effective form of fruit pre-cooling. Cold water treatment is a fast and efficient precooling approach for fruits and vegetables prior to packaging. (Baladhiya and Doshi 2016). Hydro-cooling is accomplished by spraying, or immersion. The commodity must be transferred to a cold space when precooling is done. Fruit is immersed in cold water, or the fruits are sprayed with cold water. Many chemicals are often combined with hydro-cooling water used to avoid the diseases (Jacob 2008).

12.5.1.4 Air-Cooling or Room Cooling

Air refrigeration is achieved by putting the fruits in the cold space. In order to achieve any exchange of air, fruits are stored in well ventilated containers. Air-cooling is a longer process than water or vacuum (Elansari and Siddiqui 2017).

12.5.1.5 Vacuum Cooling

Leafy vegetables such as lettuce can be cooled by vacuum easily and uniformly, rather than with water or air. Cooling is achieved by lowering the air pressure. Hermetically sealed artificial chambers are used to facilitate the same. Reducing the pressure on the atmosphere often decreases the gas of water vapor in the chamber and thereby affects the cooling (Rao 2015). It is the most expensive pre-cooling process. Pre-cooling treatment before storage at low temperatures may delay the development of chilling injury. Uniformity and speed are two advantages of vacuum cooling.

12.5.2 Modified Atmosphere Packaging (MAP)

MAP uses the latest developments in packaging technology to increase the shelf life and retain product's freshness. It normally lowers oxygen levels and increases CO_2 levels in the packaged content. Plastic film bags are used in MAP. Both the whole or fresh-cut product can be packaged either passive or active (Kerry et al. 2006). Passive MAP takes into account the O_2 and CO_2 concentrations which are the function of the weight and respiration rate of the fruit. It is also influenced by temperature, surface area, perforations, thickness, and permeability of film gases used in packaging (Costa et al. 2011). For active MAP, the desired atmosphere is applied before heat sealing in the packet headspace, but ultimately the final atmosphere will be a result of the same factors influencing passive MAP (Soltani et al. 2015). It is recommended that the packed content be high in humidity to prevent transpiration water loss and to prevent shriveling fruit (Refer Fig. 12.4).

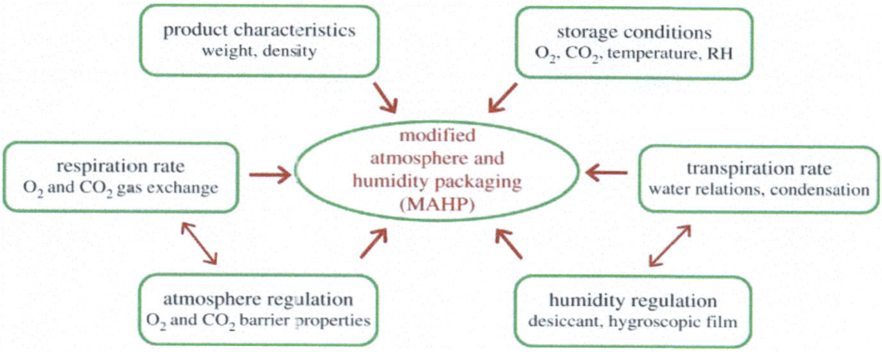

Fig. 12.4 An overview of factors affecting MAP

Correct atmosphere of equilibrium will control respiration and senescence, thereby increasing the life of the product. Active MAP also contains adsorbent technologies such as O_2, ethylene, moisture, CO_2, flavors. It facilitates release of substances such as CO_2, antimicrobials, antioxidants, and flavors (Vakkalanka et al. 2012).

12.5.2.1 Typical Gases for Modified Atmosphere Packaging

Carbon dioxide (CO_2), Oxygen (O_2), and nitrogen (N_2) are mainly used as protective gases in food packaging.

- **Oxygen** generally induces oxidation to spoil food and provides the perfect preconditions for development of aerobic microorganisms. Oxygen activates biosynthetic pathways of both respiratory and ethylene that facilitate ripening levels. As a result, oxygen is always omitted from the modified atmosphere packaging. Oxygen rates that are too high do not greatly impede breathing and should be about 12% or below when the breathing rate starts to decline. For retarding respiration levels, oxygen in bags can be as low as oxygen in bags can be as low as 3–5%. In MAP bags, less than 3–5% of oxygen results in anaerobic fermentation and distastes growth due to alcohol fermentation (Forney 2003).
- **Carbon dioxide** is colorless, odorless, and without taste. The effect on most aerobic bacteria and molds is oxidation-inhibiting and growth-inhibiting. The gas is also used to boost food shelf life of processed or stored food. Nevertheless, if the dose is too high, many items will turn sour. For containers, higher allowable CO_2 levels are 10% as more CO_2 is phototoxic. Furthermore, the gas may spread out of the packaging or be absorbed by the product-and therefore the packaging collapses. This impact can be reduced by the use of assisted or filled gases (Mullan and McDowell 2003).
- **Nitrogen** is an inert gas, which is generally relatively high purity because of its manufacturing cycle. This is commonly used in food packaging to displace carbon, especially atmospheric oxygen. This prevents food oxidation, and inhibits aerobic microorganism production (Conte et al. 2013).

- **Argon** is inert, colorless, odorless, and without taste. Argon can substitute nitrogen in many applications, due to the similarity of its properties to those of nitrogen. Some enzyme activities are thought to be inhibited, and in certain vegetable forms, argon delays metabolic reactions. Its use is very uncommon because of the marginal effects and the higher price relative to nitrogen (Pareek 2016).

In some applications hydrogen (H2) and helium (He) are also used in modified atmospheres. These gasses are not used to extend shelf life, however. They are used as trace gasses for some current on the market leak detection devices. The relatively small molecular size of the gases enables fast escape through leaks in the packaging (Singh et al. 2011).

12.5.3 Gamma-Irradiation

γ rays and e-beams (high-energy electrons) release radiant energy on food. They penetrate matter and break chemical bonds, including the DNA. Cobalt-60 or caesium-137 ionizing radiation or electron beams can prolong the shelf life of fruits. Its dose is measured in kilograys (kGy). It is secure technology and FAO/WHO well define and recognize the vital doses (WHO 1988).

Irradiation (less than 1 kGy) only interrupts cellular activity to prevent tubers, buds, and roots sprouting and delay senescence. Medium doses (1–10 kGy) lower microbial loads. High doses (over 10 kGy) kill broad range of fungi and bacteria (Lynch and Nalder 2015). Medium and high-level doses are not suitable for fresh products. These doses cause irreparable DNA and protein damage and lead to sensory deficiencies (visual, texture, and flavor) and/or accelerated senescence. Irradiation provides an important post-harvest treatment for the destruction of food spoilage bacteria, molds, and yeasts, as well as control of insect and parasite infestation. Irradiation has been commercialized for control of potato and onion sprouting, and strawberry decay (Tripathi et al. 2013).

12.5.4 Cold Chain Management (CCM)

Perishability is greatly decreased at low temperatures. Refrigeration therefore becomes significant for storage. Reportedly, the first cold store in India was founded in Calcutta in 1892. An open cold storage facility is the key to solutions for ensuring better returns (Fig. 12.5).

12.5.4.1 Status of Cold Storage and its Potential in India

India has around 6300 cold storage facilities, with 30.11 million tons capacity. However, some 75–80% of these refrigerated warehouses are only suitable for stockpiling potatoes, a product that generates just 20% of agricultural income. This was reported in a report by the UK's technical association, the Institution of

Products	Temperature°C	Humidity %	Storage Life	Freezing Point
Fruits				
Apples	1-4	90-95	1-12 months	-1.7
Pears	-1-0.5	90-95	2-7 months	-1.5
Plums & Prunes	0	90-95	2-5 weeks	-0.8
Peaches	0	90-95	2-4 weeks	-0.9
Berries				
- Blackberries	0	90-95	2-3 days	-0.8
- Raspberries	0	90-95	2-3 days	-1
- Strawberries	0	90-95	3-7 days	-0.7
Vegetables				
Peppers, sweet	4-10	90-95	2-3 weeks	-0.8
Broccoli	0	95-100	10-14 days	-0.6
Carrots, mature	0	98-100	7-9 months	-1.4
Cucumbers	10-13	95	10-14 days	-0.5
Lettuce	0	98-100	2-3 weeks	-0.1
Mushrooms	0	95	3-4 days	-0.8
Peas, green	0	95-98	1-2 weeks	-0.6

Fig. 12.5 Storage temperature, relative humidity, and shelf life of various fruits and vegetables

Mechanical Engineers (IME). The organization found that only 10–11% of Indian-produced fruits and vegetables use cold storage. In order to prevent wastage, storage space needs to be increased by 40%, the research said.

The country's total annual fruit and vegetable production is about 290 million tons which constitutes 18% of our agricultural production. Output is growing slowly due to diverse agro-climate conditions and better availability of the package of practices. The deficiency of cold storage and cold chain facilities is becoming major bottlenecks in exploiting capacity.

12.5.4.2 Storage of Foods and Storage Conditions

Low-temperature storage slows microorganisms' activities in prolonging shelf quality. Microorganisms, namely bacteria, yeasts and molds, cause heavy losses of produce. Low temperatures greatly minimize their growth, offering a safe way to conserve perishable (Lorenzo et al. 2018).

Storage time is an important factor in preservation technology. It may be short or long term. In general, there are three groups of products:

- Foods that are alive at the time of storage, distribution, and sale, e.g., fruits and vegetables,
- Foods that are no longer alive and have been processed in some form, e.g., meat and fish products, and

- Commodities that benefit from storage at controlled temperature, e.g., beer, tobacco, etc.

Living foods like fruits and vegetables have some natural defense against the microorganism activities. The only way to maintain these products is to keep the commodity alive while also preventing the development of natural enzymes that will prolong the rate of maturity or maturity. It is more difficult to maintain non-living foods because they are vulnerable to spoilage (Ramos et al. 2013). Long-term storage can only be done by freezing meat and fish product and then storing it at temperatures below -15 °C. The freezing can only help some fruits and vegetables (Abdel-Aziz et al. 2016).

Live products such as apples, tomatoes, oranges, etc. cannot be frozen. Control of storage temperature is necessary for their long-term storage. They can be benefited by storing under controlled atmosphere and modified atmosphere conditions.

Dairy products are produced from animal fats and therefore non-living foodstuff. They suffer from the oxidation and breakdown of their fats, causing rancidity. Packaging to exclude air and hence oxygen can extend storage life of such foodstuff.

A cold storage unit incorporates a refrigeration system to maintain the desired room environment for the commodities to be stored (Sharma 2018). A refrigeration system works on two principles:

- Vapor absorption system (VAS), and
- Vapor compression system (VCS).

VAS is very economical in service, although comparatively costlier, and sufficiently compensates for the higher initial investment. Wherever practicable a device of this kind should be selected to save electricity and running costs.

VCS is relatively less expensive than VAS. Depending on the cooling arrangements in the storage rooms, there are three types of VCS systems available, i.e., diffuser type, bunker type, and fin coil type. Diffuser style is comparatively more costly and is only chosen when the heights in the storage room are low. The maintenance costs are also higher for these systems. Bunker style is the cheapest and is favored when the heights of the storage room usually reach 11.5 m. The operating expenses are also low (Reque et al. 2014).

Refrigerants are used in a refrigeration device to absorb heat from the storage room. Freon used to be a common refrigerant but because it causes deterioration of the environment; its use is to be banned by 2008. Consequently, ammonia is increasingly being used and favored in cold storage units for horticultural and plantation production.

12.5.5 Chemical Treatments

12.5.5.1 1-Methylcyclopropene (1-MCP)

1-MCP is a structural analog of ethylene, a strong, gaseous inhibitor of ethylene, that attaches irreversibly to ethylene receptors and thus prevents the effects of ethylene on plants (Refer Fig. 12.6). 1-MCP negatively affects many processes of maturation and senescence, including changes in pigments, cell wall metabolism, taste, and aroma (Zhang et al. 2011).

1 MCP is approved for commercial use in more than 50 countries and continues to gain wide acceptability worldwide. 1-MCP is registered for use on a wide variety of fruits including apple, avocado, banana, broccoli, cucumber, date, kiwifruit, mango, melon, nectarine, papaya, peach, pear, pepper, persimmon, pineapple, plantain, plum, squash, and tomato. India is still behind in adopting this latest innovative chemical approach for prolonging shelf life of climacteric fruits (Watkins 2006).

Inhibitory effect of 1 MCP on climacteric fruits varies with fruit varieties. MCP treated fruits ripen slowly by delaying the process of softening, peel color, aroma, and flavor development (Zhu et al. 2015).

Apple is used worldwide to preserve consistency from storage to customer across the entire marketing chain. Under the brand name Ethyl Bloc, 1-MCP has been licensed by the US Agency for the conservation of the environment for use on ornamental crops. For cut seeds, potted flowers, nursery and foliage plants, it is used to avoid premature wilting, yellowing of the leaves, premature opening, and premature death (Li et al. 2016).

12.5.5.2 Ozone Treatment

Recent studies and industrial implementations have shown that ozone is capable of replacing conventional sanitizers. Ozone is a natural gas with oxidizing properties. For antimicrobial activity, contact times are usually four to five times lower than for chlorine (Suslow 2004). At realistic and healthy concentrations, ozone destroys rapidly bacterial cell walls. Ozone is more effective than chlorine against the thick-walled spores of plant pathogens and animal parasites. Ozone may be used

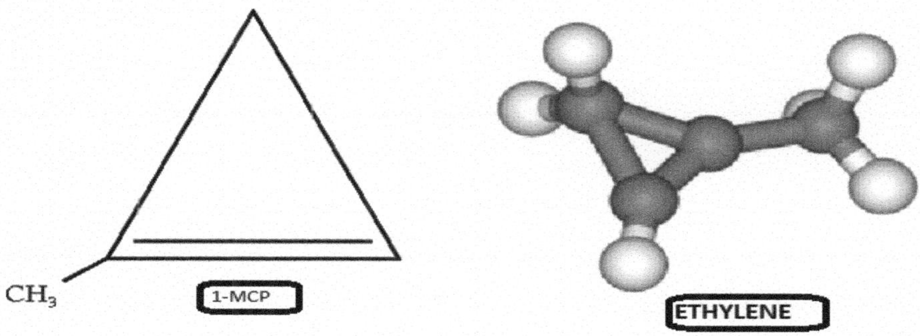

Fig. 12.6 Structural analogy between 1-MCP & Ethylene

Fig. 12.7 Breakdown of ethylene into carbon dioxide and water in the presence of ozone

as an alternative to cooling, as it disintegrates ethylene into CO_2 and H_2O, thus diluting enzymatic changes caused by ethylene during maturation (Fig. 12.7).

Treatment of fruits with 2.5 ppm ozone is reported to increase total soluble solids, ascorbic acid, β-carotene, lycopene content, and antioxidant activity. Rate of loss in weight was also delayed by 10 days in ozone treated fruits. Consumer acceptability and sensory quality of ozone treated papaya is reported to be higher. Ozone treatment may be considered as non-thermal cold food preservation (Horvitz and Cantalejo 2014).

12.6 Technology Gaps in India

Despite bounteous produce of horticultural products, India is far behind in adopting latest technologies to mitigate post-harvest losses and still its agro-based economy practices outdated practices in post-harvest management. Commercialization of methylcyclopropene technology is the basic demand for post-harvest management in India which is not yet realized. Scaling up of cold storage facilities, upgradation of mechanized infrastructure, pre-cooling facilities, and gamma-irradiation facilities need to be practiced across the country to bridge the wide gaps between the bounteous natural produce of fruits and vegetables in India and harvest that feeds the Indian population due to very poor supply chain management. Uninterrupted cold chain management from field to fork is fundamental in managing the living fruits and vegetables. Post-harvest loss reduction technology encompasses the usage of optimum harvest factors, reduction of losses in handling, packaging, transportation and storage with modern infrastructure machinery, processing into a wide variety of products, home scale preservation with low-cost technology. Containers and packaging materials confer portability as well as extend the shelf life. Adoption of these techniques could make available a large quantity of food by avoiding losses and provide better quality food and nutrition, more raw materials for processing, thus ensuring better returns to the farmers.

12.7 Success Story of Reliance Fresh from Farm to Fork

In 2006, Reliance Retail launched the first Reliance Fresh store. Today Reliance Retail operates 523 Reliance Fresh and Reliance Smart stores and sells more than 200 tons of fruits and more than 300 tons of vegetables daily. Reliance Retail explicitly demonstrates successful farm-to-fork model (Pradhan and Mangaraj

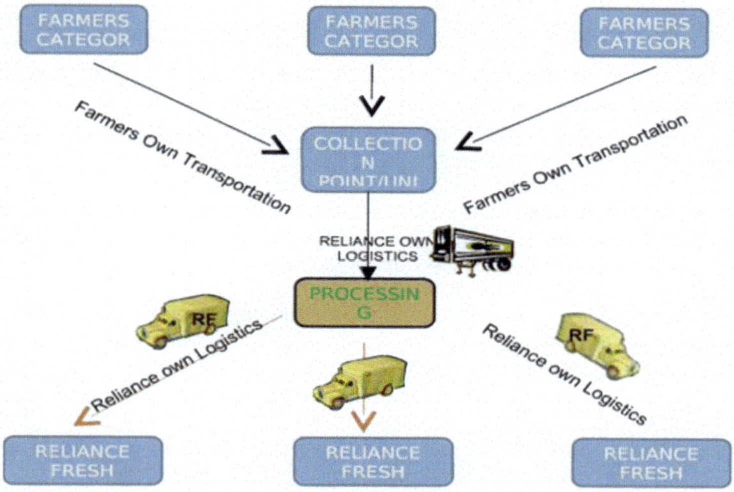

Fig. 12.8 Reliance supply chain

2008). Strong bonding with farmers from the farm level may result in dramatic improvements not only in farmers' quality of life but also the intermediaries in the supply chain thereby benefiting everyone (Fig. 12.8).

Private initiatives such as Reliance Fresh need to be set up in India in large numbers to prevent losses of post-harvest.

12.8 Conclusion

There is an earnest need for India to adopt the following to check the post-harvest losses in fruits: Establishment of Pre-cooling infrastructure at the field level, scaling up of uninterrupted Cold Chain Management from field to fork; Commercialization of MCP technology and gamma-irradiation facility; and commercialization of Modified Atmosphere Packaging at large scale across the country. Failure to adhere to these best practices has incurred in high amount of economic and nutritional loss in India. The supply chain can be structured effectively and uniformly monitored. Government of India need to undertake numerous initiatives for the supply chain management in India. Retailers like Reliance Fresh are the pioneers in establishing well developed farm to fork post-harvest operations and ensuring sustainable supply of perishable horticultural products to the consumer. Attention to the facilitation of post-harvest infrastructure, fool proof supply, and cold chain management only shall ensure sustainable supply of the bounteous horticultural produce.

References

Abdel-Aziz SM, Asker MM, Keera AA, Mahmoud MG (2016) Microbial food spoilage: control strategies for shelf life extension. Microbes Food Health 1:239–264

Baladhiya C, Doshi J (2016) Precooling techniques and applications for fruits and vegetables. Int J Process Post-Harv Technol 7(1):141–150

Baldassarre V, Cabassi G, Spadafora ND, Aprile A, Müller CT, Rogers HJ, Ferrante A (2015) Wounding tomato fruit elicits ripening-stage specific changes in gene expression and production of volatile compounds. J Exp Bot 66(5):1511–1526. https://doi.org/10.1093/jxb/eru516

Chang C (2016) How do plants respond to ethylene and what is its importance? BMC Biol 14:7. https://doi.org/10.1186/s12915-016-0230-0

Conte A, Angiolillo L, Mastromatteo M, Del Nobile A (2013) Technological options of packaging to control food quality. Food Ind:354–379

Costa C, Lucera A, Conte A, Mastromatteo M, Speranza B, Antonacci A, Del Nobile MA (2011) Effects of passive and active modified atmosphere packaging conditions on ready-to-eat table grape. J Food Eng 102:115–121

do Nascimento Nunes MC (2008) Impact of environmental conditions on fruit and vegetable quality. Stewart Postharvest Rev 4:1–14. https://doi.org/10.2212/spr.2008.4.4

Elansari AM, Siddiqui MW (2017) Recent advances in postharvest cooling of horticultural produce. In: Postharvest management of horticultural crops. Apple Academic Press, pp 25–92

El-Ramady H, Domokos-szabolcsy E, Abdalla N, Taha H, Fári M (2015) Postharvest management of fruits and vegetables storage. https://doi.org/10.1007/978-3-319-09132-7_2

Forney CF (2003) Postharvest response of horticultural products to ozone. In: Hodges DM (ed) Postharvest oxidative stress in horticultural crops. Food Products Press, New York, pp 13–54

Gibbs M, Steele P (2018) Post-harvest technology of horticultural crops. Scientific e-Resources

Horticulture Statistics (2018). http://agricoop.nic.in/sites/default/files/Horticulture%20Statistics%20at%20a%20Glance-2018.pdf. Accessed 5 May 2020

Horvitz S, Cantalejo MJ (2014) Application of ozone for the postharvest treatment of fruits and vegetables. Crit Rev Food Sci Nutr 54(3):312–339

Iqbal N, Khan NA, Ferrante A, Trivellini A, Francini A, Khan MIR (2017) Ethylene role in plant growth, development and senescence: interaction with other phytohormones. Front Plant Sci 8:475. https://doi.org/10.3389/fpls.2017.00475

Jacob P (2008) A handbook on post-harvest management of fruits and vegetables. Daya Books

Kerry JP, O'grady MN, Hogan SA (2006) Past, current and potential utilisation of active and intelligent packaging systems for meat and muscle-based products: a review. Meat Sci 74 (1):113–130

Khoo H E, Prasad K N, Kong K W, Jiang Y, Ismail A (2011) Carotenoids and their isomers: color pigments in fruits and vegetables. Molecules (Basel, Switzerland) 16(2):1710–1738. https://doi.org/10.3390/molecules16021710

Kiaya V (2014) Post-harvest losses and strategies to reduce them. Technical Paper on Postharvest Losses, Action Contre la Faim (ACF) 25

Kumar R, Khurana A, Sharma AK (2014) Role of plant hormones and their interplay in development and ripening of fleshy fruits. J Exp Bot 65:4561–4575. https://doi.org/10.1093/jxb/eru277

Li L, Lichter A, Chalupowicz D, Gamrasni D, Goldberg T, Nerya O, Ben-Arie R, Porat R (2016) Effects of the ethylene-action inhibitor 1-methylcyclopropene on postharvest quality of non-climacteric fruit crops. Postharvest Biol Technol 111:322–329. https://doi.org/10.1016/j.postharvbio.2015.09.031

Liu M, Pirrello J, Chervin C, Roustan JP, Bouzayen M (2015) Ethylene control of fruit ripening: revisiting the complex network of transcriptional regulation. Plant Physiol 169(4):2380–2390. https://doi.org/10.1104/pp.15.01361

Lorenzo JM, Munekata PE, Dominguez R, Pateiro M, Saraiva JA, Franco D (2018) Main groups of microorganisms of relevance for food safety and stability: general aspects and overall description. Innov Technol Food Preserv:53–107. https://doi.org/10.1016/B978-0-12-811031-7.00003-0

Lynch M, Nalder K (2015) Australia export programmes for irradiated fresh produce to New Zealand. Stewart Postharvest Rev 11(3):1–3. https://doi.org/10.2212/spr.2015.3.8

Maduwanthi S, Marapana R (2019) Induced ripening agents and their effect on fruit quality of Banana. Int J Food Sci 2019:2520179. https://doi.org/10.1155/2019/2520179

Mahajan PV, Caleb OJ, Singh Z, Watkins CB, Geyer M (2014) Postharvest treatments of fresh produce. Philos Trans Series A Math Phys Eng Sci 372(2017):20130309. https://doi.org/10.1098/rsta.2013.0309

McKinsey and CII Report (2001) Fruit and vegetable losses alarmingly high. The Hindu; August 01, 2001 edition

Mercier S, Villeneuve S, Mondor M, Uysal I (2017) Time–temperature management along the food cold chain: a review of recent developments. Compr Rev Food Sci Food Saf 16:647–667. https://doi.org/10.1111/1541-4337.12269

Mishra S, Guru S (2017) Physiological factors leading to post-harvest crop losses. In: Proceedings of the 35th Training on "Technological Advances to Minimize Pre-and Post-Harvest Losses in Agricultural and Horticultural Crops to Enhance Farmer's Income". 01. 68–76

Mullan M, McDowell D (2003) Modified atmosphere packaging in food packaging technology. Modified atmosphere packaging. In: Food and beverage packaging technology. CRC Press

Nascimento VDL (2017) Reviewing the functions of ethylene in growth and central metabolism in tomato

Paniagua C, Posé S, Morris VJ, Kirby AR, Quesada MA, Mercado JA (2014) Fruit softening and pectin disassembly: an overview of nanostructural pectin modifications assessed by atomic force microscopy. Ann Bot 114(6):1375–1383. https://doi.org/10.1093/aob/mcu149

Pareek S (ed) (2016) Fresh-cut fruits and vegetables: technology, physiology, and safety. CRC Press

Paul V, Pandey R, Srivastava GC (2012) The fading distinctions between classical patterns of ripening in climacteric and non-climacteric fruit and the ubiquity of ethylene-an overview. J Food Sci Technol 49(1):1–21. https://doi.org/10.1007/s13197-011-0293-4

Pradhan D, Mangaraj BK (2008) Reliance fresh stores in food retailing. London Business School UK. Pradhan D, Debasis M (eds) Reliance fresh stores in food retailing. London Business School, UK 1–30

Prasanna V, Prabha TN, Tharanathan RN (2007) Fruit ripening phenomena--an overview. Crit Rev Food Sci Nutr 47(1):1–19. https://doi.org/10.1080/10408390600976841

Ramos B, Miller FA, Brandão TR, Teixeira P, Silva CL (2013) Fresh fruits and vegetables—an overview on applied methodologies to improve its quality and safety. Innovative Food Sci Emerg Technol 20:1–15

Rao CG (2015) Engineering for storage of fruits and vegetables: cold storage, controlled atmosphere storage, modified atmosphere storage. Academic Press

Reque PM, Steffens RS, Jablonski A, Flôres SH, Rios ADO, de Jong EV (2014) Cold storage of blueberry (Vaccinium spp.) fruits and juice: anthocyanin stability and antioxidant activity. J Food Compos Anal 33(1):111–116

Sawicka B (2019) Post-harvest losses of agricultural produce. Sustain Dev 1:1–16. https://doi.org/10.1007/978-3-319-69626-3_40-1

Sharma T (2018) Design and development of a solar powered cold storage system. Doctoral Dissertation, Indian Institute of Technology, Guwahati

Sharpley AN, Bergström L, Aronsson H, Bechmann M, Bolster CH, Börling K, Tonderski KS (2015) Future agriculture with minimized phosphorus losses to waters: research needs and direction. Ambio 44(2):163–179

Singh P, Wani AA, Saengerlaub S, Langowski HC (2011) Understanding critical factors for the quality and shelf-life of MAP fresh meat: a review. Crit Rev Food Sci Nutr 51(2):146–177

Singh V, Hedayetullah M, Zaman P, Meher J (2014) Postharvest technology of fruits and vegetables: an overview. J Postharvest Technol 2(2):124–135

Slavin JL, Lloyd B (2012) Health benefits of fruits and vegetables. Adv Nutr (Bethesda, Md.) 3 (4):506–516. https://doi.org/10.3945/an.112.002154

Snowden AL (2008) Post-harvest diseases and disorders of fruits and vegetables: volume 1: general introduction and fruits. CRC Press

Soltani M, Alimardani R, Mobli H, Mohtasebi SS (2015) Modified atmosphere packaging: a progressive technology for shelf-life extension of fruits and vegetables. J Appl Packag Res 7 (3):2

Suslow T (2004) Ozone applications for postharvest disinfection of edible horticultural crops. UCANR Publications

Tripathi J, Chatterjee S, Vaishnav J, Variyar PS, Sharma A (2013) Gamma irradiation increases storability and shelf life of minimally processed ready-to-cook (RTC) ash gourd (Benincasa hispida) cubes. Postharvest Biol Technol 76:17–25

Tuomisto HL, Scheelbeek P, Chalabi Z, Green R, Smith RD, Haines A, Dangour AD (2017) Effects of environmental change on population nutrition and health: a comprehensive framework with a focus on fruits and vegetables. Welcome Open Res 2:21. https://doi.org/10.12688/wellcomeopenres.11190.2

Vakkalanka MS, D'Souza T, Ray S, Yam KL, Mir N (2012) Emerging packaging technologies for fresh produce. In: Emerging food packaging technologies. Woodhead Publishing, pp 109–133

Vanderstraeten L, Van Der Straeten D (2017) Accumulation and transport of 1-Aminocyclopropane-1-Carboxylic Acid (ACC) in plants: current status, considerations for future research and agronomic applications. Front Plant Sci 8:38. https://doi.org/10.3389/fpls.2017.00038

Watkins CB (2006) The use of 1-methylcyclopropene (1-MCP) on fruits and vegetables. Biotechnol Adv 24(4):389–409

WHO (1988). Food irradiation: a technique for preserving and improving the safety of food. World Health Organization, Geneva Switzerland hdl:10665/38544. ISBN 978-924-154240-1

Yang H, Liu J, Dang M, Zhang B, Li H, Meng R, Qu D, Yang Y, Zhao Z (2018) Analysis of β-Galactosidase during fruit development and ripening in two different texture types of apple cultivars. Front Plant Sci 9:539. https://doi.org/10.3389/fpls.2018.00539

Zhang Z, Huber DJ, Rao J (2011) Ripening delay of mid-climacteric avocado fruit in response to elevated doses of 1-methylcyclopropene and hypoxia-mediated reduction in internal ethylene concentration. Postharvest Biol Technol 60(2):83–91

Zhu X, Shen L, Fu D, Si Z, Wu B, Chen W, Li X (2015) Effects of the combination treatment of 1-MCP and ethylene on the ripening of harvested banana fruit. Postharvest Biol Technol, 107 https://doi.org/10.1016/j.postharvbio.2015.04.010

Agricultural Waste Produce: Utilization and Management

13

Deepshikha Thakur, Naleeni Ramawat, and Vineet Shyam

Abstract

Agricultural waste is the term used to denote the extensive range of waste produced as a result of several agricultural operations. Rapidly intensifying agricultural production has unsurprisingly resulted in augmented quantities of agricultural wastes that comprise crop waste, food processing waste, animal waste, and toxic agricultural waste. Presently used methods for disposal of these agricultural residues have caused prevalent concerns related to environment. Therefore, for their sustainable utilization, various approaches with respect to their utilization, reuse, and processing need to be developed. It is the need of an hour to manage these agricultural wastes in a sustainable manner so as to maintain a healthy environment and withstand the sources of energy. The chapter will discuss about the agricultural waste generation, methods of agriculture waste utilization, and agricultural waste management system.

Keywords

Agriculture waste · Sustainable utilization · Waste management

D. Thakur (✉) · N. Ramawat
Amity Institute of Organic Agriculture, Amity University, Noida, UP, India
e-mail: dthakur1@amity.edu

V. Shyam
Food Safety and Standards Authority of India, New Delhi, India

M. Thakur et al. (eds.), *Sustainable Food Waste Management*,
https://doi.org/10.1007/978-981-15-8967-6_13

13.1 Introduction

For many countries, agriculture and its associated industries are one of the most premeditated sectors. Abreast to the agricultural and related activities and operations, large volumes of solid and liquid residues are generated that is referred to as Agricultural waste (Hodaifa et al. 2019). Gontard et al. (2018) has defined agricultural waste, by-products, and co-products as plant or animal residues that are not either food or feed and that poses an additional burden to environmental and economic sectors of farming and primary processing. The term agricultural wastes encompasses crop wastes such as stalks, bagasse, fruits drops and culls, vegetables, prunings; manure, food processing waste, hazardous agricultural waste comprising of toxic chemicals such as insecticides, pesticides, and herbicides; agro-industrial waste generated from agricultural processing industries of crops, fruits, vegetables, meat, poultry, and dairy products. Agricultural waste otherwise called agro-waste can be in the form of liquids, slurries, or solids. The composition of such agro-waste depends on the system and the type of agricultural activities leading to its generation.

Increase of agricultural production resulted in increase in quantities of waste from livestock, farm activities, and from agro-industries. This increase in agricultural wastes is likely to be on a rise if all over the world, developing countries continue to exaggerate their farming systems. According to Agamuthu (2009), approximately 998 million tons of agricultural waste is produced every year. Agricultural residues generally termed as agro-wastes are produced in huge quantities every year in Indian farm fields. To an approximation, the amount of crop residues produced every year exceeds 620 million tons in India (Singh and Sidhu 2014) of which, almost 50% finds applications in various agricultural and industrial purposes like animal feedstock, paper industry, roofing material, and energy generation. This agriculture waste presents a serious environmental problem in the case of not being well treated or managed (Hodaifa et al. 2019). Major quantities of the agricultural residues generated on-farm, which is generally treated as wastes by the farmers, are burnt in the field itself or are allowed to decay in the open air creating environmental pollution. Although this is a cheap, non-labor-intensive and comparatively easy mean of agro-waste disposal, but in return this has great negative impact on the agro-ecosystem as it generates a significant amount of particulate matter in environment to create air pollution through smoke and smog and disturbs physical, chemical, and biological structure including and microflora and fauna of the soil (NMSA 2016). The disposal methods for the untreated and underutilized agricultural residues that are being followed today (burning, dumping or unplanned land filling have caused prevalent environmental concerns (Sadh et al. 2018). Jain (2014) reported that burning of agricultural crop residue contributes to the emission of greenhouse gases (CO_2, N_2O, CH_4), air pollutants (CO, NH_3, NOx, SO_2, NMHC, volatile organic compounds), particulate matter, and smoke, thereby posing risk to humans. It is therefore, imperative to make competent use of crop residues generated in the agriculture production system to help improve crop productivity, soil conditions, and environmental sustainability.

Taking into consideration the burden that the current disposal methods add to the environment and economy, approaches with respect to their utilization, reuse, and processing need to be developed to enable the sustainable utilization of feedstock and reduce pollution (Rao and Rathod 2018). There is a need to manage these crop wastes through a sustainable approach majorly through following three R technique of reducing, reusing, and recycling. Different categories of agriculture waste can be utilized efficiently for specific purposes such as energy, nutrient feed, soil amendments, adsorbents, etc., to list a few. It is imperative to use agriculture waste in a well-planned manner so that we can maintain a healthy environment for ourselves and all other living creatures.

13.2 Agricultural Waste Generation

At present, every year in India, around 960 million tons of solid waste is being generated as by-products during various industrial, mining, municipal, agricultural, and other processes. Out of this approximately 350 million tons are organic wastes from agricultural sources (Pappu et al. 2007). The nature of waste is reliant on the type of agricultural activities or processing carried out for its generation. The agricultural waste can be classified into the sources of waste generation.

The first category is the waste generated from cultivation activities. These are majorly the crop residues that encompass all agricultural wastes such as leaves, stem, straw, stalk, husk, shell, pulp, peel, stubble, etc. which originate from cereals, cotton, jute, groundnut, legumes, tea, coffee, cacao, fruits, palm oil, and all other organic materials which are produced as by-products from harvesting. Another type of waste generated is during the control of weeds and insects, i.e., the use of pesticides in order to destroy the insects and pests. Imbalanced use of these pesticides often leads to the pesticides abuse by the farmers. Majority of the bottles and packaging material containing these pesticides are discarded inefficiently into either the fields or the local water bodies. This can be threatening as it is claimed by the Plant Protection Department (PPD), that approximately 1.8% of the chemicals persist in the packaging material even after use (Dien and Vong 2006). Another serious environmental consequence is posed by the unused pesticides and pesticide packages with residue if they are either stored or buried in the erroneous way as their contents may leak into the environment via osmosis and thereby negatively impacting the environment (Obi et al. 2016). In addition, other waste generated from agricultural activities is through excess fertilizers. From this excess fertilizer, a part persists in the soil, some excess part runs off, depending upon irrigation system adopted, enters the water bodies such as ponds, lakes, rivers, another part enters the ground water, and rest of the portion either gets evaporated or de-nitrated, adding to the air pollution (Obi et al. 2016).

The second category of agricultural waste is that generated from Livestock Production. Global livestock and poultry industry producing meat, milk, and egg is growing continuously and is also generating large volumes of wastes. Livestock waste includes solid waste (manure and organic materials) produced via slaughterhouse; fluid waste (water from the animal bathing, cage washing, urination and water from slaughterhouses); pollution of air (methane and hydrogen sulfide); and odors. Also, the livestock enterprises around residential areas pose a serious threat to humans health. As reported by Ifeanyi (2012), 75 to 95% of total volume of livestock waste is water, while the rest consists of organic and inorganic matter, different species of microbes, and even parasite eggs. However, the disposal of these livestock wastes lingers to be a challenge from the stances of cost, environmental safety, and biosecurity.

The third category of agricultural waste is that generated from waste from Aquaculture. The expansion of aquaculture industry led to generation of metabolic waste that could either be dissolved or remain suspended. Approximately 30% of the feed even in a well-managed aquafarm get converted into the solid waste. Waste management in these production units can be maintained by the patterns of water flow as a proper water flow can minimize the disintegration of the fish feces and allow the speedy settling and concentration of the settleable solids (Mathieu and Timmons 1995).

The fourth category of agricultural waste is that generated from processing, i.e., process residues. Agriculture or food processing wastes are the water discharged from these operations: vegetable processing (wash water, skins, rinds, pulp, and other organic waste from fruit and vegetable cleaning, processing, cooking, and canning), Meat (Grease, fat, oils, wash water, cooking waste, dripping, hair, and feathers from slaughtering, butchering, cooking and packaging of fish, chicken, beef, and all other meat products), Dairy and Egg (Wash water and process waste from egg and milk processing, drying, bottling, and packaging), Miscellaneous food processing (grain processing and animal food production).

The types and quantities of waste production in agriculture vary between farms. Apart from the above-mentioned categories some other common agricultural wastes include packaging, redundant machinery, silage plastics, net-wrap, oils, tires, old fencing, batteries, veterinary medicines, horticultural plastics, scrap metal, and other building waste. Other uncommon wastes include spent and unused sheep dip.

13.3 Sustainable Approaches for Agriculture Waste Utilization

Recycle and reuse have emerged as major pillars for the sustainable agriculture waste utilization and for the management of resources by preventing the generation of waste and if generated then reuse, recycling, and recovery of the same is used to fulfill the objectives of sustainable developmental goals (Shulman 2011). There are many ways in which the different categories of agriculture waste can be utilized (Fig. 13.1).

13.3.1 Energy

Application of agricultural waste for energy development and utilization is becoming increasingly important due to the depletion of petroleum resources and the

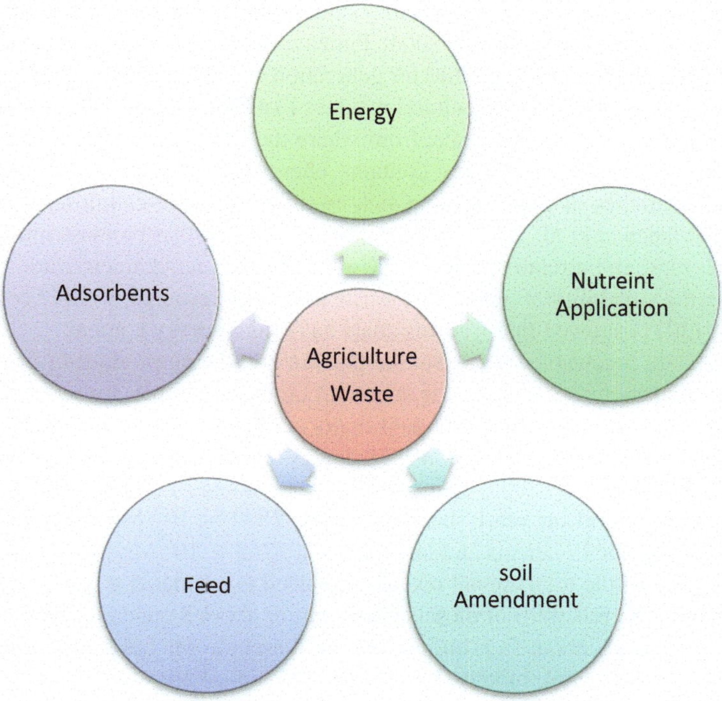

Fig. 13.1 Agriculture waste utilization

unceasing deterioration of the ecological environment (Wei et al. 2020). One of the most imperative is the utilization for the production of green energy. Due to the abundant availability of biomass throughout the world, it is essential to use it as an alternate energy resource to satisfy our needs of energy consumption. Earlier, the agricultural waste and biomass were naturally converted into organic fertilizer under favorable condition or were burnt, but at present their potential as energy source has been recognized and is thus being explored. Upadhaya and Harshvardhan (2017) have reported energy conversion as a good and effective utilization of agricultural waste.

Today, agricultural energy also known as green energy is the prime contributor in economic development of many developing countries. Variety of agricultural residues (straws, nut and fruit shells, fruit tree residues, seeds, grain dust, plant stalks and stovers, green leaves, and molasses) hold a high potential as renewable energy resource (Demirbas 2009). Energy generation from biomass holds innovative prospects for today's farmer with new possibilities to diversify agricultural activities.

The technologies such as liquefaction, pyrolysis, direct combustion, gasification are some of the technologies for conversion of biomass into energy that have potential for commercial-scale use (Demirbas 2009). In addition to the residues of the crop harvest, now forestry products are also being stressed as a substantial energy source. There is a huge list of crops and residues that are suitable for the contemporary bioenergy chains (Reijnders 2004; Pimentel et al. 2005). Apparently, power plants are using agricultural residues for generation of energy in a very cost-effective way. For power generation agricultural residues like Maize Cobs, Wheat Straw, and Rice Husks are collected at a place, from where they can easily be used for energy generation. Argo-waste feedstock contains energy in the chemical bonds of its constituent varieties of hydrocarbon (such as cellulose, hemi-cellulose, lignin, and proteins) (Mahawar et al. 2015). Wilson (2011) reported that biomass must have a high H: C ratio and a relatively low O: C ratio for thermal characterization. While analyzing the suitability of agriculture waste as energy resource, a study conducted by Nagle (2011) showed that fruit residuals have high energy content.

Biofuels are renewable energy source that are now being produced from variety of agricultural biomass. Residues from cereal crops, olive tree, tomato, and grape have been reported to provide abundant lignocellulosic waste in a study by Faraco and Hadar (2011), which can be considered as potential raw material for large-scale ethanol production. Lal (2005) estimated that the amount of crop residue generated throughout the world on yearly basis is around 2802×10^6 Mg for cereal crops, 3107×10^6 Mg for 17 cereals and legumes, and 3758×10^6 Mg for 27 food crops. The fuel value of the total annual residue produced is estimated at 1.5×10^{15} kcal, about 1 billion barrels (bbl) of diesel equivalent, or about 8 quads for the USA; and 11.3×10^{15} kcal, about 7.5 billion bbl of diesel or 60 quads for the world. Bioethanol, biomethanol, biodiesel, methane, and bio-oil are some examples of the agriculture-based biofuels. In a study by Qureshi (2010) fermentation of dilute H_2SO_4 barley straw hydrolysate by *Clostridium beijerinckii* P260 resulted in the generation of 7.09 gL^{-1} ABE (acetone, butanol, ethanol). Various agricultural waste materials like sugarcane molasses, beet molasses, corn cobs and hulls, cellulosic materials like sugarcane waste, and coffee husk hydrolyzing fermenting strains have been used for bioethanol production (Beall et al. 1992; Arni et al. 1999; Othman et al. 1992; Franca et al. 2008). Physical, chemical, and spectroscopic characteristics of the residues of an endemic marine plant, *Posidonia oceanica* were investigated by Cocozza et al. (2011). They concluded that the fibrous portion of *P. oceanica* can be utilized as a biofuel, due to its lesser heating values ranging from 13.6 to 15.7 Mg/kg.

Another technology that can treat agriculture waste is pyrolysis where agricultural waste is heated up to a temperature of 400–600 °C, in anaerobic conditions by which the material vaporizes, leaving behind a char also sometimes known as biochar. The product output of Pyrolysis of agricultural waste is oil, char, and gas (Obi et al. 2016). Others are hydro-gasification, and hydrolysis which are generally used for energy recovery and the preparation of chemicals such as preparation of alcohols for fuel, ammonia for fertilizers, glucose for food and feed.

Biogas technology provides an alternate source of energy in rural as well as urban areas as a substitute for fossil fuels. Biogas is a biofuel that is naturally produced as a by-product during the decomposition of organic waste. The agricultural waste can be used in the generation of biogas. The agricultural resources such as energy crops, grasses, agricultural residues, and aquatic weeds can be used for the production of biogas resulting in effective utilization of agri-wastes and reducing emissions from the storage of animal waste. Biogas is highly combustible, has high content of methane (50–75%), which renders it combustible, and has a good provision as an energy source.

Methane gas that has specific applications for heating purposes such as in broilers, water heating, drying of grains, etc., can be produced from agricultural wastes particularly manures. The conversion of agricultural waste to form methane-rich gas is a two-step microbial fermentation that undergoes in anaerobic conditions. The typical gas produced by this fermentation process is composed primarily of methane i.e. 50–70%, followed by 25–45% CO_2, 0.5–3% nitrogen, 1–10% hydrogen, and traces of hydrogen sulfide. The heating value of this generated gas ranges from 18–25 MJ/m^3 (Timbers and Downing 1977). There are some major disadvantages associated with the digestion system such as high capital costs and the explosive properties of the methane gas. However, these disadvantages are outweighed and overshadowed by the aforementioned advantages. This method of anaerobic digestion also makes the treatment and disposal of huge livestock waste feasible, thereby curtailing the odor problem and abreast of that produces relatively odor free sludge that still retains the fertilizer value (Obi et al. 2016).

13.3.2 Nutrient Application

After the crop is harvested, some of the crop nutrients are retained in the farm residues which are either burnt or dumped improperly. These residues if utilized efficiently, holds the potential to be the source of nutrients. The recycling of these organic agri-wastes enables the essential plant nutrients to be recovered and replenished so as to maintain organic matter, physicochemical properties of the soil, and agronomic productivity. The residual agriculture waste from the farm is rich in nutrients such as N, P, and K and can be utilized for nutrient application after conversion into composts. Numerous non-economic parts and residues of the crops are left in the field after the harvest, such as straws, stover, stubble, and haulms of various crops and even the processing wastes like groundnut shells, oil cakes, rice husks, and cobs of maize and sorghum discarded during crop processing have nutrient providing potential. However, the crop residues need composting before being used as manure (Mahimairaja et al. 2008). Composting is basically a microbiological process accomplished by the combined activity of bacteria, actinomycetes, fungi, and protozoa which are either present in the composting material or are introduced externally to speed up composting and enrich the compost. As a result, the substrate is broken down to form an amorphous brown to dark brown mixture known as compost. Compost is considered a valuable organic fertilizer, supplying nutrients for the crop and hence saving substantial amounts of mineral fertilizers (Erhart et al. 2005). It is estimated that almost 50% of the total agricultural residues are produced majorly by wheat, rice, and oilseed crops. The residues from these crops are estimated to contain almost 0.5% N, 0.2% P_2O_5, and 1.5% K_2O (Singh and Sidhu 2014). It depicts that approximately 6.5 million tons which equals to almost 30% of the total NPK mineral consumption in India, is contained in the form of these farm residues. Therefore, recycling of the agricultural farm residues or agro-wastes becomes a subject of not only major agro-ecological concern but is also well connected with the improved soil and plant health. The recycled waste products could be bio-composts, which may be fortified with microbial consortia and organic material (like amino acids, humic acid, phytostimulants and minerals, etc.) to produce bio-organic farm inputs at commercial scale which could ultimately generate economic livelihood to the rural society also (Singh and Prabha 2017).

The animal manures are the major waste generated from Livestock Production, which can be utilized for fertilizer use at the farm level. According to a report by Council for Agricultural Science and Technology, manure has the capacity to supply 19, 38, and 61% of N, P, and K in chemical fertilizer. Many studies have reported that manures contain high nutrients which can be utilized to supply nutrients to the plants (Lustosa et al. 2017). Obi et al. (2016) reported that adding manure to soil increases its fertility via improving the physical condition of the soil, increasing the nutrient retention or cation exchange capacity, enhancing the water-holding capacity and also the structure stability of the soil.

13.3.3 Adsorbents

The application of adsorbents made from agricultural waste supports the 3 Rs, i.e., reduce, reuse, and recycle rule of waste management strategy and can prove to be an efficient and revenue generating management practice for agricultural sector (Tandon and Sai 2019). Massive urbanization and industrialization have resulted in excessive release of heavy metals into the environment such as mercury, copper, zinc, cadmium, chromium, and lead. These metal ions are not susceptible to biological degradation, unlike organic pollutants. In the past decades, the contaminated effluents are being treated by the agricultural wastes through absorption, as it has been proven to be the low-cost alternative, for the treatment of heavy metals and other compounds. The low value agricultural waste such as rice husk, sugarcane bagasse, coconut husk, sawdust, neem bark, oil palm shell (Mohan and Singh 2002; Ayub et al. 2002; Tan et al. 1993; Khan et al. 2003 and Ayub et al. 2001), etc., can be adsorbents in the elimination of heavy metals as reported in many studies. Results published by Hegazi (2019) showed that low-cost adsorbents can be efficiently used for the removal of heavy metals with a concentration range of 20–60 mg/l also, rice husk was effective in the simultaneous removal of Fe, Pb, and Ni, whereas fly ash was effective in the removal of Cd and Cu from the wastewater. Kuan et al. (2011) investigated the physicochemical characteristics of residues and soluble fibrous residues produced from alkali-treated raw cereal materials (corncob and wheat straw) with high dietary-fiber content (49.87–68.65%). Cereal materials and insoluble fibrous residues were found to contain essential minerals (in contrast to soluble fibrous residues), and showed 2,2-diphenyl-1-picrylhydrazyl radical scavenging ability, good emulsification ability and emulsion stability, mineral binding capacity, water capacity, and oil holding capacity. The production of nanofibers from SF by electrospinning proved successful.

13.3.4 Feed

In most developing countries like India, the animal feed is usually deficient in nutrient sources. Agro-residues being rich in various bioactive and nutraceutical compounds, viz. carotenoids, polyphenolics, and dietary fiber, to list a few, present a potential solution to the problem of lack of nutrition in animal feed and the worldwide supply of protein and calories. The traditional method of increasing livestock production by supplementing forage and pasture with grains and protein concentrate may not meet future meat protein needs. Appropriate technologies can be used for their valorization of animal feed by nutrient enrichment. Technologies available for protein enrichment of these wastes include solid substrate fermentation, ensiling, and high solid or slurry processes. Technologies to be developed for the reprocessing of these wastes need to take account of the peculiarities of individual wastes and the environment in which they are generated, reprocessed, and used (Ajila et al. 2012).

Apart from the above-mentioned methods of utilization, there are many other methods of sustainable utilizations as well. Agriculture process residues including husks, seeds, molasses, bagasse, and plant roots have potential to be used as animal fodder. It can also be used as a soil amendment, fertilizers, and in manufacturing of various products. In a study by Lim and Matu (2015), a biofertilizer was developed using agro-wastes. Organic waste from agriculture can also be utilized for Natural fiber-polymer composites (NFPCs) (Väisänen et al. 2016). Various studies reported that different kinds of waste such as pomegranate peels, lemon peels, and green walnut husks can be used as natural antimicrobials (Adámez et al. 2017; Katalinic et al. 2013). Agro-industrial wastes are used for manufacturing of biofuels, enzymes, vitamins, antioxidants, animal feed, antibiotics, and other chemicals through solid state fermentation (SSF) (Sadh et al. 2018).

13.4 Agricultural Waste Management System

United States Department of Agriculture has defined agricultural waste management system (AWMS), as a “planned system in which all necessary components are installed and managed to control and use by-products of agricultural production in a manner that sustains or enhances the quality of air, water, soil, plant, and animal resources.” Agricultural waste management is part of the ecological cycle in which everything is cycled and recycled such that an interdependent relationship is maintained in the eco-system. By waste management, all the plant wastes are placed at the right place and right time for the best utilization in order to convert into useful products and pollution control. The Total Solids (TS) concentration of agricultural wastes is the main characteristic that determines the handling of the material. For

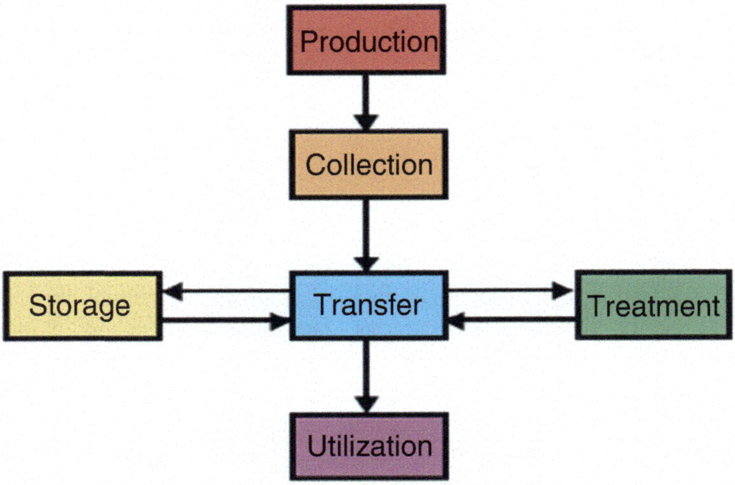

Fig. 13.2 Functions of agricultural waste management (Obi et al. 2016)

excreted manure, for example, the following factors affect the TS concentration and they include the climate, type of animal, amount of water consumed by the animal, and the feed type.

As noted by USDA, AWMS consists of following six basic functions, also depicted in Fig. 13.2:

- **Production**: Amount and nature of agricultural waste generated.
- **Collection**: Initial capture and gathering of the generated waste from the point of origin or deposition.
- **Storage**: Temporary containment or waste holding.
- **Treatment**: Function designed to reduce the pollution or toxic potential of the waste, including physical, biological, and chemical treatment and increases its potential beneficial use.
- **Transfer**: Movement and transportation of the waste throughout the system from the collection to the utilization stage either as a solid, liquid, or slurry, depending on the total solids concentration.
- **Utilization**: Application of the waste for beneficial use and it includes recycling reusable waste products and reintroducing non-reusable waste products into the environment.

Obi et al. (2016), described the "3R" Approach to agricultural waste management (AWM). The "3R" Approach to agricultural waste management (AWM) is the concept of minimizing the ill-effects of waste generation by reducing the waste quantity, reusing the wastes via using simple treatments and recycling the wastes by using it as resource for further production. The three R principle, i.e., reducing,

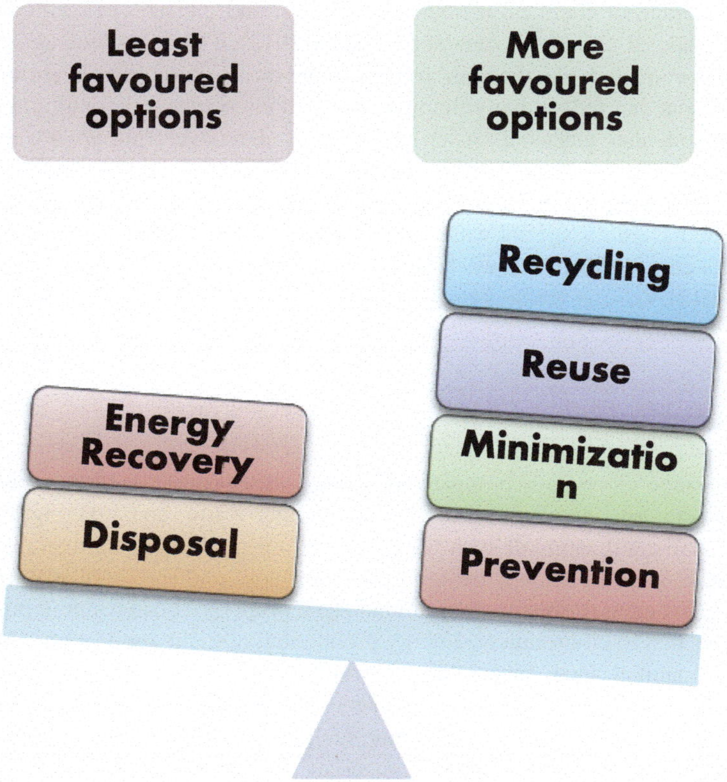

Fig. 13.3 The least and more favoured options of Agriculture waste management

reusing, and recycling of wastes, aims at achieving efficient minimization and utilization of waste generation by (Fig. 13.3)

- Reducing the quantity of waste generated by effectively utilizing the items.
- Reusing the items or parts of items which still have usable aspects.
- Recycling via use of waste itself as a resource.

13.5 Hazardous/Special Agricultural Waste

As defined by SEPA (Scottish Environment Protection Agency), special waste is defined as a hazardous waste that requires to be handled with additional care and controls for well being of humans and environment. Few examples of such waste are residual pesticides, chemicals and their containers, waste oils generated from farm machinery, asbestos roofing material, infectious waste produced during healthcare of animals, and electrical equipment containing cathode ray tubes. Many agricultural enterprises use large amounts of agricultural chemicals. With this increased usage comes the potential for surface and groundwater contamination as a result of improper storage of chemical residue, rinse water, and unused chemicals and the improper disposal of empty containers (USDA 2012). The hazardous agricultural waste can be treated via thermal, physical, chemical, or biological sorting and processing that can reduce the volume and hazardous nature of agricultural waste. The treatment also makes it easier to handle and recover (http://www.netregs.gov.uk). Disposing of hazardous/special waste should be done only in authorized sites. The transportation of hazardous waste should be strictly by a registered waste carrier. The hazardous/special waste can be disposed to an authorized landfill site. USDA has also specified the waste management system for agriculture chemicals (USDA 2012).

13.6 Conclusion

Extensive range of waste is produced as a result of several agricultural operations such as crop waste, food processing waste, animal waste, and toxic agricultural waste. Effective utilization of such waste has the potential to benefit man due to its high strength, environmentally benign nature, low cost, and ease of availability and reusability. Whereas, if this waste is not managed properly, it can result in prevalent concerns related to environment. Energy production, nutrient application, chemical adsorbents, animal feed, biofertilizer production are some of the strategies of effective waste utilization and management. The application of the knowledge of agricultural waste management systems such as the "3Rs" can be useful in transforming the waste into beneficial materials for human and agricultural usage. Appropriate waste utilization will help in promotion of world's developing agricultural sector and also the management of environment pollution.

References

Adámez DJ, Garrido M, Bote ME (2017) Chemical composition and bioactivity of essential oils from flower and fruit of Thymbra capitata and Thymus species. J Food Sci Technol 54(7):1857–1865

Agamuthu P (2009) Challenges and opportunities in Agro-waste management: an Asian perspective. In: Inaugural meeting of First Regional 3R Forum in Asia 11–12 November, Tokyo, Japan

Ajila CM, Brar K, Verma M (2012) Bio-processing of agro-byproducts to animal feed. Crit Rev Biotechnol 32(4). https://doi.org/10.3109/07388551.2012.659172

Arni S, Molinari M, Borghi M, Converti A (1999) Improvement of alcohol fermentation of a corn starch hydrolysate by viscosity raising additives. Starch-Starke:218–224

Ayub S, Ali SI, Khan NA (2001) Efficiency evaluation of neem (*Azadirachta indica*) bark in treatment of industrial wastewater. Environ Pollut Control J 4(4):34–38

Ayub S, Ali SI, Khan NA (2002) Adsorption studies on the low cost adsorbent for the removal of Cr (VI) from electroplating wastewater. Environ Pollut Control J 5(6):10–20

Beall D, Bassat A, Doran J, Fowler D, Hall R, Wood B (1992) Conversion of the hydrolysate of corn cobs and hulls into ethanol by recombinant E. coli B containing integrated genes for ethanol production. Biotechnol Lett 14:857–864

Cocozza C, Parente A, Zaccone C, Mininni C, Santamaria P, Miano T (2011) Chemical, physical and spectroscopic characterization of Posidonia oceanica (L.) Del. residues and their possible recycle. Biomass Bioenergy 35(2):799–807

Demirbas A (2009) Biofuels from agricultural biomass. Energy Sources, Part A, 31:1573–1582, Taylor & Francis Group, LLC ISSN: 1556–7036 print/1556–7230 online https://doi.org/10.1080/15567030802094011

Dien BV, Vong VD (2006) Analysis of pesticide compound residues in some water sources in the province of Gia Lai and DakLak. Vietnam Food Administrator

Erhart E, Hartl W, Putz B (2005) Biowaste compost affects yield, nitrogen supply during the vegetation period and crop quality of agricultural crops Europe. J Agron 23:305–314

Faraco V, Hadar Y (2011) The potential of lignocellulosic ethanol production in the Mediterranean Basin renew. Sustain Energ Rev 15(1):252–266

Franca A, Gouvea B, Torres C, Oliveira L, Oliveira E (2008) Feasibility of ethanol production from coffee husks. J Biotechnol 136:269–275

Gontard N, Sonesson U, Birkved M, Majone M, Bolzonella D, Celli A, Coussy HA, Jang GW, Verniquet A, Broeze J, Schaer B, Batista AP, Sebok A (2018) A research challenge vision regarding management of agricultural waste in a circular bio-based economy. Crit Rev Environ Sci Technol 48(6):614–654

Hegazi HA (2019) Removal of heavy metals from wastewater using agricultural and industrial wastes as adsorbents. HRBC J 9(3):276–282

Hodaifa G, Moya López AJ, Paraskeva C (2019) Chemical management and treatment of agriculture and food industries wastes. J Chem 2019:1–3

Ifeanyi O, Odoemelam VM, Omede A, Iloeje MU (2012) Livestock waste and its impact on the environment. Sci J Rev 1(2):17–32

Jain N, Bhatia A, Pathak H (2014) Aerosol and air quality research. Taiwan Association for Aerosol Research. Emission of Air Pollutants from Crop Residue Burning in India 14:422–430. ISSN: 1680-8584 print/2071-1409 online https://doi.org/10.4209/aaqr.2013.01.0031

Katalinic V, Mozina SS, Generalic I, Skroza D, Ljubenkov I, Klancnik A (2013) Phenolic profile, antioxidant capacity, and antimicrobial activity of leaf extracts from six vitis vinifera L. Varieties. Int J Food Prop 16(1):45–60

Khan NA, Shaaban MG, Hassan MHA (2003) Removal of heavy metal using an inexpensive adsorbent. In: Proc. UM Research Seminar 2003 organized by Institute of Research Management and Consultancy (IPPP), University of Malaya, Kuala Lumpur

Kuan CY, Yuen KH, Bhat R, Liong MT (2011) Physicochemical characterization of alkali treated fractions from corncob and wheat straw and the production of nanofibres. Food Res Int 44 (9):2822–2829

Lal R (2005) World crop residues production and implication of Its use as a biofuel. Environ Int 31:575–584

Lim SF, Matu SU (2015) Utilization of agro-wastes to produce biofertilizer. Int J Energy Environ Eng 6(1):31–35

Lustosa F, Penido J, Patricia C, Carlos S, Leonidas M (2017) Co-pyrolysis of poultry litter and phosphate and magnesium generates alternative slow-release fertilizer suitable for tropical soils. ACS Sustain Chem Eng 5(10):9043–9052. https://doi.org/10.1021/acssuschemeng.7b01935

Mahawar N, Priya G, Sunita L, Sakshi Jain S (2015) Agro waste: a new eco- friendly energy resource. Int Res J Environ Sci 4(3):47–49

Mahimairaja S, Dooraisamy P, Lakshmanan A, Rajannan G, Udayasoorian C, Natarajan S (2008) Composting technology and organic waste utilization in Agriculture. A.E Publications, P.N. Pudur Coimbatore pp 45–51

Mathieu F, Timmons MB (1995) Techniques for modern aquaculture. In: Wang JK (ed) Techniques for modern. American Society of Agricultural Engineers, St. Joseph

Mohan D, Singh KP (2002) Single and multi-component adsorption of cadmium and zinc using activated carbon derived from bagasse – an agricultural waste. Water Res 36:2304–2318

Nagle M, Habasimbi K, Mahayothee B, Haewsungcharern M, Janjai S, Muller J (2011) Fruit processing residues as an alternative fuel for drying in Northern Thailand. Fuel 90(2):818–823
National Mission on Sustainable Agriculture (NMSA), Ministry of Agriculture, Cooperation and Farmer's Welfare, Annual Report (2015–2016)
Obi FO, Ugwuishiwu O, Nwakaire JN (2016) Agricultural waste concept, generation, utilization and management. Nigerian J Technol (Nijotech) 35(4):957–964
Othman A, Othaman M, Abdulrahim A, Bapar S (1992) Cocoa, pineapples, sugarcane waste for ethanol production. Planter 68:125–130
Pappu A, Saxena M, Asolekar SR (2007) Solid wastes generation in India and their recycling potential in building materials. Build Environ 42(6):2311–2320
Pimentel D, Hepperly P, Hanson J, Douds D, Seidel R (2005) Environmental, energetic, and economic comparisons of organic and conventional farming systems. Bio Sci 55:573–582
Qureshi N, Saha B C, Bruce Dien, Ronald E, Hector M, Cotta (2010) Production of butanol (a biofuel) from agricultural residues: part I – use of barley straw hydrolysate. Biomass Bioenergy 34(4): 559–565
Rao P, Rathod V (2018) Valorization of food and agricultural waste: a step towards greener future. Chem Rec. https://doi.org/10.1002/tcr.201800094
Reijnders L (2004) Conditions for the sustainability of biomass-based fuel use. Energy Policy 34:863–876
Sadh PK, Duhan S, Duhan JS (2018) Agro-industrial wastes and their utilization using solid state fermentation: a review. Bioresour Bioprocess 5(1):2–15
Shulman VL (2011) Trends in waste management. In: Waste a handbook for management. Elsevier, pp 3–10
Singh DP, Prabha R (2017) Bioconversion of agricultural wastes into high value biocompost: a route to livelihood generation for farmers. Adv Recycl Waste Manage 2:3. https://doi.org/10.4172/2475-7675.1000137
Singh Y, Sidhu HS (2014) Management of cereal crop residues for sustainable rice-wheat production system in the Indo-Gangetic plains of India. Proc Indian Nat Sci Acad 80:95–114
Tan WT, Ooi ST, Lee CK (1993) Removal of chromium (VI) from solution by coconut husk and palm. Environ Technol 14(3):277–282
Tandon S, Sai N (2019) Recycling of agriculture waste into efficient adsorbent. In: environmentalconcerns and sustainable development. Springer, Singapore, pp 365–379
Timbers GE, Downing CGE (1977) Agricultural biomass wastes: utilization routes. Can Agric Eng 19(2):84–87
Upadhaya K, Harshvardhan K (2017) Effective utilization of agricultural waste – review paper. Int J Eng Res Technol (IJERT) 6(9)
USDA (2012) Agricultural waste management field handbook. United States Department of Agriculture, Soil conservation Service. Accessed from <http://www.info.usda.gov/viewerFS.aspx?hid=21430> on 10/06/2016
Väisänen T, Haapala A, Lappalainen R, Tomppo L (2016) Utilization of agricultural and forest industry waste and residues in natural fiber-polymer composites: a review. Waste Manag 54:62–73
Wei J, Liang G, Alex J, Zhang T, Ma C (2020) Research progress of energy utilization of agricultural waste in China: bibliometric analysis by Citespace. Article. Sustainability 12:812. https://doi.org/10.3390/su12030812
Wilson L, Yang W, Blasiak W, John GR, Mhilu CF (2011) Thermal characterization of tropical biomass feedstocks. Energ Convers Manage 52(1):191–198

Biobased Packaging from Food Industry Waste

14

Anila Zahid and Renu Khedkar

Abstract

Greener packaging and reduction in waste is the current trend in the twenty-first century. The increasing environmental concerns compel the society to look for sustainable solutions to food as well as packaging waste. Significant amount of waste is produced by the food industry, resulting from different activities which prove to be a loss of valuable biomass and nutrients and also pose disposal and pollution problems. The development of biobased packaging materials from food industry waste will lower environmental impacts compared with conventional petroleum-derived plastics. Many efforts are being directed towards commercializing biobased packaging materials with improved properties and functionalities for use in films, coatings, trays and boxes. This chapter outlines the updated research on various biobased packaging materials developed using food industry waste, their properties, limitations, applications and future trends.

Keywords

Biobased packaging · Food waste · PLA · PHA · Sustainability

14.1 Introduction

The global packaging industry has witnessed a significant transformation over the last few decades and several emerging packaging trends are expected to have a huge impact on the global packaging industry growth in this century. Today's packaging plays a far more complex role than just making containers for products during the

A. Zahid · R. Khedkar (✉)
Amity Institute of Food Technology, Amity University Uttar Pradesh, Noida, UP, India
e-mail: rdkhedkar@amity.edu

M. Thakur et al. (eds.), *Sustainable Food Waste Management*,
https://doi.org/10.1007/978-981-15-8967-6_14

process of logistics, sales, and end use. It is now also one of the most essential elements in terms of product design, branding, marketing and user experience. As the global demand for packaging continues to grow and diversify, the packaging industry is likely to be more heavily influenced by consumer preferences, industry dynamics, environmental concerns and developments in technology and manufacturing equipment.

According to a recent packaging industry report from Technavio, the global packaging industry market size will grow close to USD 278.59 billion during 2019–2023, accelerating at a CAGR of nearly 6% during the forecast period. The growth in the global packaging market is mainly driven by the increasing demand from end-user sectors such as food and beverage, retail and healthcare industries.

The packaging is any type of material that is used to hold, protect, handle, deliver and cause the presentation of goods from raw material to finished product, from farm to fork. Hence, it is usually divided on the basis of raw material which is used to produce a different type of packaging material i.e. metal, glass, polymer, paper cardboard, wood, textile, multilayered, ceramic and other types. In the twentieth century, advancement in plastic technology fuelled the growth of the packaging industry. Plastic offered benefits that other packaging materials could not. Advantages of plastic, a petroleum-based product, are: (i) economical and speedy production, (ii) good mechanical, barrier and sealing properties, (iii) appealing and convenient for the consumers. But, in recent years, huge packaging waste of over 67 million tonnes is causing environmental concerns (Song et al. 2009). Today, at the beginning of twenty-first century, great attention is given to products made from renewable sources due to their environment friendly nature. Generally, the consumers are more aware worldwide to the usage of traditional plastic products, although very demanding and useful but are not eco-friendly, therefore causes damage to the environment, hydro resources, air resources and the overall ecosystem. The principle reason behind this is the plastic material is not degradable, create problem of waste disposal and get accumulated in the environment, causes reduction of the fertile and productive land releasing poisonous gases during incineration, therefore enhances the attempt to develop biodegradable packaging (Mohatny et al. 2005). The reasons for shifting focus towards biodegradable packaging materials are: (a) decline in natural oil and gas resources; (b) increase in oil and gas prices during recent decades; (c) environmental concerns for the disposal of plastic and global warming; (d) higher cost and cross contaminations in their recycling; and (e) consumer toxicity risks about the migration of plastic monomers or oligomers to edible materials (Jamshidian et al. 2010).

Biopolymer addresses the environmental concerns as they are negligible for it, as the biodegradation process takes place in nature. Biodegradation is the degradation of a polymer in natural environments and includes changes in chemical structure, loss of mechanical and structural properties, and finally, changing into other compounds like water, carbon dioxide, minerals and intermediate products like biomass and humid materials. The rate and products of biodegradation are dependent on factors like temperature, humidity, pH, O_2, etc. (Zee 2005).

Globally, the processed food industry is valued at more than 2 trillion dollars and consists of more than 400,000 businesses (food processing: global trends). It is expected to reach 4.1 trillion dollars by the year 2024 with CAGR of 4.3% between 2019 and 2024 (Global Food Processing Market Report 2019). There is increasing pressure on the food processing industry to address environmental concerns through their company's activities. Large amount of solid and liquid waste is produced by the food industry, resulting from the manufacturing, preparation and the consumption of the food. Along with significant loss of valuable biomass and nutrients, these wastes also pose disposal and pollution problems. Waste generating from the different activities along the food chain are:

- Production and harvest: Discarded fruit or vegetable for not meeting the quality standards, left grains during harvesting, discarded fish or meat.
- Handling and storage: degraded food due to pest, microbial contamination or disease.
- Processing and packaging: Peels, trimmings, scales, fruits or vegetables unsuitable for processing and any other raw material not meeting quality standards, unintentional losses during processing or wastage due to poor order planning.
- Distribution: wastage during distribution logistics, product exceeding the 'sell by or use by' date.
- Consumption: Unintentionally not consuming the product or rejected by the consumer due to suboptimal quality, bad storage conditions, over-preparation, portioning and cooking, etc.

These different types of waste generated pose serious health and safety concerns. The waste management techniques to tackle this problem include reduction at source by plant modification, recovery, recycle or treatment of waste for value-added products and neutralization of undesirable components of the waste (Khedkar and Singh 2014, 2018).

The waste from the food processing industry contains many high-value reusable substances such as soluble sugars and fibre. Currently, the focus of finding a sustainable solution for tackling the problem of food waste lies on the valorization of these wastes and by-products, which is essential from the economic and environmental point of view. Options to produce biofuels and bio-products from these wastes offers benefits from an environmental point of view due to reduction in methane gas production from landfills and preserving the natural fossil resources and economical point of view due to cost savings as a result of surplus food production and investments in non-food crops specifically dedicated to biofuel or bioplastic production (Girotto et al. 2015).

Food wastes or by-products having high starch and protein content when treated with acids or enzymes produce simple sugars, such as maltose, glucose and amino acids (Berdanier 1998). Valuable organic acids, vitamins, antioxidants, polyphenols and colouring agents can also be extracted from these wastes.

Food industry is the largest sector which is in great demand to develop biodegradable plastic. The industries itself undergo rapid development which leads to problems with non-degradable packaging but it takes time and patience to familiarize them with biodegradable plastic (bioplastic) (Platt 2006). In addition to find the substitute for non-degradable plastic favours the development of cardboard packaging material made from renewable sources (Kolybaba et al. 2003; Narayan 2006).

14.2 Biobased Packaging Materials

In order to overcome the growing problem of waste disposal and environmental issues, leads to the development of biobased packaging materials. Biobased food packaging materials are the 'materials derived from renewable sources' (Van Tuil et al. 2000). These are the packaging materials which involves the utilization of biopolymer produced from renewable resources. It is to be noted that 'biobased' refers to the feedstock used whereas 'biodegradation' refers to how a material degrades during disposal. Biopolymers are manufactured either from plant raw material or animal raw material. Their important property is biodegradability.

Biopolymers are different from natural polymers on behalf of their synthesis which is induced intentionally. Conventional/Traditional polymers are not eco-friendly as they are non-biodegradable because of long chains of molecules that are too big and connected to each other to make them able to separate the microorganisms to break. Biopolymers are produced naturally from wheat, whey, potato or corn starch, etc. that are capable of degrading by microorganisms. For 1 kg of bioplastic, 1–2 kg of maize and 5–10 kg of potato starch are required. Therefore 500,000 tons of bioplastic are produced per year by the utilization of 50,000–100,000 ha of soil. Additionally, it causes the demolition of large areas for the production of natural biodegradable material (Goodship and Ogar 2004).

They are classified on the basis of chemical structure, origin, method of production, cost-effectiveness, and their application, etc. (Davidovic and Savic 2010).

Traditionally, the classification of biobased packaging materials has been done based on their developments through the years and have been divided into three generations (Fig. 14.1).

14.2.1 First Generation

The first generation biobased packaging materials included shopping bags made of synthetic polymers such as LDPE (Low-density polyethylene) mixed with 5–15% starch fillers and additives promoting oxidation. Since these materials did not biodegrade, but fragmented into smaller molecules, gave a poor image to biodegradable packaging.

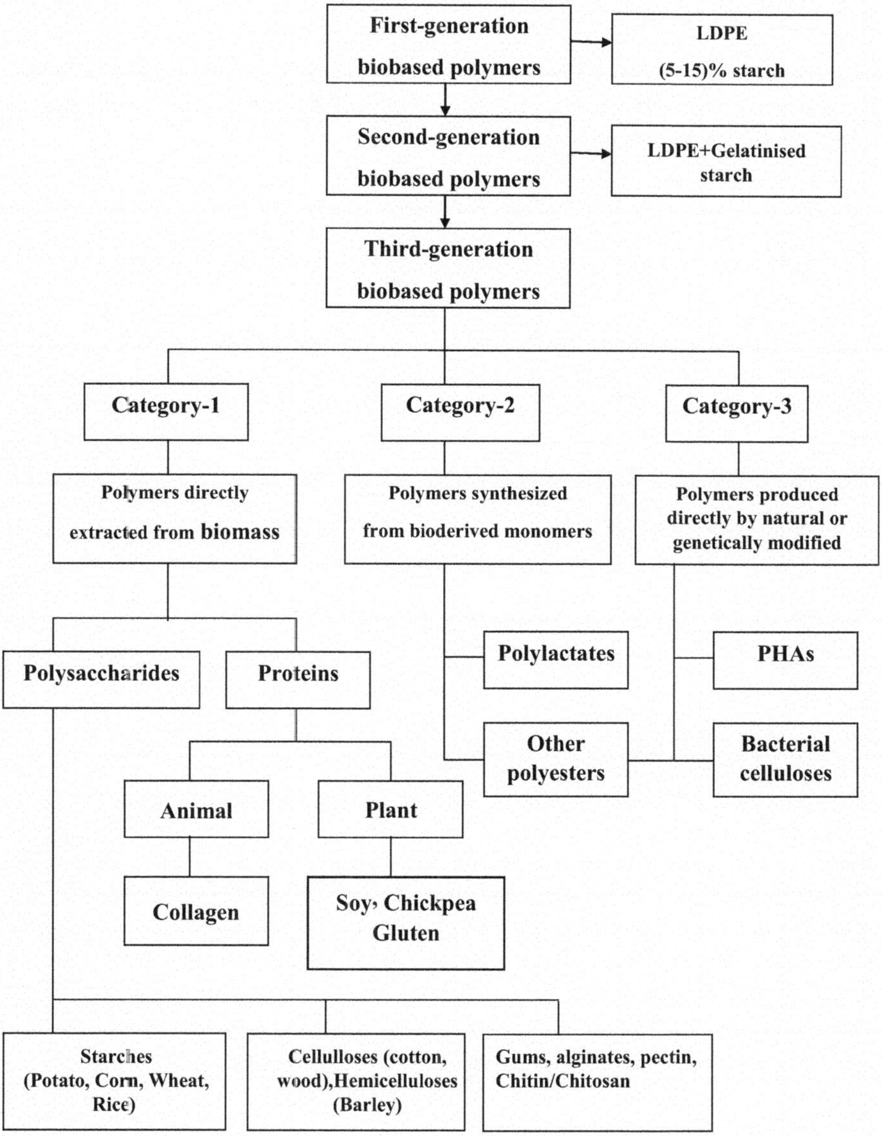

Fig. 14.1 Classification of biobased packaging materials (modified from Robertson 2008)

14.2.2 Second Generation

The second generation biobased packaging materials included films made of mixture of gelatinised starch (40–75%) and LDPE with ethylene acrylic acid, polyvinyl alcohol (PVOH) and vinyl acetate. Complete biodegradation of the entire film took a minimum 2–3 years.

14.2.3 Third Generation

The third generation biobased packaging materials consist of completely biodegradable materials. They are further classified according to their origin and method of production (Robertson 2008)

- Polymers are directly extracted from biomass.
- Polymers are produced by chemical synthesis from bio-monomers.
- Polymers obtained from natural or genetically modified organisms.

14.2.3.1 Polymers Extracted/Isolated Directly from Biomass

This category includes biopolymers obtained from plants, marine and domestic animals. Examples are polysaccharides such as cellulose and its derivatives, pectin, carrageenan, chitin and starch, whey protein casein, collagen, soy protein, myofibrillar proteins of animal muscle, etc., can be used alone or as a mixture with synthetic polyesters such as polylactic acid (PLA). Cellulose-based paper and hemicellulose are the most widely used plant polymers in food packaging (Grondahl et al. 2006). The advantages and limitations of these polymers are shown in (Table 14.1).

Some of the commonly used biopolymer sources are:

14.2.3.1.1 Starch

Green plants such as potato, corn, maize, wheat, rice, etc. are rich in starch which is utilized for the synthesis of biopolymer. The component is potentially obtained from renewable resources and its advantages in food packaging industries are due to its renewability, low cost, good oxygen barrier in the dry state, abundant, biodegradability, etc. Starch is a natural polymer consisting two units of glucose, namely amylose and amylopectin. The development of starch-based biopolymer has witnessed the consumption of different forms of starch such as native starch, modified starch, plasticized starch, thermoplastic starch(TPS) and is mixed with many synthetic polymers whether they are biodegradable or non-biodegradable. In

Table 14.1 Advantages and limitations of biopolymers extracted from biomass

Category	Name	Advantages	Limitations
Polysaccharides	Starch, celluloses, hemicelluloses, alginates, pectin, chitin, chitosan	Low cost, easily and widely available, non-toxic, thermoprocessable	Poor barrier properties against moisture
Proteins	Casein, gluten, zein, collagen, gelatin, soy proteins	Good gas barrier properties, easily available	Poor water vapour barrier properties, brittle
Lipids	Beeswax, Carnauba wax, Candelilla wax, rice bran wax, edible oils, paraffin oils, triglycerides	Excellent barrier against moisture	Waxy taste and texture, greasy surface, risk of rancidity

order to prepare biopolymer, the crystal structure of grains is distorted by the application of heat, pressure, mechanical work and plasticizers such as water, glycol and other polyols to synthesize thermoplastic starch (Bastioli 2005). TPS is obtained by the combination of starch, plasticizer and thermomechanical energy. TPS achieves equilibrium properties after some days due to its low mechanical properties and high sensitivity to water vapour (Averous and Baqquillon 2004). Many studies have been conducted to improve the properties of starch-based bioplastics. Various plasticizers have been added to reduce the problem of the brittleness of the packaging material. Zeng et al. (2005) noted that clay particles, e.g. saponite, hectorite when used as nanocomposites significantly improve the mechanical performance and moisture resistance. Starch is also used in combination with Polyvinyl chloride (PVC) for the synthesis of the starch film (Adeodato et al. 2011).

Partal et al. (2010) studied the development of albumen/starch (corn and potato) based bioplastics resulting in transparent packaging comparable to oil-derived plastics.

14.2.3.1.2 Chitin and Chitosan

Chitin is the structural material of crustaceans, insects and fungi whereas chitosan is obtained by de-acetylation of chitin (Clarinval and Halleux 2005). It is potentially versatile biopolymer, differs from cellulose only by the OH group. Due to the presence of amino acids, it is consumed in industrial applications. Biodegradable films made of chitin and chitosan are used as a coating for fresh fruits and vegetables (Zhao and Mc Daniel 2005). Composite films are also prepared from the combination of chitin, chitosan and starch for prolonging the shelf life of foods (Chiellini 2008).

14.2.3.1.3 Soy Proteins

Soy proteins are isolated for the production of biodegradable films, in combination with stearic acid to improve its tensile and thermal properties and reduced moisture sensitivity (Lodha and Nteravali 2005). Soy protein concentrate, isolate or flakes could be compounded with synthetic biodegradable plastics such as polycaprolactone or polylactic acid for the synthesis of moulded products and edible films of packaging to make the environment cleaner and greener. In recent researches, it is used together with other polysaccharides such as carrageenan, glycerol, gellan gum, etc. for the synthesis of soybean-based packaging trays (Mohareb and Mittal 2007). Soy proteins have different functional properties such as cohesiveness, adhesiveness, Dough emulsification, water and Fat absorption. Soy protein isolates (SPI) are used for the synthesis of soy protein films that contain at least 90% protein in it. The methods that are involved in the synthesis of soy protein films are heating, extruding, spinning, casting and compaction. SPI together with polyethylene oxide (PEO), low-density polyethylene (LDPE) are used to produce protein-based films. The combination of glycerol and SPI at 10 MPa and compaction at 150–160 °C are consumed to manufacture protein-based material.

14.2.3.1.4 Gluten

Gluten, a by-product of the wheat starch industry, is soluble in aqueous alcohols. It comprises two groups, i.e. gliadin (soluble) and glutenin (insoluble). On the basis of cohesiveness and elasticity, gluten is used in the formation of packaging film. Basically, two methods are involved in the generation of gluten films: One is casting into a thin layer then drying and the other is boiling of protein solutions, after that collecting the protein films on the surface of the solution by thermos pressing. The structure of films is affected by the process involved in its synthesis and is explained by stress–strain relationship. Water ethanol is the most commonly used solvent for the formation of films. The uniformity of films is regulated by the acidic or basic conditions. The packaging materials produced from basic solutions have more tensile strength as compared to films produced from acidic conditions. The process like mechanical mixing and heating enhances the properties of films. The substances like non-polar hydrophobic such as mineral oil are added into the film-forming solutions in order to reduce its water vapour permeability by 25%. The gluten film surfaces are shiny and act as an active layer in edible films. The substances like Thyme essential oil (TO) are to be added to improve its in vitro antioxidant and antimicrobial properties.

14.2.3.1.5 Zein

Zein is a corn protein that is obtained as a by-product during corn processing and is present at about 45–50% of corn proteins. It contains non-polar amino acids due to which it is soluble in alcohol helps to improve water vapour barrier properties. Wet or solvent methods are employed for the formation of protein (zein) based packaging material. Firstly, a solvent is chosen which can dissolve zein in it. Secondly, the solution is cast on the flat and non-stick surface. At last, the film-forming solution is evaporated and the film is peeled off from the surface. The solutions that can be used for the formation of zein based films include water, methanol, ethanol and acetone. In order to cure the zein films, several physical treatments such as UV irradiation and ɣ- radiations are employed.

14.2.3.1.6 Gelatin

Gelatin is produced from collagen by partial degradation and their differences in the degree of hydrolysis leads to a range of different molecular weights of gelatine, from 65,000 and 300,000 g/mol. Proline, 4-hydroxyproline and glycine are the main compounds in gelatin. It has all good film-forming properties like low cost, biocompatibility and biodegradability. The casting of gelatin films is carried out by aqueous solutions. On the basis of synthesis temperature, it is classified into cold-cast and hot-cast films. The cold-cast films are prepared at approximately room temperature whereas hot-cast films are prepared at a temperature above 35 °C.

14.2.3.1.7 Casein

Casein is a milk protein and is comprised of four subunits: Kappa casein, beta-casein, alpha casein and alpha s2-casein in the composition of 13%, 36%, 38% and 10%, respectively. These four fractions build characteristics of the film. Casein molecules are arranged in a number of strong intermolecular hydrogen, hydrophobic and electrostatic bonds. Casein films have certain film-forming properties such as biodegradability, high thermal stability, non-toxicity and also capable of forming micelle to bind small molecules and ions. Casein is used for the manufacturing of edible films due to high nutritional value, water solubility, emulsification capability. Casein is insoluble due to the presence of sodium caseinate which is useful to prepare packaging material apart from traditional packaging. Edible plasticizers like sorbitol or glycerol can be added to casein films in order to prevent its structure from shrinking during the drying process (Chen et al. 2019).

14.2.3.2 Whey Proteins

Packaging films made from whey, a by-product of cheese manufacturing, has shown better mechanical and barrier properties compared with other protein-based films such as zein, gluten and SPI or polysaccharide-based films, e.g. starch, cellulose, pectin, carrageenan. But, to improve the resistance to moisture barrier and also to reduce the brittleness, plasticizers are required to be added. The formation of whey-based films involves heat denaturation of the whey proteins in aqueous solution. Heating modifies the three-dimensional structure of the protein, promoting the intermolecular S-S bonding and hydrophobic reactions. The final step in the formation of whey-based protein films is dehydration of the heat or cold-set gel. The films can be used to produce pouches, bags, casings, wraps, capsules, etc. (Ramos et al. 2012).

14.2.3.3 Lipids

Lipid molecules comprise glycerides to impart hydrophobicity, therefore, added to film to act as a water vapour barrier. The functional property of film depends on the structure, length of chain, degree of saturation, state as well as dimension of crystal. Beeswax, Carnauba wax, oils, free fatty acids are components of lipid-based films. Lipids play a key role to form a moisture barrier in order to prevent microbial and physicochemical deterioration. It has two components: one is polar and the other is non-polar (waxes). Waxes are the most effective moisture barrier and are completely insoluble in water. Lipids are classified into different classes. Class I of polar lipids have triglycerides, hence forms a stable monolayer as they are insoluble in water. Class II or III have monoglycerides depending on its chain length. Monoglycerides act as an emulsifier which stabilizes the edible lipid-based film. The classification of different lipids used in films is shown in Table 14.2.

14.2.3.4 Polymers Produced by Chemical Synthesis from Bio-Monomers

Polymers produced by classical chemical synthesis from bio-monomers such as aliphatic polyesters, aliphatic-aromatic copolymers, polylactide aliphatic copolymer (CPLA), Polycaprolactone (PCL).

14.2.3.4.1 Polylactic Acid (PLA)

It is the most famous biopolymer of this group. The properties of PLA are similar to the thermoplastic polystyrene. It is colourless, glossy, brittle, and has a tensile

Table 14.2 Classification of biologically active lipids

Class	Surface properties	Bulk properties	Examples
Non-polar	Not spread to form monolayer	Insoluble	Paraffin oil, waxes
Polar			
Class I (insoluble, non-swelling, amphiphiles)	Spread to form stable monolayer	Low solubility	Triglycerides, cholesterol
Class II (insoluble, non-swelling, amphiphiles)	Spread to form stable monolayer	Insoluble but swell in water	Phospholipids, monoglycerides
Class IIIA (soluble amphiphiles with lyotropic mesomorphism)	Spread but form unstable monolayer	Soluble, form micelles above a critical micellar concentration at low water concentrations	Detergents, Lysolecithin, fatty acids, gangliosides
Class IIIB (soluble amphiphiles, no lyotropic mesomorphism)	Spread but form unstable monolayer due to solubility in aqueous substrate	Form micelles but not liquid crystals	Bile salts, rosin soaps, saponins

Source: Aydin et al. (2017)

strength of 50–62 MPa. It has a melting temperature of 175 °C and glass transition temperature in the range of 55–60 °C. It has Young's modulus of 384–481 MPa (Chen and Patel 2012; Garlotta 2001).

It has two principal monomers i.e. lactic acid and lactide. The most common way to synthesize PLA is the polymerization of lactide involving metal as catalysts (especially tin (II)-ethyl hexanoate) in solution or as suspension. PLA is also prepared by direct condensation of lactic acid monomer at a temperature lower than 200 °C.

PLA is prepared with a different range of properties because lactic acid exists in four different molecular forms (Mehta et al. 2005) and is produced from polymer from different molecular weight ranging over thousands to millions. Lactic acid (2-hydroxy propionic acid), the single monomer of PLA, is produced via fermentation or chemical synthesis. The optically active L(+) and D(−) stereoisomers are produced by bacterial (homofermentative and heterofermentative) fermentation of carbohydrates. The lactic fermentation process is preferred rather than synthesis because of the limitations of the latter in the production of desirable L-lactic acid stereoisomer, and high manufacturing costs among others (Datta and Henry 2006).

The homofermentative method is preferably used for industrial production because its pathways lead to greater yields of lactic acid and to lower levels of by-products. The general process consists of using species of the *Lactobacillus* genus such as *Lactobacillus delbrueckii, L. amylophilus, L. bulgaricus, and L. leichmannii*, a pH range of 5.4–6.4, a temperature range of 38–42 °C, and a low oxygen concentration. Generally, pure L-lactic acid is used for PLA production (Mehta et al. 2005).

The raw material for obtaining the lactic acid is obtained by fermentation of glucose or starch from other sources. The source of carbohydrate corn, wheat or alternatively whey and molasses (Wackett 2008). Disposal of whey, a by-product of milk processing and cheese production processes is a major pollution problem for dairy industries. It is a suitable raw material for lactic acid production (Panesar et al. 2007). Many studies have been conducted to find other sources of carbohydrates for lactic acid production. Some agricultural by-products, which are potential substrates for lactic acid production include, cassava starch, lignocellulose/hemicellulose hydrolysates, cottonseed hulls, Jerusalem artichokes, corn cobs, corn stalks, beet molasses, wheat bran, rye flour, sweet sorghum, sugarcane press mud, cassava, barley starch, cellulose, carrot processing waste, molasses spent wash, corn fiber hydrolysates and potato starch (Reddy et al. 2008). Other sources of carbohydrate for lactic acid production include kitchen wastes (Kim et al. 2003; Zhang et al. 2008) and fish meal wastes (Huang et al. 2008).

PLA is usually processed into thermoformed pads, containers and bags for the packaging of fresh foods as it has excellent water vapour permeability and moisture retention capability which is the most appropriate feature of packaging material. Unlike polypropylene bags, PLA bags are in high demand as it maintains optimal gas composition i.e. 10–20% $C0_2$ and 5–10% O_2 for the storage of raspberries (Seglia et al. 2009).

PLA can be easily moulded into film, fiber and spun bond. It has a practical application as absorbable sutures for drug delivery and bone fracture internal fixation devices. Another application includes as a growth promoter in plant at agricultural sector, in textiles, and non-woven applications such as fibrefill, crop covers, geotextiles, wipes, diapers and binder fibers. Conventional plastic material like PE, PVS, PE,PET, etc. are replaced by PLA in packaging as candy wraps, films and shrink labels. In order to process PLA for mass production lines for thermoforming, and extrusion to possess thermal stability of polymer.

A biodegradable packaging made of PLA plasticized with lactic acid, D-Lactide, L-Lactide and other derivatives of oligomers of Lactic acid has been a substitute of thermoplastic polymers such as polyethylene (Sinclair, 1993). The slow biodegradation, majorly dependent on hydrolysis is the main limitation of PLA (Stloukal et al. 2015).

Several studies have been conducted on the use of nanocomposites for improving the functionality of biobased packaging materials. Chow and Lok (2009) observed an increase in thermal stability of PLA/OMMA (Organo montmorillonite) nanocomposites on the addition of maleic anhydride grafted ethylene-propylene rubber (EPMgMA). Nakayama and Hayashi (2007) studied the effect of TiO_2 nanoparticles on the degradability of PLA-TiO_2 nanocomposites and concluded that it aids in promoting photodegradability of the nanocomposite material. Blending of 4% kaolinite nanofillers with amorphous PLA improved the oxygen barrier property by 43% (Cabedo et al. 2006).

14.2.3.4.2 Polylactide Aliphatic Copolymer (CPLA)

As the name suggests, this material is a mixture of lactide and aliphatic polyesters like dicarboxylic acid or glycol, with the addition of both the hard and soft flexible properties. Due to this, it is easy to process and is thermally stable at 200 °C. The rate and amount of carbon dioxide generated during the combustion of this polymer are approximately half as compared to other commercial polymers. Naturally, it starts degrading in 5–6 months following by a complete decomposition after 12 months. If the polymer is composted with food garbage, it begins to decompose after 2 weeks.

14.2.3.4.3 Polyethylene Furanoate (PEF)

Polyethylene Furanoate (PEF) is a 100% biobased polymer synthesized from 2,5-furandicarboxylic acid (FDCA) and mono-ethylene glycol (MEG) and has a potential to replace PET. FDCA is a monomer derived using bio-fermentation of renewable sources of sugars from corn, wheat or waste, wood and bagasse (Eerhart et al. 2012). The barrier, mechanical and thermal properties of PEF are superior to PET. It has a high glass transition temperature of 86 °C and lower melting temperature of 235 °C (Poulopoulou et al. 2019).

14.2.3.5 Polymers Obtained from Natural or Genetically Modified Organisms

These polymers are synthesized by microbes and are biodegradable and can be used for packaging purposes (Chiellini 2008). Under imbalanced growth conditions, some bacteria like *Bacillus Azotobacter, Clostridium, Thiothrix,* etc. Shift from their original physiological pathways and synthesize different carbon reserve compounds such as PHA (Poly-b-Hydroxyalkonates). Majorly, the bacteria synthesizing PHAs can be broadly classified into two types. One group produces SCL-PHA (short-chain- length) having monomers ranging from 3 to 6 carbons, while another group synthesizes MCL-PHA (medium-chain-length) with 6 to 16 carbon atom monomers.

Among the group of PHAs, PHB (Poly-b-Hydroxy Butyrate) can become a very good substitute for synthetic polymers. Depending on the bacteria and the carbon source, the polyhydroxyalkanoate (PHA) may be manufactured from rigid brittle to plastic to rubber-like polymer and can have similar properties such as propylene and polyethylene, elastic and the thermoplastic (Zivkovic 2009). Mechanical properties such as excellent strength and toughness, barrier properties of nearly complete resistance against moisture and low oxygen permeability of PHA make it an excellent option for the production of bottles and water-resistant films.

The main limitation of using bacterial PHAs for biopolymers is the production cost, which is two and a quarter times their capital equipment cost (Van-Wegen et al. 1998).

14.2.3.5.1 PHA from Whey

In the last two decades, a broad number of studies were related to the production of biodegradable plastics from milk whey. For the European dairy industry, it is estimated annual production of 75 million tons of whey from cheesemakers, and about 40% of this amount is discarded and managed as waste. From an economical and environmental point of view, whey disposal represents a serious problem, whereas its use in fermentation processes (e.g. bioplastic production) may be advantageous not only for environment but also for economy (Jambunathan and Zhang 2016; Lenczak et al. 2013). Biotechnological production of polyhydroxyalcanoates (PHA) from lactose was described in different works (Marangoni et al. 2002; Nikel et al. 2005; Lenczak et al. 2013); because of few wild type microorganisms were able to directly convert lactose to PHA (Koller et al. 2007), different strains of *E. Coli* were used on whey. The use of recombinant strain was reported in many works: Kim (2000) described PHA production with *E. coli* GCSC6576 in fed-batch oxygen-limited culture; Ahn et al. (2001) reported polyhydroxybutyrate (PHB) production with *E. Coli* GCSC 4401 in highly concentrated whey solution and a pH-stationary fed-batch solution; Nikel et al. (2005) applied a statistical optimization on cultural medium containing whey powder and corn steep liquor in presence of different *E. Coli* strains. In all the aforesaid examples of PHA production on whey, the use of a pure culture, recombinant or not, required aseptic conditions with additional operations and costs. The opposite approach to PHA production was the use of mixed cultures, e.g. activated sludge derived from food processing wastewater treatment plant (Suresh et al. 2004). In the work of Khardenavis et al. (2007), wastewater from the food processing industry (potato cheeps, wafers and sweets) was used as a substrate for PHB production and the value obtained after 48 h of incubation in filtered wastewater was 39.1% (on a dry weight basis).

14.2.3.5.2 PHA from Starch

Starch is a glucose polymer produced from renewable resources such as potato, corn, maize, wheat and cassava. Corn starch comprises 12% of starch production as industrial waste. Maize is used as a bulk source of starch production and act as major contributors of raw material in an industrial scale (Cerquiglini et al. 2016). Although starch is readily consumed by humans, hence is consumed by PHA producing microorganisms. Cassava starch hydrolysates are used to synthesize PHA followed by the isolation of *Cupriavidus sp.* (Poomipuk et al. 2014). This species produces a high biomass concentration of 5.97 g/l with PHA content of 61.6%.

14.2.3.5.3 PHA from Waste Oil

The oil from food waste is obtained by both the food industry as well as households, rich carbon source and capable of producing PHAs. The easiest method to use these oils is to incorporate them directly into media as carbon substrate. The functionally active microorganisms are *Cupriavidus necator* which converts the waste oil and produce PHA (Taniguchi et al. 2003). The commonly used oils are palm oil and lard utilized as carbon substrate attain a dry cell weight of 6.8 g/l and PHA content is about 83%.

14.2.3.5.4 PHA from Lignocellulosic Waste

Lignocellulosic materials are renewable sources developed from plant material comprises of cellulose, hemicelluloses, pectin and lignin. These types of waste are obtained from the food industry including wheat straw, rice straw, bagasse and bran. Lignocellulosic materials can hydrolysed and converted to fermentable sugars, then undergo detoxification to remove inhibitory compounds (Obruca et al. 2015). These materials can be used without pre-treatment but produce low level of cell growth. Therefore, an ammonia fibre expansion (APEX) phenomenon is used as pre-treatment followed by enzymatic hydrolysis of wheat straw to produce glucose, xylose and arabinose. The hydrolysate undergoes Fed-batch fermentation by the role of *Burkholderia sacchari* achieving biomass concentration of 146 g/l a PHA content is 72% (Cesario et al. 2014).

14.2.3.5.5 PHA from Recombinant Microbes

Food-based carbon substrates are consumed for the production of PHAs by the action of microbial strains. The use of recombinant microbes, however, could be advantageous as it is cost-effective in the sense that it does not require any stressed condition either nitrogen or phosphorus starvation, therefore potentially active in a given environment for the production of PHA. Native PHA producer cannot utilize certain substrates like bacteria, transformed to PHA producing genes to synthesize PHA from waste oils, whey and starch as a carbon source. The standard organism to produce PHA is *Escherichia coli.* Many PHA producing organisms are unable to metabolize, hence, high lactose-containing whey uses *E. coli* with PHA producing gene (the pha gene operon). Conventional laboratory *E. coli* strains like XL1-Blue, JM OR DH5α, etc. are incapable of utilizing lactose as a nutrient source, therefore, wild type *E. coli* strains were developed to produce PHAs. Documents revealed that the microbial strains GCSC4401 and GCSC6576 transformed using self-replicating plasmid pSYl107 containing *A. eutrophus* for the synthesis of PHA (Nielson et al. 2017).

14.3 Properties of Bio-Based Packaging Materials

The food industry is one of the greatest packaging disposal producers and causes the development of biodegradable packages for foodstuffs in order to sort out environmental issues. Due to their physical and mechanical properties, the utilization of these materials is in high demand.

14.3.1 Barrier Properties

Barrier property refers to the permeability of water vapour, oxygen and other gases. Conventional/traditional packaging material has poor barrier properties. Some most commonly used biomaterials like cellophane, cellulose paper, cellulose films, etc. also have low humidity resistance, therefore, used in combination with synthetic polymers with the aim to attain the appropriate barrier capability for the packaging of foodstuff. Biomaterials prepared from starch are 4–6 times higher moisture vapour transmission rate as compared to the conventional polymer. The basic factor on which the barrier property rely on is ambient humidity. The materials made up of arabinoxylan have low permeability for O_2 and CO_2 but highly permeable for water vapour. Some of the biopolymer derives from chemical synthesis like PLA have 3–5 times more moisture transmission rate as compared to PET (Polyethylene terephthalate), LDPE (Low-density polyethylene), HDPE (High-density polyethylene) and OPS (Oriented polystyrene) (Table 14.3).

14.3.2 Mechanical Properties

The mechanical property refers to the stiffness, strength, elongation at break, etc. which get affected by the molecular weight and crystallinity, both vary with degradation. The properties of any biomaterial are determined by the molecular weight of open chain structure (linear or branched) and degree of crystallization. The mechanical properties of most of the biobased packaging materials compete well with the conventional synthetic polymers (Table 14.4). PLA has improved its mechanical strength and heat stability. PLA has a melting temperature in the range of 130–180 °C. The physical properties of PHA depend on the molecular structure and the composition. It is generally rigid and brittle, hence, increases its mechanical properties, due to which used in the various food packaging industry. PHA has a melting temperature of 50–180 °C.

14.4 Biodegradable Packaging Forms

Various different forms of biopolymers packaging are used in the food packaging industry to package different food products for their storage and presentation to the consumer. These are available in the form of biodegradable gels, films, bags, containers, boxes with lids and trays.

14.4.1 Biodegradable Gels

The biodegradable gels are prepared from natural or renewable resources. The gels are commonly used to prevent the foodstuff from microbial contamination such as hydrogels (Farrisa et al. 2009). Gels do not show a positive effect on each and every product, hence, depends on the material from which it is synthesized to protect which components such as pectin, glucosinolate, gelatin, etc. *Solanum muricatum* is a fruit used in the preparation of gel for the protection of beta-carotene (Schreiner et al. 2003). Starch-based radish gel coat proved to be effective to protect pectin content while the same gel had no effect on Glucosinolate content. The incorporation

Table 14.3 Comparison of barrier properties of conventional and biobased packaging materials

Polymer	Oxygen transmission Rate[a]	Water Vapour transmission Rate[b]	Temp. (°C)	Thickness (mm)	Source
OPLA	56.33	15.30	22	4.6	Auras et al. (2005)
PLA	200	66[c]	23	0.1	Plackett et al. (2006)
PLLA-M[d]		210[e]	25	0.25	Tsuji et al. (2006)
P (LLA-DLA) (50:50)		214[e]	25	0.25	Tsuji et al. (2006)
PLLA[d]		31[e]	27	1	Uemura et al. (2006)
PLLA+SiO_x		18[e]	27	1	Uemura et al. (2006)
PLA/NC[f]	84–99	34–40[c]	23	0.1	Plackett et al. (2006)
PLA-PCL[g]	105	118[c]	23	0.1	Plackett et al. (2006)
PLA(IM)[h]	233	118	23	0.1	Plackett et al. (2006)
PLA-PCL/ NC[f]	183	146[c]	23	0.1	Plackett et al. (2006)
PHB		1.16	30	1	Miguel and Iruin (1999)
PHBV (14% HV)		1.39	30	1	Miguel and Iruin (1999)
PET	9.44	3.48	22	4.6	Auras et al. (2005)
OPS	532	5.18	22	4.6	Auras et al. (2005)
LDPE/ APET	33	6[c]		0.1	Plackett et al. (2006)
LDPE		7.9	38[e]	0.75	Jagannath et al. (2006)
LDPE+ 5% starch		36.85	38[e]	0.75	Jagannath et al. (2006)

Source: Robertson (2008)

OPLA oriented PLA, PLLA-M poly(l-lactic acid) middle molecular weight, P(LLADLA) poly (l-lactic acid) and poly(d-lactic acid), NC nanoclay, PLA (IM) impact modified PLA, PCL poly (caprolactone), PHB poly(hydroxybutyrate), PHV poly(hydroxyvalerate), PHBV poly (hydroxybutyrate-*co*-hydroxyvalerate), PET poly(ethylene terephthalate), OPS orientedpolystyrene, LDPE/APET low-density polyethylene/amorphous poly(ethylene terephthalate)

[a]Units: mL m^{-2} day^{-1} at 0% relative humidity (RH)

[b]Units: g m^{-2} day−1 at 100% RH

[c]37.8 °C

[d]Poly(l-lactic acid)
[e]90% RH
[f]5% nanoclay
[g]3–4% caprolactone
[h]Impact modified

of white ginseng together with biodegradable films proved to be functionally active in maintaining the composition of antioxidants (Rico et al. 2007).

14.4.2 Biodegradable Films

The biodegradable films are basically designed to replace plastic films for the utilization of organic industrial waste. Such biomaterials are eco-friendly as they are degradable. They are more flexible and resistant to moisture for several weeks or even months. They are potentially active for the replacement of conventional films used in storage, transport, packaging of the product and are completely biodegradable. The permeability of biodegradable films is probably decreased as compared to polyphenol foils. A comparative study has been carried out on two different types of films applied on fresh cuts of pineapple for the determination of microbial quality during storage at 10°. One film is a conventional plastic film and the other one is methylcellulose film that contains vanilla as an antimicrobial agent. Fresh-cut pineapple without any foil is used as a control. The results show that the cuts wrapped in methylcellulose films show an inhibitory effect against *E. coli* and *yeast* and the intensity of the yellow colour is also enhanced due to vanillin and a decrease of ascorbic acid by 90% is observed. The plastic film has a larger amount of ethanol (Sungsuwan et al. 2008).

14.4.3 Biodegradable Bags

The composition of raw materials builds up the properties of packaging material. Hence, Biodegradable bags are strong, flexible, resistant to breakage and damage and resistant to temperature change. These bags are totally eco-friendly and safe for the storage of foodstuff as they are decomposable to CO_2 for the storage of several weeks. The usage of biodegradable bags extends to many food industries. These bags are potentially different from conventionally used plastic or polyethylene (Nampoothri et al. 2010).

14.4.4 Biodegradable Boxes with Lids

Biodegradable boxes are considered as eco-wares which maintains the quality of foodstuff. Box with the lids is prepared from polystyrene synthesize from corn. Such materials are biodegradable, hence it does not release harmful substances or gases

Table 14.4 Comparison of the Mechanical properties of conventional and biobased packaging materials

Polymer	Melting temperature, T_m (°C)	Glass transition temperature, T_g (°C)	Young's modulus (Gpa)	Tensile strength (Mpa)	Elongation at break (%)	Source
Starch[a]			0.2–0.4	24–30	200–1000	Bastioli (2005)
Starch[b]			0.2–2.0	20–30	20–500	Bastioli (2005)
Starch	110–115	–	0.6–0.85	35–80	580–820	Clarinval and Halleux (2005)
PLA	130–180	40–70	3.5	48–53	30–240	Clarinval and Halleux (2005)
PHA	70–170	−30 to 10	0.7–1.8	18–24	3–25	Clarinval and Halleux (2005)
PHB	140–180	0	3.5	25–40	5–8	Clarinval and Halleux (2005)
PHBV	100–190	0–30	0.6–1	25–30	7–15	Clarinval and Halleux (2005)
PHB	180	4	3.5	43	5	Sudesh and Doi (2005)
PHBV[c]	145	1	1.2	20	50	Sudesh and Doi (2005)
PET	245–265	73–80	2.8–4.1	48–72	30–300	Clarinval and Halleux (2005)
PS	100	70–115	2.3–3.3	34–50	1.2–2.5	Clarinval and Halleux (2005)
LDPE	98–115	−100	0.3–0.5	8–20	100–1000	Clarinval and Halleux (2005)
LDPE	110	−30	0.2	10	620	Sudesh and Doi (2005)
PP	176	0	1.7	38	400	Sudesh and Doi (2005)

Source: Robertson (2008)
[a]Film grade
[b]Injection moulding grade
[c]20% HV

into the environment after 47 days. Due to these crystalline structures, the visibility of food contents is improved and is resistant to grease and tolerates temperature range from −60 °C to +80 °C. Blueberries wrapped in PLA containers are more functional to maintain fruit quality as compared to conventional containers at temperature 10–23 °C (Alemar et al. 2008).

14.4.5 Biodegradable Trays

It is used for the protection and storage for fruits and vegetables such as salad, and sliced broccoli, tomatoes, sweet corn and blueberries. They are usually kept in trays in wrapped form in the food packaging industry. The trays are resistant to moisture and brittle due to its composition. The positive impact of these materials is to maintain their structural properties (Makino and Hirata 1996).

14.5 Biodegradable Versus Biobased Plastics

A packaging material is either biodegradable or non-biodegradable. A biodegradable is considered as biopolymer which is renewable. Some partially biodegradable biopolymers are also developed, for example, renewable resource is blended with synthetic one (Table 14.5). A bioplastic material such as Polylactic acid (PLA), Thermoplastic starches (TPS), and Polyhydroxyalkanoates (PHA), etc. are developed from natural resources and exerts biodegradability under various conditions. Other materials such as biobased polyamides and bio-polyethylene are obtained from bio-derived products but are not biodegradable. Some other products like Polybutylene terephthalate (PBT) and Polybutylene succinate (PBS) are developed from petrochemicals but are biodegradable. Bioplastics decrease carbon dioxide emission by 30–70% as compared to traditional plastics so as to the production cost (Table 14.6).

14.6 Challenges and Future Trends for Biobased Food Packaging

Numerous studies have been conducted on biobased packaging materials in the last few decades which show the potential applications. But its cost, processing and performance are the major limitations for food packaging (Sorrentino et al. 2007). The sorting of polymers from a mixed waste during processing is a critical factor since even a low amount of cross-contamination can ruin the entire batch. Obtaining the desired shelf life, prevention of premature biodegradation and insect infestation followed by efficient biodegradation after disposal is also a challenge for the successful use of biobased packaging materials.

Globally, the ever-increasing food and energy demand will give a boost to the success of biobased packaging materials for food products. Stricter environmental

Table 14.5 Type of biodegradable and non-biodegradable polymers

Type of polymer	Biodegradable	Non-biodegradable
Biobased	Cellulose acetate, cellulose acetate butyrate, cellulose acetate propionate, cellulose nitrate, starch, chitosan	Polyethylene, Aminoundecanoic acid-derived polyamide, Laurolactam-derived polyamide, Polyethylene terephthalate, Polytrimethylene terephthalate
Partially biobased	Polybutylene succinate, poly butylene adipate-co-terephthalate, starch blends, polylactic acid blends	Polybutylene terephthalate, polyethylene terephthalate, polytrimethylene, poly vinyl chloride, epoxy resin
Petroleum based	Polybutylene succinate, Polybutylene succinate-co-adipate, Polybutylene succinate-co-lactide, poly €-caprolactone, Polyglycolide, polyvinyl alcohol	Polyethylene, polypropylene, polystyrene, poly vinyl chloride, acrylonitrile-butadiene-styrene, Polybutylene terephthalate, synthetic rubber

Source: Niaounakis (2013)

Table 14.6 Comparison of sustainability of biobased packaging materials with petroleum-based plastics

Biobased packaging	Sustainability
Polyhydroxyalkanoates (PHA)	Highly biodegradable
Polylactic acid (PLA)	Production uses 30–50% less energy, 50–70% reduction in CO_2 emissions
Thermoplastic starch (TPS)	Production uses 68% less energy, lower CO_2 emission, recyclable and biodegradable
Biourethanes	Production requires 23% less energy 36% less greenhouse gases
Cellulose and lignin	The biological degradation of lignin is lower than cellulose and compostable
Polytrimethylene terephthalate	Production requires 26–50% less energy, no additives are used, recyclable and biodegradable
Corn protein and soy protein	Biodegradable and compostable

Source: Alvarez-Chavez et al. (2012)

legislations and political willingness will also play a role in it. Biobased packaging materials with improved functionality and decreasing costs will ensure the growth of this industry. The use of nanotechnology, improvements in the processing steps, blending with different polymers and addition of various fillers has resulted in improving the barrier properties of the biobased packaging materials e.g. PLA. The use of the injection moulding process has shown less process losses compared with other techniques such as extrusion moulding.

14.7 Conclusion

Food waste produced from agricultural and food processing is abundant and location specific. It has the potential of industrial utilization by supplying it as raw material for other sectors. For global sustainability, conversion of food waste into value-added products can be the desired end use of it. Production of Biobased packaging is an ideal strategy for food waste disposal and a solution for the environmental burden of petroleum-derived plastics, producing smaller carbon footprint. Biobased packaging materials should ideally be produced from the waste or by-products of agricultural or food processing, which are the focus of the industry and are not competing with primary food production.

Biobased packaging materials, such as PHAs and PLA have good strength and barrier properties and the global demand for such eco-friendly packaging will definitely increase due to improved properties and decreasing cost as a result of technological developments. The properties of these biobased packaging materials can be improved with the use of blends of the polymers and nanocomposites. Large scale production of the biobased materials with desired properties will help them compete with the conventional packaging materials and also address the environmental concerns.

References

Adeodato VMG, Altenhofen da-Silva M, Oliveira dos-Santos L, Beppu MM (2011) Natural-based plasticizers and bio -polymer films: a review. Eur Polym J 47(3):254–263

Ahn WS, Park SJ, Lee SY (2001) Production of poly(3-hydroxybutyrate) from whey by cell recycle fed-batch culture of recombinant Escherichia coli. Biotechnol Lett 23:235–240

Alemar E, Samsudin H, Auras R, Harte B, Rubino M (2008) Postharvest shelf life extension of blueberries using a biodegradable package. Food Chem 110:120–127

Alvarez-Chavez CR, Edwards S, Moure-Eraso R, Geiser K (2012) Sustainability of bio-based plastics: general comparative analysis and recommendations for improvement. J Clean Prod 23 (1):47–56

Auras RA, Singh SP, Singh JJ (2005) Evaluation of oriented poly(lactide) polymers vs. existing PET and oriented PS for fresh food service containers. Packag Technol Sci 18:207–216

Averous L, Baqquillon N (2004) Biocomposite based on plasticizes starch: thermal and mechanical behaviours. Carbohydr Polym 56:111–122

Aydin F, Kahve HI, Ardic M (2017) Lipid based edible films. J Sci Eng Res 4(9):86–92

Bastioli C (2005) Starch-based technology, handbook of biodegradable polymers. Rapa Technology Ltd., Shawbury, pp 257–286

Berdanier CD (1998) CRC desk reference for nutrition. CRC Press, Boca Raton

Cabedo L, Feijoo JL, Villanueva MP, Lagaron JM, Gimenez E (2006) Optimization of biodegradable nanocomposites based on a PLA/PCL blends for food packaging applications. Macromol Symp 233:191–197

Cerquiglini C, Claro J, Giusti AM, Karumathy G, Mancini D, Marocco E, Mascianá P, Michetti M, Milo M (2016) Food Outlook June 2016. Food Agric. Organ. United Nations 14

Cesario MT, Raposo RS, De Almeida MCMD, Van KF, Ferreira BS, Da Fonseca MMR (2014) Enhanced bioproduction of poly-3-hydroxybutyratefrom wheat straw lignocellulosic hydrolysates. New Biotechnol 31:104–113

Chen GQ, Patel MK (2012) Plastics derived from biological sources: present and future: a technical and environmental review. Chem Rev 112:2082–2099
Chen H, Wang J, Cheng Y, Wang C, Liu H, Bian H, Pan Y, Sun J, Hen W (2019) Applications of protein based films and coatings for food packaging: a review. Polymers 11:2039
Chiellini E (2008) Environmentally compatible food packaging. Woodhead Publishing Limited, Cambridge, pp 8–10
Chow WS, Lok SK (2009) Thermal properties of poly(lactic acid)/organo-montmorillonite nanocomposites. J Therm Anal Calorim 95:627–632
Clarinval AM, Halleux J (2005) Classification of biodegradable polymers. In: Smith R (ed) Biodegradable polymers for industrial applications, 1st edn. CRC Press, Boca Raton, pp 3–31
Datta R, Henry M (2006) Lactic acid: recent advances in products, processes and technologies: a review. J Chem Technol Biotechnol 81:1119–1129
Davidovic A, Savic A (2010) Microbial production of biodegradable polymer. Tehnologica Acta 3:3–13. (in Croatian)
Eerhart AJJE, Faaij APC, Patel MK (2012) Replacing fossil-based PET with bio based PEF; Process analysis, energy and GHG balance. Energy Environ Sci 5:6407
Farrisa S, Schaich KM, Liu LS, Piergiovanni L, Yamb KL (2009) Development of polyion-complex hydrogels as an alternative approach for the production of bio-based polymers for food packaging applications. Trends Food Sci Technol 20:316–332
Garlotta D (2001) A literature review of poly-lactic acid. J Polym Environ 9(2):63–84
Girotto F, Alibardi L, Cossu R (2015) Food waste generation and industrial uses: a review. Waste Manag 45:32–41
Global Food Processing Market report, 2019. Retrieved from https://www.businesswire.com/news/home/20190904005488/en/Global-Food-ProcessingMarket-Report-2019-Trends. Accessed 20 Mar 2020
Goodship V, Ogar EO (2004) Polymer processing with supercritical fluids. Rapra Rev Rep 15(8)
Grondahl M, Gustafsson A, Gatenholm P (2006) Gas-phase surface fluorination of arabinoxlyan films. Macromolecules 39:2718–2712
Huang L, Sheng J, Chen J and Li N (2008) Direct fermentation of fishmeal wastewater and starch wastewater to lactic acid by Rhizopus oryzae. In: 2nd International Conference on Bioinformatics and Biomedical Engineering, iCBBE 2008
Jagannath JH, Nadanasabapathi S, Bawa AS (2006) Effect of starch on thermal, mechanical, and barrier properties of low density polyethylene film. J Appl Polym Sci 99:3355–3364
Jambunathan P, Zhang K (2016) Engineered biosynthesis of biodegradable polymers. J Ind Microbiol Biotechnol 43:1037–1058
Jamshidian M, Tehrany EA, Imran M, Jacquot M, Desobry S (2010) Polylactic acid: production, applications, Nanocomposites and release studies. Compr Rev Food Sci Food Saf. https://doi.org/10.1111/j.1541-4337.2010.00126.x
Khardenavis AA, Suresh KM, Mudliar SN, Chakrabarty T (2007) Biotechnological conversion of agro-industrial wastewaters into biodegradable plastic, poly β-hydroxybutyrate. Bioresour Technol 98:3579–3584
Khedkar RD, Singh K (2014) New approaches for food industry waste utilization. In: Biologix, ISBN no. 81-88919-15-2, Chap. 3, 51–65
Khedkar RD, Singh K (2018) Food industry waste: a panacea or pollution hazard?. Chapter 3 In: Paradigms in Pollution Prevention, ED: Tanu Jindal, Springer Briefs in Environmental Science, ISBN 978-3-319-58414-0, 35–48
Kim BS (2000) Production of poly(3-hydroxybutyrate) from inexpensive substrates. Enzym Microb Technol 27:774–777
Kim KI, Kim WK, Seo DK, Yoo IS, Kim EK, Yoon HH (2003) Production of lactic acid from food wastes. Appl Biochem Biotechnol 107:637–648

Koller M, Hesse P, Bona R, Kutschera C, Atlic A, Braunegg G (2007) Potential of various archae- and eubacterial strains as industrial polyhydroxyalkanoate producers from whey. Macromol Biosci 7:218–226

Kolybaba M, Tabil LG, Panigrahil S, Crerar WJ, Powell T, Wang B (2003) Biodegradable polymers: past, present, and future. In: American Society of Agricultural Engineers Annual Meeting, Fargo, North Dakata, USA

Lenczak JL, Schmidell W, de Aragao GMF (2013) High-cell-density culture strategies for polyhydroxyalkanoate production: a review. J Ind Microbiol Biotechnol 40:275–286

Lodha P, Nteravali AN (2005) Thermal and mechanical properties of environmental-friendly green plastics from stearic acid modified-soy protein isolate. Ind Crops Prod 21:49–64

Makino Y, Hirata T (1996) Modified atmosphere packaging of fresh produce with a biodegradable laminate of chitosan-cellulose and polycaprolactone. Postharvest Biol Technol 10:247–254

Marangoni C, Furigo AJ, de Aragão GMF (2002) Production of poly(3-hydroxybutyrate-co-3-hydroxyvalerate) by Ralstonia eutropha in whey and inverted sugar with propionic acid feeding. Process Biochem 38:137–141

Mehta R, Kumar V, Bhunia H, Upadhyay SN (2005) Synthesis of poly(lactic acid): a review. J Macromol Sci Part B Polym Rev 45:325–349

Miguel O, Iruin JJ (1999) Water transport properties in poly(3-hydroxybutyrate) and poly (3-hydroxybutyrate-*co*-3-hydroxyvalerate) biopolymers. J Appl Polym Sci 73:455–468

Mohareb E, Mittal GS (2007) Formulation and process conditions for biodegradable/edible soy-based packaging trays. Package Technol Sci 20:1–15

Mohatny AK, Misra M, Drzal LT, Selke SE, Harte BR, Hinrichsen G (2005) Natural fibers, biopolymers, and biocomposites: an introduction. In: Mohatny AK (ed) Natural fibers, biopolymers and biocomposites. CRC Press. Chapter 1

Nakayama N, Hayashi T (2007) Preparation and characterization of poly(l-lactic acid)/TiO_2 nanoparticle nanocomposite films with high transparency and efficient photodegradability. Polym Degrad Stab 92:1255–1264

Nampoothri NM, Nair NR, John RP (2010) An overview of the recent developments in polylactide (PLA) research. Bioresour Technol 101:8494–8501

Narayan R (2006) Biobased and biodegradable polymer materials: rationale, drivers and technology exemplars. In: Khemani K, Scholz C (eds) Degradable polymers and materials: principles and practice, vol 939. American Chemical Society, pp 282–306

Niaounakis M (2013) Biopolymers: reuse, recycling, and disposal. William Andrew Publishing

Nielson C, Rahman A, Rehman AU, Walsh MK, Miller CD (2017) Food waste conversion to microbial polyhydroxyalkanoates. Microb Biotechnol 10(6):1338–1352

Nikel PI, Pettinari MJ, Mendez BS, Galvagno MA (2005) Statistical optimization of a culture medium for biomass and poly(3-hydroxybutirate) production by a recombinant Escherichia coli strain using agro industrial byproducts. Int Microbiol 8:243–250

Obruca S, Benesova P, Marsalek L, Marova I (2015) Use of lignocellulosic materials for PHA production. Chem Biochem Eng Q 29:135–144

Panesar PS, Kennedy JF, Gandhi DN, Bunko K (2007) Bio utilisation of whey for lactic acid production. Food Chem 105:1–14

Partal P, Gonzalez-Gutierrez J, Garcia-Morales M, Gallegos C (2010) Development of highly-transparent protein-starch based bioplastics. Bioresour Technol 101(6):2007–2013

Plackett DY, Holm K, Johansen P, Ndoni S, Nielsen PV, Sipilainen-malm T, Sodergabrd A, Verstichel S (2006) Characterization of l-polylactide and l-polylactone– polycaprolactone co polymer films for use in cheese-packaging applications. Packag Technol Sci 19:1–24

Platt DK (2006) Biodegradable Polymers Market Report, Shawnbury, UK. Rapa Technology Ltd. 2006

Poomipuk N, Reungsang A, Plangklang P (2014) Poly-b-hydroxyalkanoates production from cassava starch hydrolysate by Cupriavidus sp. KKU38. Int J Biol Macromol 65:51–64

Poulopoulou N, Kasmi N, Siampani M, Terzopoulou ZN et al (2019) Exploring next generation engineering bioplastics: poly(alkylene furanoate)/poly (alkylene terephthalate) (PAF/PET) blends. Polymers 11:556

Ramos OL, Fernandes JC, Silva SI, Pintado ME, Malcata FX (2012) Edible films and coatings from whey proteins: a review on formulation, and on mechanical and bioactive properties. Crit Rev Food Sci Nutr 52:533–552

Reddy G, Altaf M, Naveena BJ, Venkateshwar M, Kumar EV (2008) Amylolytic bacterial lactic acid fermentation—a review. Biotechnol Adv 26:22–34

Rico D, Martin-Diana AB, Barat JM, Barry-Ryan C (2007) Extending and measuring the quality of fresh-cut fruit and vegetables: a review. Trends Sci Technol 18:373–386

Robertson G (2008) State of the art biobased food packaging materials. Chapter 1 In: Chiellini E (ed) Environmentally compatible food packaging. Woodhead Publishing, CRC Press, pp 3–28

Schreiner M, Huyskens-Keil S, Krumbein A, Prono-Widayat H, Ludders P (2003) Effect of film packaging and surface coating on primary and secondary plant compounds in fruit and vegetable products. J Food Eng 56:237–240

Seglia D, Krasanova I, Heidemane G, Kampuse S, Dukalska L, Kampus K (2009) Packaging technology influence on the shelf, life extension of fresh raspberries. Acta Hort 877:433–440

Sinclair RG (1993) Biodegradable packaging thermoplastics from lactides. US Patent no.5180765

Song JH, Murphy RJ, Narayan R, Davies GBH (2009) Biodegradable and compostable alternatives to conventional plastics. Phil Trans R Soc B364:2127–2139

Sorrentino A, Gorassi G, Vittoria V (2007) Potential perspectives for bio-nanocomposites for food packaging applications. Trends Food Sci Technol 18:84–95

Stloukal P, Pekarova S, Kalendova A, Mattausch H, Laske S (2015) Kinetics and mechanism of the biodegradation of PLA/clay nanocomposites during thermophilic phase of composting process. Waste Manag 42:31–40

Sudesh K, Doi Y (2005) Polyhydroxyalkanoates. In: Bastioli C (ed) Handbook of biodegradable polymers. Rapra Technology Ltd, Shawbury, pp 219–256

Sungsuwan J, Rattanapanone N, Rachanapun P (2008) Effect of chitosan/methyl cellulose films on microbial and quality characteristics of fresh-cut cantaloupe and pineapple. Postharvest Biol Technol 49:403–410

Suresh KM, Mudliar SN, Reddy KMK, Chakrabarty T (2004) Production of biodegradable plastics from activated sludge generated from a food processing industrial wastewater treatment plant. Bioresour Technol 95:327–330

Taniguchi I, Kagotani K, Kimura Y (2003) Microbial production of poly(hydroxyalkanoate)s from waste edible oils. Green Chem 5:545–548

Tsuji H, Okino R, Daimon H, Fujie K (2006) Water vapor permeability of poly(lactide)s: effects of molecular characteristics and crystallinity. J Appl Polym Sci 99:2245–2252

Uemura Y, Maetsuru Y, Fujita T, Yoshida M, Hatate Y, Yamada K (2006) The effect of coatings formed by low temperature tetramethoxysilane plasma treatment on water-vapor permeability of poly(l-lactic acid) film. Korean J Chem Eng 23:144–147

Van Tuil R, Flower P, Lawther M, Weber CJ (2000) Properties of biobased packaging materials. In: Weber CJ (ed) Biobased packaging materials for the food industry- status and perspectives. The Royal Veterinary and Agricultural University, Department of Dairy and Food Science, Copenhagen, pp 13–44

Van-Wegen RJ, Ling Y, Middelberg APJ (1998) Industrial production of polyhydroxyalkonates using Escherichia coli: an economic analysis. Trans I Chem E 76:417–426

Wackett LP (2008) Polylactic acid (PLA) an annotated selection of world wide web sites relevant to the topics in environmental microbiology. Microb Biotechnol 1(5):432–433

Zee MV (2005) Biodegradability of polymers–mechanisms and evaluation methods. In: Bastioli C (ed) Handbook of biodegradable polymer, 1st edn. Rapra Technology Limited. Shropshire, pp 1–22

Zeng QH, Yu AB, Lu (Max) GQ, Paul DR (2005) Clay based polymer nanocomposites: research and commercial development. J Nanosci Nanotechnol 5:1574–1592

Zhang B, He PJ, Ye NF, Shao LM (2008) Enhanced isomer purity of lactic acid from the nonsterile fermentation of kitchen wastes. Bioresour Technol 99:855–862

Zhao Y, Mc Daniel M (2005) Sensory quality of foods associated with edible film and coating systems and shelf-life extension. In: Innovations in food packaging. Elsevier, San Diego, pp 434–453

Zivkovic N (2009) Polyhydroxyalkanoates, green chemistry (seminar). Food Technology Zagreb, (in Croatian)

Emerging Opportunities for Effective Valorization of Dairy By-Products 15

Amrita Poonia

Abstract

The huge and ever-growing number of dairy by-products has become a major concern throughout the whole world. These by-products produced by the dairy industry are a great loss of valuable compounds and also increases severe environmental problems. However, numerous of these by-products have the potential to be reused into the development of other food products. Dairy waste valorization is one of the recent areas of research that has attracted a great deal of attention as a potential alternative to the utilization of these by-products. By-products like skim milk, buttermilk, ghee residue and whey arise in significant amounts and are consequently of high relevance. Due to their composition with various beneficial ingredients, the by-products can be utilized by using different techniques leading to economic and environmental advantages. Particularly the development of environment friendly and innovative strategies to process such type of by-products is of great importance in our society. New researches have also been oriented on recent technologies to utilize these by-products for food and non-food products. This chapter focuses on various by-products waste generation during the processing of milk and milk products and reviews state-of-the-art technologies for their valorization. The main uses of functional ingredients obtained from these by-products are presented and discussed, highlighting major mainstream food areas of application, e.g. in the functional and bakery and meat industry. Furthermore, their applications, various strategies to be adopted for successful valorization, research gap, current options and future scenario are discussed.

A. Poonia (✉)
Department of Dairy Science and Food Technology, Institute of Agricultural Sciences, Banaras Hindu University, Varanasi, Uttar Pradesh, India

M. Thakur et al. (eds.), *Sustainable Food Waste Management*,
https://doi.org/10.1007/978-981-15-8967-6_15

Keywords

Valorization · Dairy products · Bovine serum albumin · Ghee residues

15.1 Introduction

Valorization is a process of changing by-products of any industry into products at a much greater value. The results can include quality chemicals, materials, fuels and energy as well as many other products beneficial for a local economy. The scenario of food industries adopting the concepts of valorization for reusing the large quantity of food processing by-products is high. There is enough scope for the preparation of a range of value-added products including many biomolecules of economic importance that may ultimately contribute to sustainable development. It is very important that the new technology for valorization of the food processing by-products is available so that the by-products are appropriately subjected to the value addition process and adequately utilized. Only few industries have adopted and applied valorization of by-products. This chapter will provide useful information about the valorization of dairy by-products to the policymakers.

To fulfill the consumer demands in the context of growing interests in convenient foods, the food industry research and developments conducting intensive research for the development of modified food with additives. The application of by-products for the development of food products may not only solves environmental problems but also reduce the wastage of the by-products. Apart from this, it will also increase their nutritive value and consumer acceptance. However, lots of research work is required on the quality and food safety, risk of contamination, foodborne diseases, development of various processes and their control.

15.2 Need of Valorization

15.2.1 Promote Sustainable Development

There is a need to find out the sustainable ways of using dairy by-products. Proper processing of dairy by-products is very essential. According to United Nations (2015) worldwide, this sustainable development is a vital matter and every industry is concentrating on sustainability issues in various segments including dairy industry waste as in its sustainable development areas. Valorization of dairy by-products with different recent technologies and in various products is vital to develop different aspects of sustainability, i.e. environmental, economy and society.

15.2.2 To Enhance Value

Valorization is a process of changing residues into products at a much greater value. This can include the latest technologies, quality chemicals, machines, materials, fuels and energy as well as many other products also.

15.2.3 To Compensate Growing Environmental Problems

In dairy industry during the processing of milk that generates voluminous by-products and wastes, valorization can help sole the environmental problems.

15.2.4 Maximum Use of Dairy Produce

By using the dairy by-products in food and other uses it will boost global competitiveness, promote sustainable economic growth and generate employment.

15.2.5 Boost Economy

A contribution towards sustainability, valorization makes maximum use of dairy products while employing low-energy and cost-effective processes.

15.3 Scope of Valorization in Dairy Industry

Mirabella et al. (2014) reported that total milk production is more than 801 million tons. Out of the total production, more than 37% is used for the manufacturing of cheese and other coagulated products. About 30% of total production is used for the preparation of butter. During the processing of all these products, only 10–20% of milk od total production is recovered as the final product and the residual 80–90% liquid part is a by-product, i.e. whey. The total milk production in India is likely to double in the next 10 years and there is an opportunity for large investments in food and food processing technologies, skills and equipment. The demand of consumers for various food groups like confectionery, chocolates, high protein food, soft beverages, functional dairy food, designer foods, etc. along with the health food and health food supplements is another rapidly rising segment of this industry which is gaining vast popularity day by day.

By 2030, the population in India is expected to be about 590 million people and there is an enormous demand for foods other than the demand from the exports in urban parts of the country. The changing consumption patterns, working culture of the family members rising income levels among the middle-class and changing lifestyles, are some of the factors providing the demand for these kinds of food products. Furthermore, the central government has given important rank to dairy as well as all agro-processing businesses.

15.4 Classification of Dairy Industry By-Products

A by-product may be defined as a product of commercial value produced during the manufacture of the main product. Whey, buttermilk, skim milk and ghee residue are the major dairy by-products. Each of these by-products has unique nutritional value and hence must be processed judiciously into the edible as well as non-edible value-added products. The disposal of these by-products has now become a challenge and danger to the environment. The major products of the dairy industry with their by-products are listed in Fig. 15.1.

15.5 Major Components of Dairy By-Products

15.5.1 Lactose

Lactose is the only milk sugar accounts approximately for two-third of the total solids present in whey Yadav et al. (2015). It can be used in various food products to enhance colour, flavour, aroma, texture, anti-caking agent, reducing sweetness to extend the shelf—life of food products. During fermentation of whey, lactose is used as an important ingredient for producing single-cell protein, amino acids, ethanol, enzymes, organic acids, vitamins, bioplastics and insecticides.

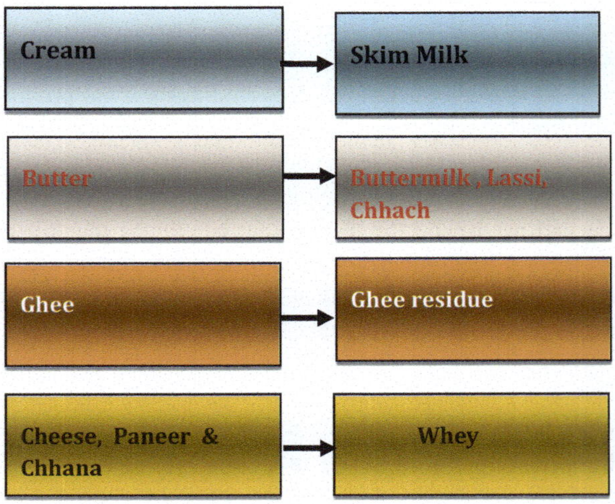

Fig. 15.1 Major products of the dairy industry with their by-products

15.5.2 β-Lactoglobulins

β-Lactoglobulins is the major whey protein that accounts for 50–55% of the total whey proteins. It is rich in amino acids, bind long-chain fatty acids, retinoids and steroids and increase bioavailability of amounts of nutrients.

15.5.3 α-Lactalbumin

α-Lactalbumin constitutes 20–25% of the protein in whey. It is a calcium-binding protein and has the ability to bind metal cations and may help in the absorption of essential minerals. The unique properties of this protein are its richness in essential amino acids especially tryptophan. Tryptophan is the precursor of serotonin which controls sleep, appetite, improves mood, cognitive performance and also controls stress and regulates melatonin. Lopez-Exposito and Recio (2008) reported that α-lactalbumin has bacteriocidal activity and causes apoptosis of tumour cells.

15.5.4 Glycomacropeptide

It is a C-terminal peptide of the kappa casein released by chymosin in the first stage of the enzymatic coagulation of milk during cheese manufacturing. It constitutes about 0–15% of the whey proteins. It is found maximum in sweet whey derived by enzymatic coagulation of milk. It is known for its various functional properties like foaming and gel formation and emulsification. Its biological properties include anti-carcinogenic effects, reduction of gastric secretion, growth factor for *Bifidobacterium*, hemagglutinin inhibition, modulation of immune response and nutritional management.

15.5.5 Bovine Serum Albumin (BSA)

Bovine Serum Albumin is a globular protein having a good amino acid profile. It also maintains homeostatic colloidal pressure binds lipids, acts as a carrier for fatty acids and catecholamines. BSA constitutes about 5–10% of the total whey proteins present in whey.

15.5.6 Lactoferrin

It is iron-binding glycoprotein and constitutes 1–2% of the total whey proteins. It has immunomodulatory, antimicrobial anti-tumoral properties. It is used as a functional food ingredient as a natural solubilizer of iron in foods due to its iron-binding ability.

15.5.7 Immunoglobulins

Immunoglobulins constitute about 10–15% of the total whey proteins. Cheese whey has three types of immunoglobulins, i.e. IgG, IgM and IgA. Their intake is a boon to the peoples who are immunocompromised.

15.5.8 Lactoperoxidase

It constitutes about 0.5% of the total whey proteins. It is a glycoprotein with bactericidal and bacteriostatic effects.

15.5.9 Lysozyme

It is a hydrolytic enzyme that hydrolyzes the peptidoglycan of the bacterial cell wall. It also has antibacterial activity against the gram-positive bacteria. It is mostly found in tears, egg white and cheese whey. Lysozyme is mainly used as a food additive in the preservation of many foods.

15.5.10 Minerals

Almost all the minerals that are present in milk are found in whey. Macrominerals like K, Ca, P, Na and Mg are found in whey. Trace minerals like Zn, Cu and Mn are also found in whey. It was found that the concentration of minerals is more in acid whey as compared to sweet whey.

15.5.11 Vitamins

Whey contains all the vitamins, i.e. A, C, D, E and B complex vitamins in appropriate amounts.

15.6 Recent Extraction Methods Applied to Dairy By-Products

15.6.1 Affinity Chromatography (AC)

The main basis of separation of proteins in AC is a reversible interaction between protein and a precise ligand attached to a chromatographic matrix. The unique features of this technique are its high selectivity, frequently high capacity and high resolution for the target protein. It is mostly applied as the initial step in a two-step purification process and then a chromatographic or polishing step to remove the residual impurities. Affinity chromatography is helpful for polishing or completing the protein purification process. Due to the high affinity of proteins towards the specific chemical groups, they attach covalently and fix to the column matrix through non-bound proteins pass through the column (Kumar and Sharma 2015).

15.6.2 Supercritical Fluid Extraction (SCF)

Supercritical Fluid Extraction is a process that consists of separating one component from another using supercritical fluids, most commonly carbon dioxide. The main principle of this is based on the fluid dissolving power, which under specific conditions above its critical temperature and pressure results in different dissolving properties. Supercritical fluids are used widely to modify different foods. Mainly this technique is applied in various dairy foods, i.e. cheese whey, whey cream, buttermilk and butter. SCF has the potential for extending the shelf life of the product by decreasing the microbial population. Application of SFE on buttermilk also is possible for obtaining concentrates of polar lipids, as phospholipids, from milk fat globule membrane (MFGM). Supercritical fluid extraction with CO_2 has a great prospective in the dairy industry. SCF is very useful in commercial applications for providing functional food products and ingredients with an increased economic and bioactive value.

15.6.3 Ultrasound Processing

Ultrasound is a type of energy produced by sound waves of frequencies 20–500 kHz. These sound waves are very high in nature and cannot be detected by the human ear. Ultrasound when conveyed through a biological structure forms compressions and depressions of the particles of medium and a high amount of energy can be reported. Ultrasonic waves with high intensity are used in the dairy industry for the inactivation of bacteria and enzymes, homogenization of milk, purification and increased production of cheese. Low-intensity waves were used to determine the composition of milk and dairy products, that is, fat cells concentration, fat-free solids and the whole solid (Yasaman 2014). Like many other methods, ultrasounds have considerable advantages over other methods of food analysis. In the food industry ultrasound is applied to quality control, process control, homogenization, cleaning and sterilization. This method has its advantages and disadvantages. Its non-destructive feature and less time consuming are the main advantages. In the dairy industry, it is used to inactivate microorganisms and ultrasonic homogenization.

15.6.4 Ion-Exchange Chromatography

The basis of separation of proteins in Ion-exchange chromatography is by alterations in their surface charge to result in a very high-resolution separation with high sample loading capacity. Mostly the separation is founded on the reversible interaction between a charged protein and an oppositely charged chromatographic medium. Proteins bind while they are loaded into the column. After this, the conditions are changed so that bound substances are eluted differentially. Elution is done by increasing the salt concentration or by altering pH. During this protocol, samples are diluted with NaCl using gradient elution. Target proteins are concentrated during binding and collected in a purified and concentrated form (Kumar and Sharma 2015).

15.7 Valorization of Dairy By-Products

15.7.1 Whey and Its By-Products

Whey is a yellowish watery portion and by-product of cheese and paneer industry. Biological Oxygen Demand (BOD) and Chemical Oxygen Demand (COD) of whey are very high, e.g. acid whey has a BOD value of 35,000–45,000 mg per litre and COD values of 55,000–70,000 mg per litre. Macwan et al. (2016) reported the World whey production about 180 million tones which contains about 1.5 million tonnes of high-value protein and 8.6 million tonnes of lactose. About 40% of whey is discarded as a waste around the world which causes a vast valuable nutrients loss. It shows that high amount of milk nutrients is going waste and not used for human consumption. This will also lead to serious sustainability consequences. Whey has long been recognized as the best source of protein supplementation to repair tissue and to build muscle (Krissansen 2007). Whey and whey components are used by the food industry in a wide variety of applications on the basis of their excellent nutritional and functional properties (Foegeding et al. 2002; Tunick 2008). It is increasingly recognized as a functional food with a number of health benefits, has received growing interest as functional ingredients in dietary and healthy foods such as slimming foods, diets for the elderly and clinical foods. Poonia and Arti (2018) developed *Rabadi* by fermenting pearl millet (*Pennisetum typhoideum* (L.) PM flour with buttermilk and then the standardized product was packed in indigenous pouches and stored at 4 °C and 10 °C. The shelf life of the product was 8 days at 4 °C and 5 days at 10 °C, respectively. Bioactive whey ingredients such as bioactive proteins, which exert an additional health benefit for the consumer, are increasingly used in pharmaceuticals as well as nutraceuticals (Foegeding and Luck 2002). During the last 30 years, various efforts have been done to utilize these by-products for edible purpose. Whey can be utilized in various dairy products for manufacturing cheese and coagulated dairy products. In the case of acid whey, due to its high mineral content and low pH, the considerable difficulty can be faced in its utilization and mostly remain unused. This problem is solved by condensing, drying and demineralization of whey. Further, it can be used for the manufacturing of whey drinks, whey protein concentrates (WPC), lactose production, whey powder and whey-based non-fat beverages and powder, fruits whey beverages, production of mineral biomass protein, ethanol, whey drinks, infant formulas and designer dairy foods.

15.7.2 Whey Proteins

There are two subclasses of whey proteins, i.e. major proteins and minor proteins. Major whey proteins are β-lactoglobulin (β-Lg) 65%, α-lactalbumin (α-La) 25%, immunoglobulins (Ig), bovine serum albumin and proteose peptones. Minor whey proteins include Lactoferrin, lysosome, glycomacropeptide, lactoperoxidase and phospholipoproteins are the minor whey proteins. They have a high protein efficiency ratio (3.6), net protein use (95) and biological value (110) that makes them comparable with egg protein and greater than soy protein and even casein, gluten and beef. Whey protein contains all of the 20 amino acids and each of the nine essential amino acids. Sulphur-containing amino acids, i.e. methionine and cysteine, are also found in large amounts (Macwan et al. 2016). Most of the methods reported for the production of biomaterials from milk-derived proteins have been developed by using caseins or only whey proteins. Nanofiltration (NF), microfiltration (MF), reverse osmosis (RO) and ultra-filtration (UF), hydrolysis, electrodialysis and ion exchange are the separation processes for the recovery of solids from whey.

Different products obtained from why are: whey concentrate and whey cream, sweetened condensed whey and dried whey powder, Whey protein concentrates and isolates, Whey lactose, whey beverages, Fruit-flavoured whey-based beverage, carbonated beverages, fermented beverages, whey-based herbal beverage and probiotic whey beverage. Various types of whey-based beverages have been developed, i.e. alcoholic, fruit-flavoured, plain and fermented beverage, carbonated and new products (Table 15.1). Whey and whey proteins are also playing an important role in the manufacture of biomaterials (Table 15.2).

15.7.3 Skim Milk and By-Products

Skim milk is obtained during the manufacture of cream. It is a rich source of solid-not–fat (SNF) and high nutritional value. The main use of skim milk is in standardization of dairy products or skim milk powder by spray drying. The major by-products of skim milk are casein and related products, co-precipitates and proteins hydrolysates. Edible casein and caseinates are used in dairy and food products. Now a day's casein has various applications for the edible and non-edible purposes. Co-precipitates have several advantages over casein i.e. high yield and flexible functional properties and higher nutritional value.

15.7.4 Casein and Caseinates

Casein is used in industries for producing paper, rubber, textile and paints for since long time. Edible casein and caseinates are also used in the development of various dairy and food products. Most of the casein is used for nutraceuticals and for the manufacture of imitation cheese. The major drawbacks in its production are

Table 15.1 Utilization of whey for the development of various food products

Type of products	Form used	References
Low-fat products		
Soups, sausages, salad dressings, mayonnaise, meat, yogurts, and ice creams	Whey proteins	Dua et al. (2018), Yilsay et al. (2006), Young (2007)
Bakery and confectionary products		
Breads, cakes, cookies, biscuits, crackers, muffins, and icing	Whey	Ceglinska et al. (2007), De Wit (2001), Stoliar (2009)
Meat and meat products		
Frankfurters, sausages, mortadellas, luncheon meat or surimi	Sweet whey powder, whey protein concentrates (WPCs) whey protein isolate (WPI) whey with reduced lactose content, and demineralised whey	De Wit (2001)
Dairy products		
Yoghurt and ice creams	Sweet whey powder, WPC	Hugunin (2009)
Ice creams and sundaes	Whey powder, demineralised whey powder, WPCs and WPI	Young (2007)
Cheese and their analogues	Sweet whey powder WPCs and WPI	Young (2007)
Different beverages		
Probiotic beverage	*Lactobacillus acidophilus, Bifidobacterium* and *Streptococcus thermophilus*	Faisal et al. (2017)
Fruits and vegetables beverage	Papaya and whey	Panghal et al. (2017a, b)
Carbonated beverage	Mango,pineapple and whey	Katke and Patil (2017)
Fermented beverages	*Streptococcus thermophilus* and whey	Saha et al. (2017)
Herbal beverage	Basil, mint, ginger, aloe vera, lemon grass	Maya and Ritu (2016)
Alcoholic beverages	Whey	Baldissera et al. (2011), Smithers (2008)
(Ready to drink) beverages (RTD)	Whey protein isolates and concentrates	Rittmanic (2006)
Other uses		
Edible coatings	Whey	Galus and Kadzińska (2016)
Baby foods	Whey	Chung and Yamini (2012)
Infant formulas	Whey	De Wit (2001)

production cost, using spray drying, low bulk density and high packaging, storage and transportation costs. The manufacturing process of casein has shown in Fig. 15.2.

Table 15.2 Application of casein and whey proteins in the production of biomaterials

Name of product	Form of protein used	References
Reconstituted fibres	Casein	Roff and Scott (1971)
Gels/hydrogels and emulsions	Mixture of casein and whey proteins	Garcia-Moreno (2014)
Foams	Whey protein isolates (WPI)	Rouimi (2005)
Bioplastics	Whey protein isolates (WPI)	Sothornvit (2007)
Films and coatings	Whey proteins	Anker (2000)
Adhesives	Casein	Kuckova et al. (2007)
Biocomposites	Casein	Aberg (2004)
Bioemulsifiers	Whey proteins	Nitschke and Costa (2007)
Natural nanomaterials (Nanocapsules, nanoparticles, nanoemulsions, nanocoatings, nanofibres and nanoceuticals)	Casein and whey proteins	Poonia (2017)

15.7.5 Co-Precipitates

Co-precipitates have many benefits like their functional properties, higher nutritional value and yield as compare to casein. The major disadvantage is its poor solubility especially of the more calcium co-precipitates. Co-precipitates can be used as an ingredient in the preparation of different food/dairy products viz. confectionary, snack, meat, baked products and animal and pet foods.

15.7.6 Milk Protein Hydrolysates

Milk protein hydrolysates have been extensively used in nutritional, dietetic and formulated foods. It is very beneficial to the people who are suffering from protein allergy and those who need easily digestible foods. Casein can be used for casein glues, as paints, in the leather industry, rubber products, textile fibres, rennet casein plastics. In agriculture, casein is used in insecticide sprays, fungicides, fertilizer and for coating the seeds as an adhesive. Casein hydrolysates can be manufactured by using acid, alkali and enzymatic hydrolysis of casein. The functionality of milk proteins can be further modified by changing its physico-chemical properties by various treatments like heat treatment, acidification, homogenization, shear, use of chemicals, enzymatic modifications and high-pressure processing. The advantages

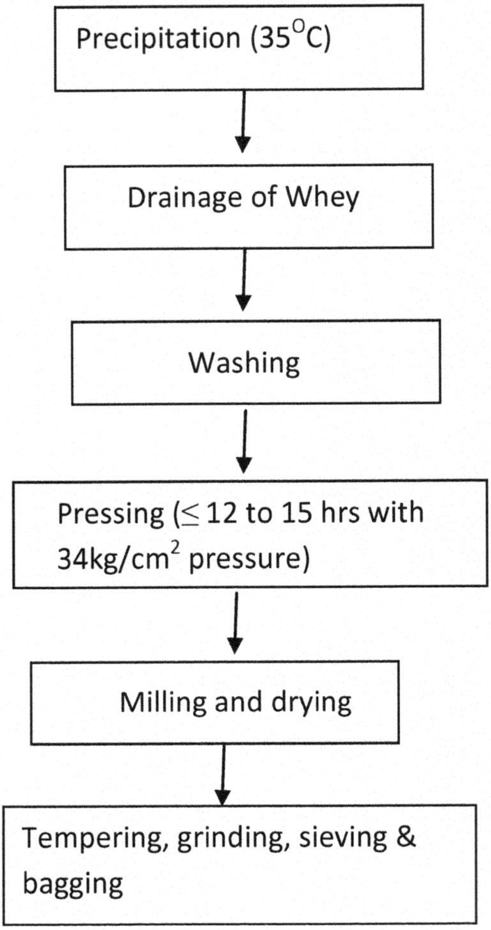

Fig. 15.2 Manufacturing Process of Casein

of casein hydrolysates have been discussed in Box 15.1. The functional properties of casein and whey proteins are:

- Solubility
- Heat stability
- Water binding capacity
- Viscosity
- Gelation
- Film formation
- Oil binding
- Emulsifying properties
- Foaming properties

Box 15.1 Advantages of Casein Hydrolysates

Available in Predigested Form: Predigested form helps in faster intestinal absorption of the hydrolysates and beneficial to the human beings suffering from impaired digestive functions. Here the amino acids are more easily available as compare to whole proteins. α-amino (N) absorption was more from the casein hydrolysates and lactalbumin hydrolysate then respective free amino acid mixture (Silk et al. 1980).

Reduced Allergenicity: Food allergies are very common and generally associated with proteins mainly glycoproteins. But milk proteins are hydrolyzed to reduce their allergenicity. The thermal treatment provided to hydrolyze the proteins also destroys the main allergenic determinants.

Improve Weight Gain and Nitrogen Balance: Whey protein hydrolysates containing diet improve weight gain and higher nitrogen retention when compared with whole protein and amino acid mixture diets.

Prevent Bone and Dental Disorder: Hydrolysates of casein, especially casein glycomacropeptide (CGMP) is a bioavailable form that retain the ability of CGMP to prevent bone loss and help in absorption and calcification.

15.7.7 Buttermilk

Buttermilk is the by-product of butter in the liquid phase obtained during churning of cream in the butter making. Generally, buttermilk has been considered as the invaluable by-product of the dairy industry. To counter the increasing solids disposed of as waste, buttermilk is used in animal feed or dried to be incorporated in dairy or bakery products as an emulsifying agent. Buttermilk is very comparable to skim milk in many ways, i.e. more than 80% of its proteins are major milk proteins, like caseins and whey proteins. However, its fat composition differs significantly from that of skim milk. The main reason for this difference in amount and composition is due to the presence of milk fat globule membrane (MFGM) derived substances. Near about 20% of buttermilk proteins are of MFGM origin. Alike skimmed milk and whey, buttermilk contains lactose, minerals, caseins and serum proteins which can be extracted, purified and valorized in very distinct food or non-food applications. Due to its high content of residual MFGM, ghee residue is of great interest to food industry. The MFGM residues in buttermilk are responsible for the unique nutritional and technological properties of this dairy by-product This biological membrane protects and stabilize it in the aqueous phase during churning. Additionally it contains specific proteins and unique polar lipids (PL) closely associated with a complex structure. Due to its rich composition, buttermilk opens a broad range of new products in valorization as emulsifiers, stabilizers and health promoters in food or non-food products. Buttermilk contains lactose, minerals and skimmed milk proteins (caseins and whey proteins) in the same proportion as skimmed milk.

In the case of buttermilk, the majority of the studies have focused on either fractionating or concentrating various MFGM components, primarily minor lipids. But the studies on biological effects of it are very few and poorly understood. Baumgartner et al. (2013) conducted well-controlled clinical trials provide support for the buttermilk health benefits hypothesis. They studied the clinical evidence of the cholesterol-lowering properties of buttermilk consumption Vanderghem et al. (2010) studied that MFGM contains about 60–70% of all milk polar lipids and these are largely responsible for the stability of fat globules in the milk oil/water emulsion, due to their amphiphilic nature. Short-term consumption of buttermilk significantly reduced plasma cholesterol and triglyceride concentrations in a double-blind, randomized, placebo-controlled crossover study on healthy subjects ($n = 34$) with mild-hypercholesterolemia (Conway et al. 2013).

15.7.7.1 Types of Buttermilk

15.7.7.1.1 *Sweet Cream Buttermilk (SCBM)*

This type of buttermilk is obtained by churning fresh/pasteurized cream and has very little developed acidity.

15.7.7.1.2 *Sour Buttermilk*

This buttermilk is obtained by churning the sour cream/milk and this is generally prepared in the organized dairy sector.

15.7.7.1.3 *Chhachh/Lassi*

This is obtained by churning curd during the manufacture of *makkhan*. *Lassi* is generally prepared at the household level for household use only.

15.8 Valorization of Buttermilk

15.8.1 Buttermilk Powder

Generally spray drying is used for the preparation of buttermilk powder. The major difference between buttermilk powder and skim milk powder is that the sour cream buttermilk powder contains high total lipids including phospholipids and low bulk density as compared to skim milk powder. Applications of buttermilk in different food products are shown in Fig. 15.3.

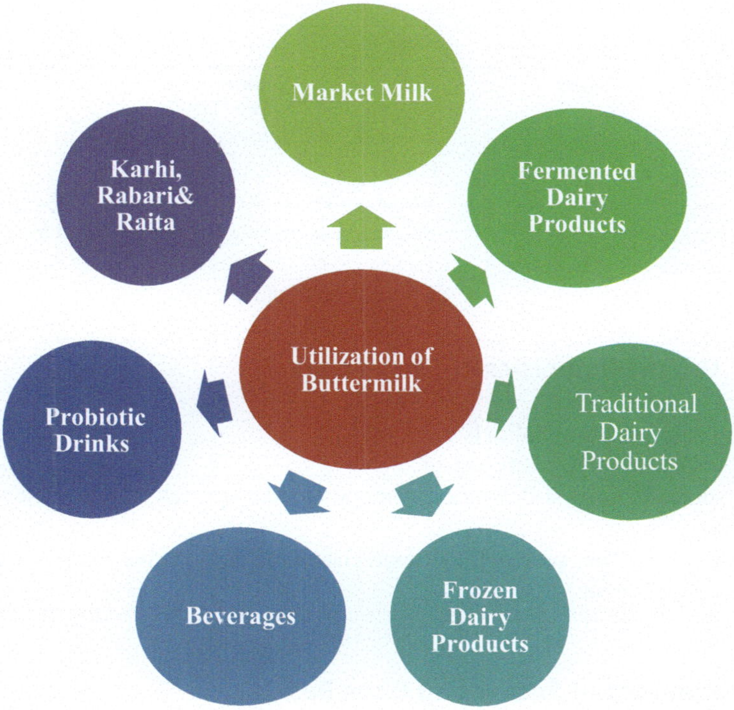

Fig. 15.3 Applications of buttermilk in food products

15.9 Ghee Residue

Ghee residue (GR) is a solid mass brownish in colour found as a by-product during ghee manufacturing. Ghee residue is one of the largest by-product of the milk industry. It is a good source of fat, proteins and minerals. Ramesh et al. (2018) reported that ghee residue is a rich source of fat, protein, unsaturated fatty acids and amino acids. Ghee residue can also be a good source for overcoming the problem of protein-energy malnutrition (De Wit 2001). Ghee residue is generally obtained during the manufacturing of indigenous butter, creamery butter, sweet cream, sour cream and washed sweet cream. The yield of ghee residue depends on the method of preparation of ghee and this variation is due to the difference of non-fatty serum constituents of the various raw materials used for the preparation of ghee. Direct-creamery (DC) has the highest amount of ghee residue, i.e. (12%) followed by almost the same yield in creamery butter (CB) and desi-butter (DB) methods, i.e. (3.7%). Santha and Narayanan (1978) reported that ripening of cream prior to clarification of cream, reduce the amount of ghee residue. This nutritious by-product can be utilized as a food supplement in a variety of foods, food-spreads, soups and so on (Table 15.3).

Table 15.3 Application of ghee residue in food Industry

Name of product	References
Confectionary	Prahlad (1954)
Candy	Galhotra and Wadhwa (1993)
Chocolate	Reddy and Khan (1978), Wadhwa (1997)
Edible pastes	Prahlad (1954)
Burfi	Verma and De (1978)
Bakery products	Borawake and Bhosale (1996), Bajwa and Kaur (1995)
Flavour enhancer	Tamine (2009)
Food-spreads, soups	Galhotra and Wadhwa (1993)

15.9.1 Unique Properties of Ghee Residue

- Good source of phospholipids
- Rich source of natural antioxidants
- Rich source of natural flavouring compounds like free fatty acids, lactones and carbonyls
- Rich source of fat, proteins and appreciated for its nutritive value for human dietary supplement

15.9.2 Methods of Recovering Ghee Residue

15.9.2.1 Centrifugal Process

In this process GR is heated at 65 °C and ghee is recovered by centrifugation. The yield of ghee by this method is 25% with 46% efficiency.

15.9.2.2 Pressure Technique

In this technique, GR is heated at 65–70 °C and there a limited pressure is applied by hand, screw or hydraulic press. The maximum yield of this method is about 45%.

15.10 Strategies to Be Adopted for Successful Valorization

It is very important that appropriate strategies be adopted for a successful implementation of the valorization of dairy by-products. However, several strategies have been invented and adopted by the dairy and food industries. The following strategies are used due to their merits and importance in the dairy industry.

- The shelf life of by-products of the dairy industry
- Recent methods of waste and dairy by-products disposal
- Development of alternative technologies and methods for the utilization of dairy by-products
- Scope for the utilization of dairy by-products in the development of food and non-food products

- Identification of potential applications
- Good manufacturing practices (GMP)
- Development of greener valorization strategies
- Quality management systems under the International Standards Organization (ISO 9000)
- Information of quality of dairy by-products along with their physico-chemical and biological properties to judge their potential as raw materials for developing value-added products
- Hazard analysis and critical control points (HACCP)
- Awareness programmes on the dissemination of knowledge about the importance of valorization
- Human resources development toward capacity building in valorization programme relevant to food industries

15.11 Research Gaps to Be Investigated

Firstly, excessive disposal of dairy wastes especially the whey is deteriorating the environment due to high BOD and COD in many parts of the World. By-products utilization is an attractive concept that has gained increasing popularity in India and other parts of the country due to the rapid increase in the generation of such types of wastes. Researchers are now focusing on the development of valorization strategies along with its proper disposal. Secondly, there is a need for different types of innovative solutions and alternatives to utilize these dairy by-products by using advanced valorization strategies. These will need collaboration from a range of disciplines, i.e. engineering, biotechnology, environmental sciences, legislation, biochemistry and economics to come up with innovative alternatives. This type of effort will lead the way toward a more sustainable bio-based society and economy.

15.12 Current Options and Future Scenario

Interdisciplinary approach in biotechnology, biochemistry is the key approach for the valorization of dairy by-products. There is a need to adopt eco-friendly technologies and conservation of the environment and sustainable utilization of dairy products. Initiative should be taken that would transform dairy by-products significantly contributing to meet the growing demand of healthy food products with environmental issues and socio-economic considerations. The major market drivers for this initiative are the huge amount of by-products, consumer demands, new technologies and production costs.

Various valuable compounds/ingredients can be processed by different techniques from dairy by-products. Furthermore, the processes and technologies used for this purpose can be improved with regard to yield, reduced labour, costs and various application opportunities. Hence more and more research might be necessary and optimization and on valorization technologies. Rather than limiting the research in laboratory only, it should focus on their relevance for industrial applications.

15.13 Conclusions

Various government schemes of strict natural direction are making waste management as a compulsory. Many regulatory bodies are making the consumers aware of it and also encouraging the food industries for proper management and utilization of these by-products. Mainly, dairy-based by-products, i.e. whey contains a large amount of protein, lactose and minor nutrients. High nutritional, functional and therapeutic value of these by-products are their major advantages, which not only enhance the other nutrients to the body but also increase diversity to the flavour and taste. Future valorization of these by-products lies in the nutraceutical health benefits, probiotics, bioactive compounds, consumer awareness and industrial venture. This chapter focuses on sustainable technologies, applications and processes of different methods to use this large number of dairy by-products in environmental, social and economic aspects along with the way forward.

The excessive production of dairy by-products leads to the necessity to utilize these waste products and also to develop further advanced processing technologies for their beneficial utilization. This chapter summarized the valorization perspective and various technologies of dairy waste products.

Glossary

Biological Oxygen Demand Amount of dissolved oxygen required for aerobic biological microorganisms to break down organic material found in a given water sample at a certain temperature over a specific time period.

By-Products Product of commercial value produced during the manufacture of a main product.

Chemical Oxygen Demand Amount of the capacity of water to consume oxygen during the decomposition of organic material and the oxidation of inorganic chemicals.

Milk Fat Globule Membrane Exceptional and complex structure contains mainly lipids and proteins that backdrops milk fat globule secreted from the milk-producing cells of humans and other mammals.

Nanocomposites Basically provide a highly versatile chemical functionality and consequently they are used for the development of high barrier properties.

Nanoparticles Aids in improving the food's flow property, colour and stability.

Supercritical Fluid Extraction Process which separates one component from another using supercritical fluids and mainly CO_2.

Sweet Cream buttermilk Buttermilk is obtained by churning fresh/pasteurized cream.

Valorization Process of changing by-products of any industry into products at a much greater value.

Whey Yellowish watery portion and by-product of the cheese and paneer industry.

References

Aberg CM, Chen T, Olumide A, Raghavan SR, Payne GF (2004) Enzymatic grafting of peptides from casein hydrolysate to chitosan. Potential for value-added byproducts from food-processing wastes. J Agric Food Chem 52:788–793

Anker M, Stading M, Hermansson AM (2000) Relationship between the microstructure and the mechanical and barrier properties of whey protein films. J Agric Food Chem 48(9):3806–3816

Bajwa U, Kaur A (1995) Effect of ghee (butteroil) residue and additives on physical and sensory characteristics of cookies. Chemie-mikrobiologie-Technologie-der-Lebensmittel 17 (5/6):151–155

Baldissera AC, Della Betta F, Penna ALB, Lindner JD (2011) Functional foods: a new frontier for developing whey based protein beverages. Semin-Cienc Agrar 32:1497–1511

Baumgartner SER, Kelly S, van der Made TT, Berendschot C, Husche D, Lütjohann D, Plat J (2013) The influence of consuming an egg or an egg-yolk buttermilk drink for 12 weeks on serum lipids, inflammation, and liver function markers in human volunteers. Nutrition 29:1237–1244

Borawake KN, Bhosale DN (1996) Utilization of ghee residue in preparation of nankatai type cookies and sponge cakes. Indian J Dairy Sci 49(2):114–119

Ceglinska A, Pluta A, Skrzypek J, Krawczyk P (2007) Study on the application of nanofiltrated whey-derived mineral components in the production of bread. Zywnosc Nauka Technologia Jakosc 6(55):234–241

Chung CS, Yamini S (2012) FDA's Health Claim Review: Whey-protein partially hydrolyzed infant formula and atopic dermatitis. Pediatrics 130:e408–e414

Conway VP, Couture SF, Gauthier Y, Pouliot Y, Lamarche B (2013) Impact of buttermilk consumption on plasma lipids and surrogate markers of cholesterol. Nutr Metab Cardiovasc Dis 23(12):1255–1262

De Wit JN (2001) Lecturer's handbook on whey and whey products. European Whey Products Association. Brussels, Belgium. Available at: [http://ewpa.euromilk.org/publications.html]

Dua S, Kumar S, Kaur S, Ganai AW, Khursheed I (2018) Chemical and sensory attributes of ghee residue burfi supplemented with corn flour. J Pharmacogn Phytochem 7:3818–3822

Faisal S, Chakraborty S, Devi WE, Hazarika MK, Puranik V (2017) Sensory evaluation of probiotic whey beverages formulated from orange powder and flavor using fuzzy logic. Int Food Res J 24 (2):703–710

Foegeding EA, Luck PJ (2002) Whey protein products. In: Roginski H, Fox PF, Fuquay JW (eds) Encyclopedia of dairy sciences. Academic Press, London

Foegeding EA, Davis JP, Doucet D, Guffey MK (2002) Advances in modifying and understanding whey protein functionality. Trends Food Sci Technol 13(5):151–159

Galhotra KK, Wadhwa BK (1993) Chemistry of ghee-residue, its significance and utilization-a review. Indian J Dairy Sci:142–146

Galus S, Kadzińska J (2016) Whey protein edible films modified with almond and walnut oils. Food Hydrocoll 52:78–86

Garcia-Moreno PJ, Horn AF, Jacobsen C (2014) Influence of casein–phospholipid combinations as emulsifier on the physical and oxidative stability of fish oil-in-water emulsions. J Agric Food Chem 62(5):1142–1152

Hugunin A (2009) Whey products in yogurt and fermented dairy products. U.S. Dairy Export Council, Applications Monographs. Yogurt, pp 151–154

Katke SD, Patil PS (2017) Studies on development of carbonated fruit flavoured shrikhand whey beverage. Contemporary Research India, Vol. 7

Krissansen GW (2007) Emerging health properties of whey protein and their clinical implication. J Nutr 26(6):7135–7235

Kuckova S, Hynek R, Kodicek M (2007) Identification of proteinaceous binders used in artworks by MALDI-TOF mass spectrometry. Anal Bioanal Chem 388(1):201–206

Kumar P, Sharma SM (2015) Magnetic separation: an overview of purification methods for proteins. Int J Appl Res 1(12):450–459

Lopez-Exposito I, Recio I (2008) Protective effect of milk peptides: antibacterial and antitumor properties. Adv Exp Med Biol 606:271–293

Macwan SR, Dabhi BK, Parmar SC, Aparnathi KD (2016) Whey and its utilization. Int J Curr Microbiol Appl Sci 5(8):134–155

Maya D, Ritu P (2016) Formulation of fruit (orange juice) and whey based beverages flavoured with different herbs using natural sweetener as stevia. Int J Adv Res 4(10):2183–2187

Mirabella N, Castellani V, Sala S (2014) Current options for the valorization of food manufacturing waste: a review. J Clean Prod 65:28–41

Nitschke M, Costa SGVAO (2007) Biosurfactants in food industry. Trends Food Sci Technol 18:252–259

Panghal A, Kumar V, Dhull SB, Gat Y, Chhikara N (2017a) Utilization of dairy industry waste-whey in formulation of papaya RTS beverage. Curr Res Nutr Food Sci 5(2):168–174

Panghal A, Virkar K, Kumar V, Dhull SB, Gat Y, Chhikara N (2017b) Development of probiotic beetroot drink. Curr Res Nutr Food Sci 5(3)

Poonia A (2017) Potential of milk proteins as nanoencapsulation materials in food industry. In: Nanoscience in Food and Agriculture, vol. 5. Springer, pp 139–168

Poonia A, Arti K (2018) Development and shelf-life evaluation of functional Rabadi (a fermented pearl millet product) by incorporation of whey. Indian J Nutr Diet 55(3):318–333

Prahlad SN (1954) Studies on the technological aspects of processing and storage on dairy by-products, ghee residue. M.Sc. Thesis, Banaras University, Banaras, 1954. (Cited by Sukumar, D.E., 2007. In: Outlines of Dairy Technology. Oxford University Press, Madras)

Ramesh P, Valavan SE, Gnanaraj PT, Omprakash AV, Varun A (2018) Nutrient composition of ghee residue. J Pharmacogn Phytochem 7:3316–3319

Reddy PVS, Khan AQ (1978) Recovery of ghee from ghee residue and utilization of the same residue for chocolate preparation. In: 20th International Dairy Congress, France, pp 984–985

Rittmanic S, U.S (2006) Whey proteins in ready-to-drink beverages. U.S. Dairy Export Council, Applications Monograph. Beverages. pp 1–8

Roff WR, Scott JR (1971) Section 21 – regenerated proteins. In: Roff WJ, Scott JR (eds) Fibres, films, plastics and rubbers: a handbook of common polymers, 1st edn. Butterworth, London, pp 197–204

Rouimi S, Schorsch C, Valentini C, Vaslin S (2005) Foam stability and interfacial properties of milk protein-surfactant systems. Food Hydrocoll 19(3):467–478

Saha P, Ray PR, Hazra T (2017) Evaluation of quality and stability of chhana whey beverage fermented with lactic acid bacteria. Asian J Dairy Food Res 36(2):112–116

Santha IM, Narayanan KM (1978) Composition of ghee-residue. J Food Sci Tech-Mys 6:24–27

Silk DB, Fairclough PD, Clark ML, Hegarty JE, Marrs TC, Addison JM, Burston D, Clegg KM, Matthews DM (1980) Use of a peptide rather than free amino acid nitrogen source in chemically defined “elemental” diets. JPEN J Parenter Enteral Nutr 4(6):548–553

Smithers GW (2008) Whey and whey proteins – from ‘gutter-to-gold’. Int Dairy J 18:695–704

Sothornvit R, Olsen CW, McHugh TH, Krochta JM (2007) Tensile properties of compression-molded whey protein sheets: determination of molding condition and glycerol- content effects and comparison with solution-cast films. J Food Eng 78(3):855–860

Stoliar M, U.S (2009) Whey ingredients in bakery products. U.S. Dairy Export Council, Applications Monographs. Bakery. pp 1–8

Tamine AY (2009) Dairy fats and related products. Blackwell Publishing Ltd

Tunick MH (2008) Whey protein production and utilization In: Onwalata CL, Huth PJ (abstract) Whey processing functionality and health benefits. Blackwell Publishing Press, Ames, Lowa, pp 1–13

United Nations (2015) The Sustainable Development Agenda, available at: www.un.org/sustainable/development/development-agenda

Vanderghem CP, Bodson S, Danthine M, Paquot C, Deroanne C, Blecker C (2010) Milk fat globule membrane and buttermilks: from composition to valorization. Biotechnol Agron Soc Environ 14:485–500

Verma BB, De S (1978) Preparation of chocsidu burfi from ghee residue. Indian J Dairy Sci pp 370–374

Wadhwa BK (1997) Functional properties of ghee residue. In: Technological advances in dairy by-products. NDRI Publication, pp 141–146

Yadav JSS, Yan S, Pilli S, Kumar L, Tyagi RD, Surampalli RY (2015) Cheese whey: a potential resource to transform into bioprotein, functional/nutritional proteins and bioactive peptides. Biotechnol Adv 33:756–774

Yasaman E (2014) Ultrasound effect on the preservation of dairy products. J Appl Environ Biol Sci 4(1s):82–86

Yilsay TO, Yilmaz L, Bayizit AA (2006) The effect of using a whey protein fat replacer on textural and sensory characteristics of low-fat vanilla ice cream. Eur Food Res Technol 222:171–175

Young S (2007) Whey products in ice cream and frozen dairy desserts [on line]. U.S. Dairy Export Council, Applications Monograph, Ice cream, pp 1–12

Advances in Sugarcane Industry: By-Product Valorization

16

Narendra Mohan and Anushka Agarwal

Abstract

Agro-industrial production and processing generate huge amount of by-product and waste resulting in significant environmental impact and a great deal of investment being incurred during its storage and disposal. Therefore, proper management and efficient waste utilization are essential for environmental sustainability and human health. Sugar industry is said to be the second largest agro-processing industry next to cotton textiles, playing significant role in the growth and development of India's economy. During the processing of sugar, sugar mills generate several waste such as molasses, bagasse, filter cake, waste water, and boiler ash. Utilization of these waste generated from the sugar industry and converting them into value-added by-products is of paramount importance. Since the market price of sugar remains volatile and at times fall below the production cost, innovations and diversification to produce more value-added products i.e. converting "waste to resource" may help the sugar industry to have a better economic sustainability as compared to being a stand-alone sugar factory. The paper elaborates on innovative ways to utilize the by-products and waste of the sugar industry thereby adding value to the existing product line and lowering dependency on revenues from sugar. Improvement in product quality and such innovative management of waste may accelerate the growth of the sugar industry as well as help in maintaining clean and healthy surroundings and providing quality product as per consumer need.

Keywords

Agro-industry · Value-addition · By-product · Diversification · Sugar industry

N. Mohan · A. Agarwal (✉)
National Sugar Institute, Kanpur, India

M. Thakur et al. (eds.), *Sustainable Food Waste Management*,
https://doi.org/10.1007/978-981-15-8967-6_16

16.1 Introduction

The advent of modernization along with technological innovation and advancement in the field of agriculture has significantly increased the agricultural produce. The effect of reformation in the agriculture sector has indeed resulted in quick and swift development of various food processing industries all across globe in order to cater to the needs of the growing population. While on one hand revolution in the food sector resulted in generation of huge quantities of food products as desired by its consumer base, employment opportunities to a large number of people, boosting the economic status, on the other hand, it also resulted in generation of large amount of waste causing sustainability and environmental issues.

Agriculture forms the backbone of the Indian economy supporting more than two-third of the total population of the country and also contributing to the GDP of the country. Therefore, agriculture holds a significant stature in social, political, and economic affairs. Sugarcane is one of the major cash crop in India where more than 50 million farmers are directly or indirectly dependent on the sugar or sugar based industries. India in amongst the top ranked country for production and consumption of sugar of around 33 MMT (crushing season 2018–19) (Indian Sugar 2020) next to Brazil. In spite of many bigs, this industry often encounters issues of economic sustainability to the extent that the issues of pending cane prices areas become a burning issue attracting the government. As it is rightly said, "experience is the mother of all," sugar industry too, through its tireless experience have learnt a lesson and is therefore in the process of transformation from being a stand-alone sugar factory into being an integrated complex having facilities for power export and ethanol production.

The importance of by-product utilization in a more attractive fashion has been realized by the industry as merely depending upon revenues from sugar alone to a greater extend does not sound as a profitable venture in times to come and hence better management of by-products could be a promising asset to the sugar industry as a whole. The by-products of the sugar industry whose potential has not been explored to its maximum, have huge scope to be converted into value-added products through innovative approach and thereby would yield more income to the industry through sale and marketing of such innovative and attractive products rather than sugar. There is a long way to go as it is well said that Rome was not built in a day; similarly changes do not just happen overnight, but a thought process is required to be initiated so as to see a self-sustainable sugar industry in times to come. The potential of revenue generation through utilization of various by-products viz. bagasse, filter cake, and molasses is to be taken up in an innovative manner besides developing technologies for converting huge amount of surplus water into good quality water to meet human needs and earn revenue. In fact, sugar industry is to be converted from a single product factory to multiproduct factory.

16.2 Products and By-Products of Sugar Industry: Existing Scenario and Way Forward

As evident from past 2–3 years, India has grown to be the highest sugar producing geography with sugar production of around 33 MMT (Indian Sugar 2020). Sugar is highly fungible with comparatively small portion of production being amenable to special sugar market. Several policy interventions in regard to facilitating sugar export, and fixation of "Minimum Selling Price" of sugar by the government have helped the sugar industry to be economically viable but still there is a long way to go for achieving economic stability and sustainability in times to come. Therefore, in order to be sustainable, The need of the hour is for better allocation of the present resources in making assorted product line that provides the industry as a cushion for its better survival. Figure 16.1 gives a brief idea of how the by-products of the sugar industry are being utilized in traditional manner. Primarily sugarcane is processed to obtain sugar, the fibrous residue known as bagasse is majorly used for cogeneration,

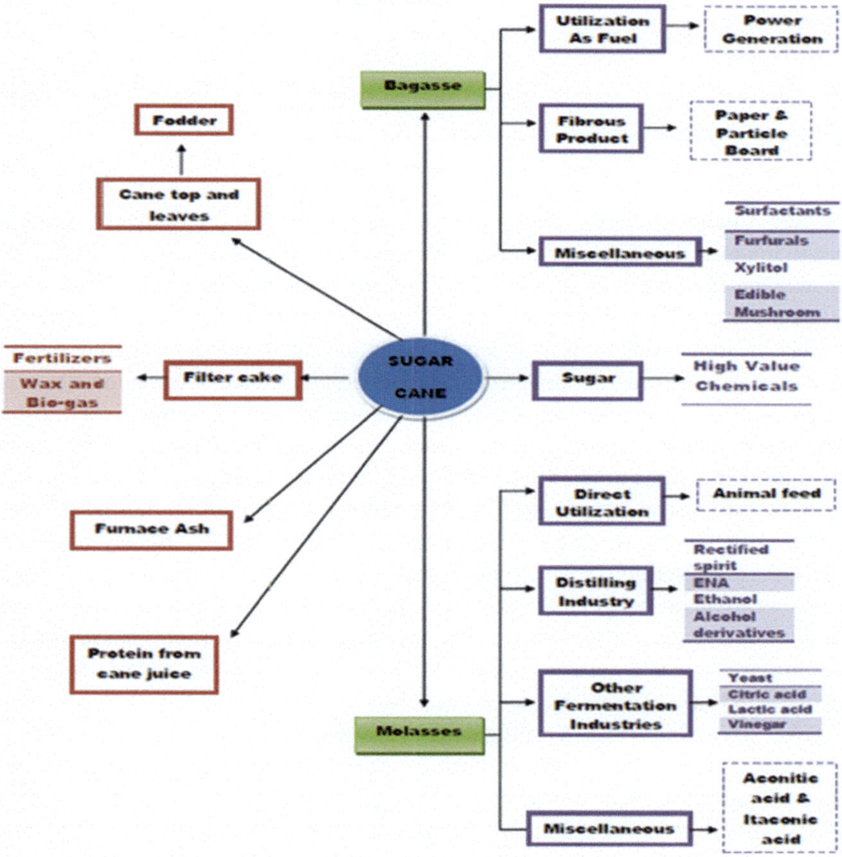

Fig. 16.1 Utilization of products and by-products of sugar industry (existing scenario)

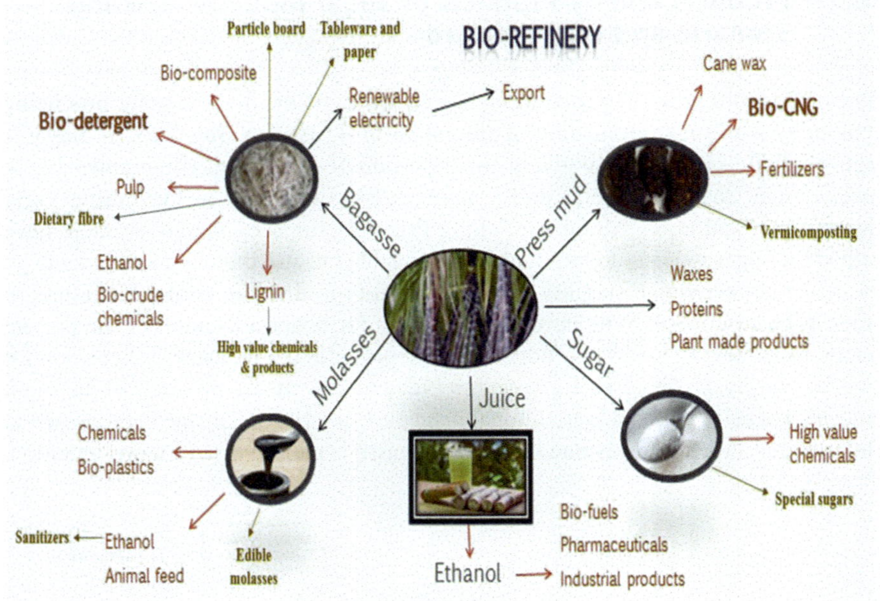

Fig. 16.2 Bio-refinery concept for sustainable sugar sector

molasses is mainly diverted for animal feed or used by distilling industry, and filter cake is being utilized for production of cane wax or is being used as fertilizers. Indian sugar industry is an industry with immense potential. It is an industry where every product and by-product can serve as a potential and economic raw materials for other manufacturers for production of numerous value-added, bio-based products that may have great market prospect and acceptability amongst consumers as well. Some countries for instance have maintain competitiveness through diversification for example, Mitr Phol's total revenue generation from sugar alone stands for around 42% only, while remaining 58% revenue generation is through by-product diversification. In a similar manner, Kaset Thai International Sugar Corporation (KTIS) aims high in reducing its revenues from sugar to 50% from 80% by investing into co-products. Therefore, Indian sugar industry too should start thinking outside of the box, bringing a change in the way the industry looks at the market as well as at its customer is necessary for its survival in present times. There is plethora of possibilities and challenges for sustainable innovations that the sugar industry can look upon in times to come whose potentials are yet to be harnessed to the maximum possible extent (Rao 1997). Keeping this in view Fig. 16.2 illustrates the possibilities and futuristic approach for better utilization of product and by-product that the sugar industry may think upon for better survival in the market. The model as given in Fig. 16.2 gives an idea about the bio-refinery concept the industry must adopt. Such model would help the industry to explore food sector, as well as healthcare sector and also the nutraceutical market as well.

16.3 Innovative and Sustainable Approach Towards By-Product Utilization of Sugar Industry

16.3.1 Bagasse

During the processing of sugarcane for manufacture of sugar, the fibrous biomass left after the extraction of cane juice is termed as bagasse. This fibrous biomass is considered as one of the most important by-product of the sugar industry in terms of volume (28–32% on cane). Usually sugarcane bagasse is used as a fuel for production of steam and electricity which is further utilized by the sugar factories itself and to some extent even exported to the grid. With the passage of time, the charm of bagasse based cogeneration and export of power to grid was fading away as result of reduced power tariffs due to competition with tariffs offered by power generated through non-conventional energy resources and therefore a more comprehensive attitude was required for proper disposal of such huge voluminous by-product. Bagasse, a lignocellulosic biomass is generally composed of cellulose 23%, hemicellulose 12.3%, lignin 9.9%, fat and wax 1.8%, and many other elements such as carbon 48.7%, hydrogen 4.9%, nitrogen 1.3%, phosphorus 1.1%, and ash 2.4% (Pandey et al. 2000). This by-product being widely available, cheap, and also being environment friendly can be a potential source to a huge number of products and therefore its potential is to be fully harnessed by the industry in order to add benefits to the present system. As bagasse is primarily used for fuel, it also finds its application for several other purposes likewise paper and pulp making or it is also seen to be used in cattle feed as well (Mohan and Kanaujia 2019). With rising consciousness and concerns of consumers towards environment, there has been a shift in the use of disposables and non-plastic materials for packaging that are biodegradable, low cost incurred for their manufacturing. This trend has led an altogether different sector for efficient utilization of bagasse. Present times have shown various manufacturers jumping into the production of disposables, tablewares, bottling, packaging materials, particle board, etc. that are processed using bagasse (Poopak and Roodan 2012). The examples as given under Figure 16.3a–h given a broad perspective of attractive use of bagasse for various consumer friendly products.

Sugarcane bagasse contains complex lignocellulosic material which can be used as low cost energy and carbon source for fungal cultures. Generally, fungi are grown on potato dextrose agar (PDA), Sabouraud dextrose agar (SDA), or Cornmeal agar (CMA) which is very expensive. Basically every fungus requires carbon, nitrogen, and energy source to grow and survive. Studies have shown that sugarcane bagasse may meet these requirements and work as a fungal growth medium and can replace expensive media in the market. One quintal of bagasse could yield 87 kg of oyster mushroom (Pleurotus sajor caju) and 79.6 kg of white button mushrooms (Agaricus bisporus). To add to the shelf, sugarcane bagasse may be seen as a rich source of dietary fibre but its major limitation is its low digestibility which is due to association of lignin with cellulose and hemicellulose. Lignin is said to reduce the digestibility of cellulose and hemicellulose by physically protecting them against enzyme degradation. Efforts are being made to overcome this difficulty by using number of

Fig. 16.3 (**a**) Multiple bagasse products: tableware, paper, packaging, bottles, MDF board, plastic composite resins. (**b**) Bagasse tableware/dinnerware-disposable plate, bowl, tray, cutlery. (**c**) Bagasse paper: a4 copier paper, kraft, printing, tissue, toilet, wallpaper, bags, cardboard. (**d**) Bagasse packaging: food, fruits, eggs, beer, perfume, deo, cosmetics. (**e**) Bagasse bottles: mineral water, milk. (**f**) Bagasse bottles: milk, water, wine, oil, medicine. (**g**) Bagasse board: MDF, HDF, panel, walls, designer panel, furniture. (**h**) Bagasse resin composites: (bio plastic) resin for multiple industrial products: automobile parts, chair, dustbins, pen, mobile, computer

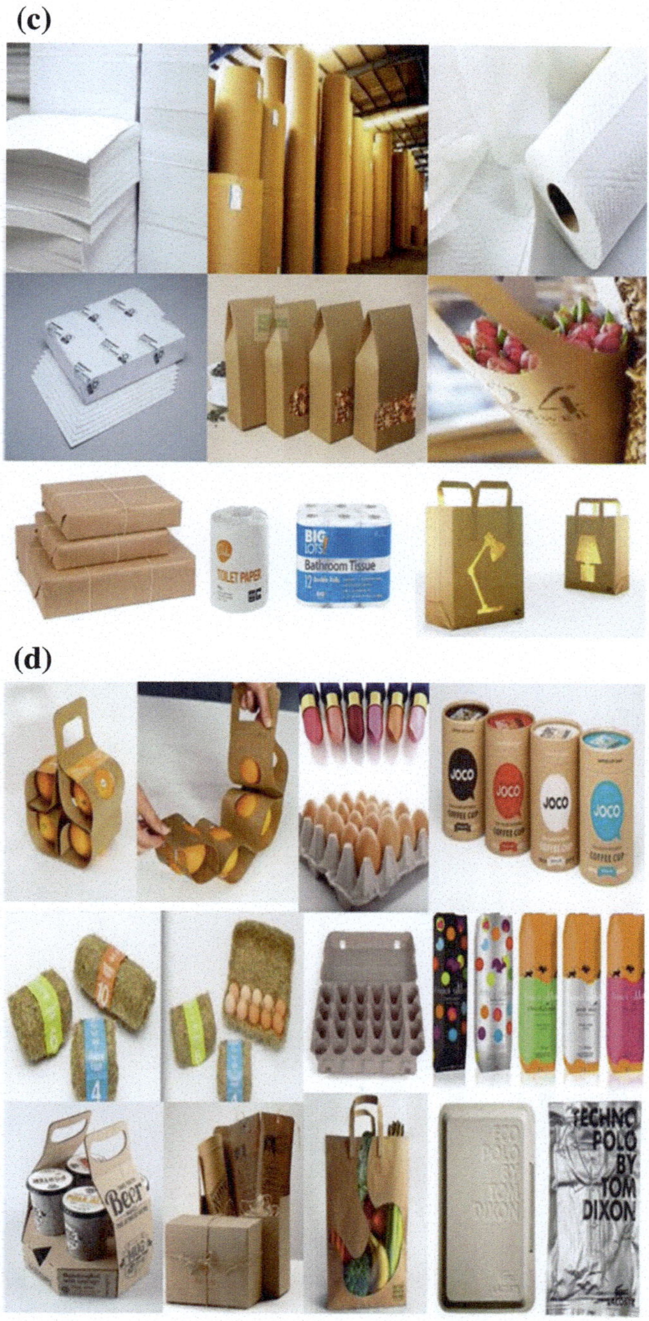

Fig. 16.3 (continued)

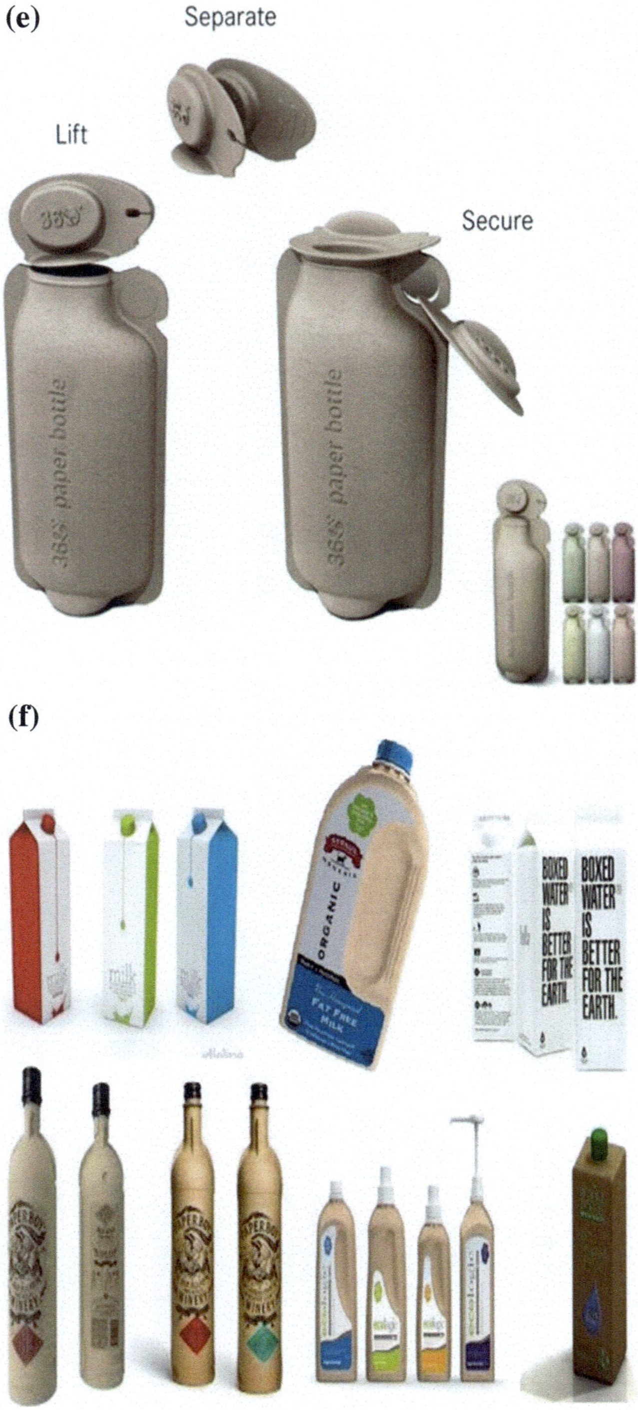

Fig. 16.3 (continued)

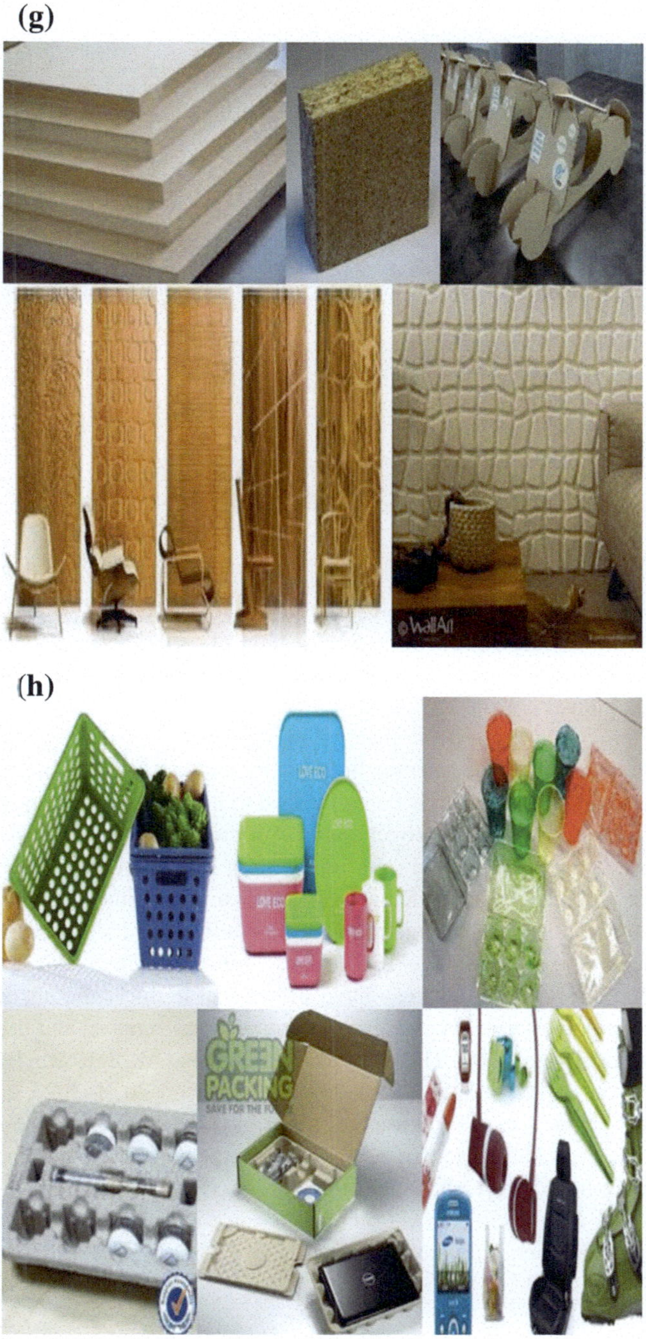

Fig. 16.3 (continued)

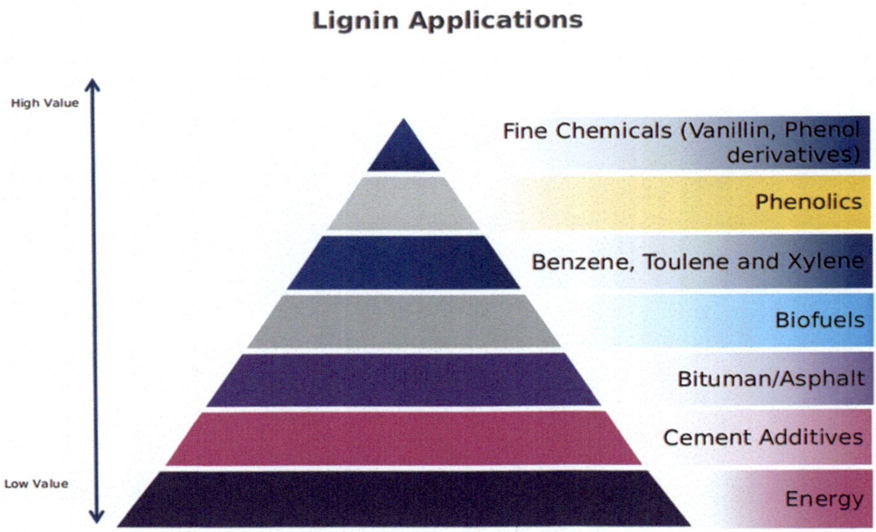

Fig. 16.4 Utilization of lignin for high-value chemicals

chemical and biological treatments towards delignification. Therefore, in times to come, sugarcane bagasse fibre may serve as one of the most popular nutrients in the nutraceuticals industry.

Bagasse being a lignocellulosic material comprises cellulose, hemicellulose, and lignin. While the cellulose and hemicellulose component of this by-product is being exploited for value-addition and is being utilized to its maximum extent possible for production of products viz-a-viz cogeneration, production of dietary fibre, paper and particle board, paper etc. the lignin portion is often left unexplored and untouched. Several literatures have shown that while cellulose and hemicellulose portion of bagasse are of great use for making different value-added products, the lignin portion of bagasse also serves as a base for various high-value chemicals as shown in Fig. 16.4. Studies and market survey have shown that lignin is the upcoming name which has huge potential for manufacture of numerous aromatics and different kind of polymers (Dhiman 2019). The market for lignin-based high-value products is expected to grow at the rate of 3.42% during the period 2018–2023 from current 974.6 million, where it may form a major share in the cement industry as a cement additive. From a global perspective, Europe is expected to be the major market player for lignin-based products which at present accounts for about 34% of the total market share. In a nutshell, innovation and advance research in the field of lignin bio-refinery hold immense potential in near future.

16.3.2 Molasses

Molasses is one of the most valued by-products of the sugar industry. Several value-added products like organic acids, enzymes, etc. can be produced by utilizing molasses in a better and efficient way (Rao 1997). In present times molasses is used primarily by the distillation industry. Although due to the recent pricing policy announced by the Government of India, production of ethanol has gained momentum, but a part of the molasses can still be used for converting it to a value-added product earning higher revenues as compared to those with ethanol production (Gopal and Kammen 2009). Wide range of molasses as obtained from different sources exists in the market such as cane molasses, beet molasses, and refinery molasses. Various MNCs are branding and selling edible molasses @ Rs. 1000/kg or so which is much higher than the normal price of molasses sold by the sugar factories. This edible molasses has wide application ranging from bakery product to enhancing flavouring properties in meat, herbs, chocolate, spices, etc. and also helps in fortifying sweets, savoury, or spice flavours. Many innovative technologies and ideas can be explored to make this product more nutritious and more easily available in the market which therefore opens a new field for the sugar industry to prosper.

Increasing energy demand, depleting natural resources, and the need to reduce carbon dioxide emission have greatly increased interest for the use of biomass as alternative energy source. Bio-ethanol is most abundant bio-fuel for automobile transportation. The Table 16.1 given below gives an idea about the growing population of vehicles been added in the country over the years and hence the possible increase in emission.

Although there are many factors which contribute to air pollution, the quantum of vehicular emission cannot be ignored or underestimated. Unfortunately, out of the 50 most polluted cities of the world, 25 cities are in India as reflected from their poor air quality index (AQI). Bio-ethanol, being a clean and green fuel can play a dominant role in pursuit of reducing emission levels providing a relatively cleaner environment. Example of Brazil is before us which speak for drastic improvement in the air quality after adoption of the EBP 27 programme. In a report published by Indian Sugars (Indian Sugars 2020) União da Indústria de Cana-de-Açúcar UNICA mentioned that Brazil has seen success cases in its large cities with improved pollution levels (particulate matter) close to the World Health Organization levels. Such levels were achieved mainly due to adoption of ethanol by Brazil in vehicle fuel. Therefore, India too can improve its air quality levels by introducing a mandatory ethanol blend and diverting a part of its sugarcane production to produce ethanol fuel. More diversion of cane towards ethanol rather than sugar would help reduce the glut in the sugar market and greater use of the bio-fuel would reduce India's dependence on crude oil imports, ensure energy security, reduce pollution, and generate local jobs. Studies and reports reveal that India is determined for an ethanol blending target of 10% by 2022.

Table 16.1 Automobile domestic sales trends

Category	2013–14	2014–15	2015–16	2016–17	2017–18	2018–19
Passenger vehicles	25,03,509	26,01,236	27,89,208	30,47,582	32,88,581	33,77,436
Commercial vehicles	6,32,851	6,14,948	6,85,704	7,14,082	8,56,916	10,07,319
Three wheelers	4,80,085	5,32,626	5,38,208	5,11,879	6,35,698	7,01,011
Two wheelers	1,48,06,778	1,59,75,561	1,64,55,851	1,75,89,738	2,02,00,117	2,11,81,390
Quadri-cycle[a]			0	0	0	627
Grand Total	*1,84,23,223*	*1,97,24,371*	*2,04,68,971*	*2,18,63,281*	*2,49,81,312*	*2,62,67,783*

[a]Only Aug 18–March 2019 data is available for 2018–19

16.3.3 Filter Cake

Amongst the by-products of the sugar industry, filter cake in particular is second main solid waste generated from the sugar industry with annual production of around 9–10.5 million metric ton (Awasthi et al. 2019). Proper disposal of this by-product is of great concern for the industry. Its management, handling, storage, and transportation become difficult due to high water content of filter cake and also its peculiar smell which causes insect and pest infestation. Filter cake is largely being utilized as bio-fertilizers and is used as compost in sugarcane fields. Since direct use of filter cake poses risk and several limitations, it is used with other fertilizers to improve the fertility of the soil, pH balance in soil, improve drainage, and also to promote growth of healthy microflora to enhance soil quality for better crop management (Prado et al. 2013).

An emerging trend to utilize the press cake for production of Bio-gas or Bio-CNG is gaining momentum in coming times. The press cake contains appreciable proportion of biodegradable organic matter which has very good potential for the production of bio-gas. The bio-gas produced as a result of anaerobic reaction of various degradable substrates including press cake serves as clean energy and may also prove to be a value-added product from the by-product of the sugar industry (Awasthi et al. 2019). Considering filter cake (% on cane) being around 3.5 and limited option available for its commercial exploitation, this by-product having substantial amount of organic material can be used for production of bio-gas, compressed bio-gas/ bio-CNG. With proper investment and planning, this untapped potential can be harnessed which will also help in value-addition for the sugar factories. Although a thorough study relating to the cost of production and the quality of the raw material is to be carried out.

As the sugar industry generates surplus amount of waste, these can be utilized as bio-resource for production of ethanol, animal feed, paper, and many more resourceful products as detailed in the chapter. However, still appreciable amount of the by-products is leftover for disposal and could not be put to better use. For efficient disposal and sustainable utilization of such waste or by-product of the industry, vermicomposting is one such method which as per several studies have found to be an effective method to enrich the soil and farms with organic matter (Hanc and Pliva 2013).

Vermicomposting is a decomposition process wherein coupled action of earthworms and microorganisms on the waste and cow dung mixture takes place. Vermicompost is a peat-like substance or material which is highly porous and is enriched with plant nutrients that facilitate sustainable agriculture by supporting as essential nutrient growth media along with being a hub for several microbes. Using sugar waste alone does not support vermicomposting as the earthworms cannot survive in pure sugar waste. However, when mixed with cow dung, sugar waste such as filter cake or bagasse may serve the purpose for successful vermicomposting process. Cow dung being a rich source of microorganisms enhances the overall nitrogen content of the Vermicompost (Nagavallemma et al. 2006). Figure 16.5 illustrates the use of sugarcane bagasse and filter cake for vermicomposting. In a

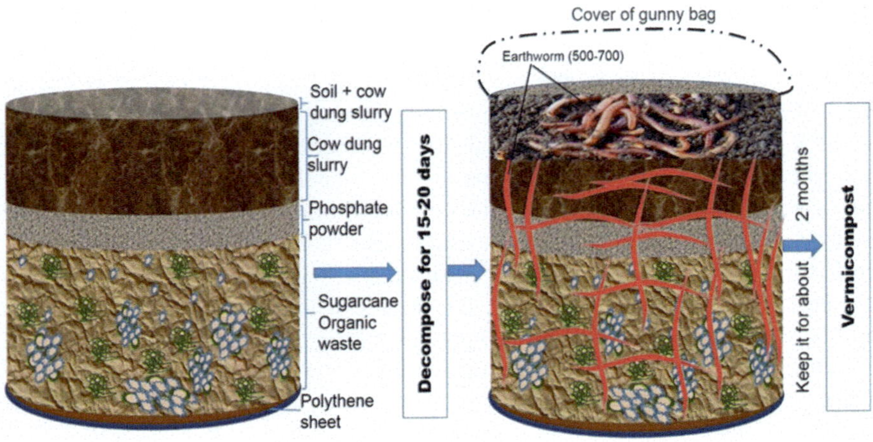

Fig. 16.5 Vermicomposting using sugarcane filter cake and sugarcane bagasse

nutshell, vermicomposting is said to be a friendly, sustainable, and economically viable alternate for utilizing organic waste such as filter cake from sugar industry (Saranraj and Stella 2014).

16.3.4 Water

Sugar industry is responsible for being the major water consumer and effluent generator. Approximately, 20–30 tonnes of water is required in order to cater to the process requirements for production of around one tonne of sugar. Around 10–20% of water is rendered surplus as waste water. The waste water originating from different segments of the sugar factory viz. milling house, process house as well as boiler house all contribute together to form a great volume of waste water that the industry generates which holds potential to cause environment imbalance. It is said that the amount of effluent generated from the industry depends on the cane crushing capacity of the individual industry and the manner in which water usage is managed within the industry. For example, in India, around 1 kL of waste water is generated from one tonne of sugarcane processed by the sugar industry. So, an industry with 2500 tons crushed per day (TCD) capacity would result in generating around 450 ML of waste water for an average crushing season of 6 months (Ingaramo et al. 2009). The waste water generated from the sugar industry is characterized by high biochemical oxygen demand (BOD), chemical oxygen demand (COD), and total soluble solids. Proper disposal of this waste water from the industry is an area of concern worldwide. Studies have shown several methods adopted for wastewater management of sugar industry, for instance, a review published by Kushwaha revealed various advancements and upgradation in aerobic, anaerobic, and physico-chemical treatment techniques (Kushwaha 2015). The studies give a comprehensive idea about the primary, secondary, and tertiary treatment techniques of waste water wherein physico-chemical treatment methods such as ultrasonic (US) and nanoparticle exposure were adopted. Advance methods for example electrocoagulation, ozone treatment, ultrasonic membrane anaerobic system could be promising approached where the sugar industry could innovate and evolve in times to come. In a nut shell, different other innovative technologies could be adopted in the long run with a view to help sugar industry mitigate the problem of waste water disposal as well as stand as an industry that could return certain portion of the water to the society which in turn could be in any form viz-a-viz for irrigation purpose, water suitable for other human needs, and even for drinking purposes. Therefore, the sugar industry can be a hub for "bio-water".

16.4 Conclusion

An integrated strategy towards sustainable waste management and value-addition and utilization of by-products in an innovative manner is an effective approach for economic feasibility and sustainability of the sugar industry altogether. It is anticipated that sustainable waste management strategies will contribute to efficient management of the triple bottom line components of environmental responsibility, economic return, and social development for the overall growth and development of the sugar industry. The essence of success will be "attitude" to invest in a judicious way rather than building capacities. While quality raw material development shall be the priority, diversifications and integrations are going to be the key to success.

References

Awasthi S, Paroha S, Mohan N (2019) Bioenergy from filter cake. In: Proceedings of 77th annual convention of SATI 505-515

Dhiman G (2019) Lignin bio-refinery: an effective biomass conversion to value added product. Patent Blog and Patent News for Legal Services

Gopal AR, Kammen DM (2009) Molasses for ethanol: the economic and environmental impacts of a new pathway for the lifecycle greenhouse gas analysis of sugarcane ethanol. Environ Res Lett 4:044005. https://doi.org/10.1088/1748-9326/4/4/044005

Hanc A, Pliva P (2013) Vermicomposting technology as a tool for nutrient recovery from kitchen bio-waste. J Mater Cycles Waste Manag 15:431–439. https://doi.org/10.1007/s10163-013-0127-8

Indian Sugar - The Complete Sugar Journal, January 2020, volume XX, issue (11)

Ingaramo A, Heluane H, Colombo M, Cesca M (2009) Water and wastewater eco-efficiency indicators for the sugar cane industry. J Clean Prod 17:487–495

Kushwaha JP (2015) A review on sugar industry wastewater: sources, treatment technologies and reuse. Desalin Water Treat 53:309–318

Mohan N, Kanaujia AK (2019) Biomass energy for economic and environmental sustainability in India. Sugar Tech 21(2):197–201

Nagavallemma KP, Wani SP, Lacroix S, Padmaja VV, Vineela C, Babu RM, Sahrawat KL (2006) Vermicomposting: recycling wastes into valuable organic fertilizer, vol 2. International Crops Research Institute for the Semi-Arid Tropics, Hyderabad, pp 1–16

Pandey A, Carlos RS, Nigam P, Soccol TV (2000) Biotechnological potential of agro-industrial residues. I: sugarcane bagasse. Bioresour Technol 74:69–80. https://doi.org/10.1016/S0960-8524(99)00142-X

Poopak S, Roodan A (2012) Environmental benefit of using bagasse in paper production—a case study of LCA in Iran. InTech, London. https://doi.org/10.5772/51553

Prado R, Caione G, Campos C (2013) Filter cake and vinasse as fertilizers contributing to conservation agriculture. Appl Environ Soil Sci 2013:581984. https://doi.org/10.1155/2013/581984

Rao PJM (1997) Industrial utilization of sugarcane and its co-products. ISPCK Publishers & Distributors, New Delhi

Saranraj P, Stella D (2014) Composting of sugar mill wastes: a review. World Appl Sci J 31:2029–2044. https://doi.org/10.5829/idosi.wasj.2014.31.12.546

Part III

Sustainable Food Waste Management Technologies

Mushroom: A Potential Tool for Food Industry Waste

17

Shweta Kulshreshtha and Monika Thakur

Abstract

The generation of food waste is a global issue. Food waste is generated at various levels such as pre- and post-harvest level, consumer level and due to administrative factors and unavoidable weather conditions and pests' attack. Mushroom is nutritious food due to the presence of proteins, vitamins and several nutraceuticals which can fight with nutrient deficiency diseases. Therefore, mushroom is a boon to convert food wastes into food. Besides, it can remediate the pollutants when spent mushroom substrate is introduced in the soil. This leads to the improvement in soil quality. Mushroom waste can also be used for the recovery of nutraceuticals. These approaches are discussed in this chapter in order to make the people aware about this magical zero waste technology. This chapter covers different aspects of mushroom cultivation and removal of waste from different levels and their conversion into food along with the waste removal. In near term, mushroom cultivation will be a boon for the society and human beings.

Keywords

Mushrooms · Nutraceuticals · Zero waste · Food waste

S. Kulshreshtha (✉)
Amity Insttitute of Biotechnology, Amity University Rajasthan, Jaipur, India

M. Thakur
Amity Institute of Food Technology, Amity University Uttar Pradesh, Noida, Uttar Pradesh, India

M. Thakur et al. (eds.), *Sustainable Food Waste Management*,
https://doi.org/10.1007/978-981-15-8967-6_17

17.1 Introduction

The sustenance of human beings depends on food and water resources. Food is an important commodity without which life is not possible. Food resources provide sufficient food which can satisfy the need of every living being. However, food wastes and unequal distribution of the food lead to the crisis and increase the hunger index. Moreover, it increases the cost of the food for the poor people. Wastage of food is associated with different levels from its production to its consumption (Muth et al. 2019). Like, it occurs at harvesting stage, after harvesting due to lack of storage facilities, and throughout the distribution and consumption chain. Food wastage can be seen in any street, municipal waste, garbage waste, and landfill area. According to food wastage footprints, about 1.3 billion tonnes of food is wasted worldwide which is approximately one-third of the total production. United Nations Development Programme reported that food waste in India is 40% of the total food produced. Wastage of food not only affects the ecosystem, human and natural resources but also impairs the economy of the country (Aragie et al. 2018). Therefore, the management of the food resource is the need of the hour in order to prevent food wastages and to feed a huge population.

Food wastes can be disposed in landfill area which gives no financial output. Moreover, it involves the cost of transportation from the houses to landfill area. When accumulated in the landfill area, this waste produces enormous odour and flies and insects which pose harmful effects on human beings. Even, this food cannot be consumed by the animals, therefore giving no benefit or financial return. If food waste is collected separately and used to feed the animals, it gives little return by reducing the demand of fodder. Anaerobic digestion of food waste generates biogas for using as fuel and biogas slurry as nutrient-rich organic supplement for soil. Therefore, this contributes to energy supply, financial benefit along with positive effect on environment (Paritosh et al. 2017). The conversion of food waste into mushroom provides useful products by solid-state fermentation like protein-rich mushroom fruit bodies, vitamins and enzymes (Papoutsis et al. 2020). This is beneficial aspect of food waste management as food waste is giving a new food, i.e. mushroom (Kulshreshtha et al. 2013), which is a viable solution to the problem of protein deficiency among children. Besides, mushrooms cultivation on food waste provides many enzymes and vitamins. Therefore, the bioconversion of food waste into useful products is an affordable technology which gives high return with monetary benefits and protein- and vitamin-rich food along with bioremediation of food waste from the environment.

This paper comprehensively reviews the food or agricultural waste management through mushroom cultivation and production of various nutritional compounds. This will develop the awareness among the people about the management of food waste. The bioconversion approaches will develop food or nutritional compounds from the food waste. This chapter will cover the aspects of bioremediation or bioconversion of food waste into food for protecting the environment and improving human health. It provides a framework for making new policies on waste management.

17.2 Mushroom Cultivation: One Technology with Plethora of Applications

Food wastes are generated at various levels such as pre-harvest level, post-harvest level, consumer level and unconsumed food due to administrative, management and storage issues. Pre-harvest level is associated with on-going food wastage due to unharvested crops, unsatisfactory appearance, rotten and damaged by insects and not meeting customers' expectations (Johnson et al. 2018). Second wastage occurs at post-harvest level on agricultural farms due to uncertainty in weather conditions and attack of insects, improper agricultural operations such as threshing and drying of grains, unavailability of transport system, and imperfect packaging technique. Third, the huge wastage of food occurs at consumer level during transfer from storage to bulk trader, from bulk traders to retailers, from retailers to consumers in the form of unconsumed food, improperly stored food and loss due to improper cooking methods. Fourth part of food loss is associated with the poor administration and laws, insufficient storage centres and political issues (Fig. 17.1).

The prevention of food wastage can satisfy the need of poor people and conquer nutritional deficiency-related problems. There is an urgent need for implementation of strategies to increase the food safety and maintaining its quality. However, implementation of such technologies is time-consuming process. The solution which can be implemented with ease to prevent wastages of food at every level is adopting the method and technology of mushroom cultivation. Mushroom cultivation is widely accepted technology worldwide because it is a sustainable solution with numerous benefits. The fruit bodies of mushroom can be consumed as the source of nutrients like protein (Prandi et al. 2019) and vitamin which help in fighting the problem of malnourishment. During cultivation, mushroom produces various enzymes of industrial and pharmaceutical importance. The fruit bodies certainly considered as functional food due to its pharmaceutical application (Jebapriya et al. 2013). At this instant, the question arises about the disposal of post-harvest spent mushroom substrate. Again, the disposal of spent mushroom substrate provides several benefits. For instance, this can be applied in the soil for improving the soil quality, water retention ability and maintaining moisture content. After the harvesting of mushroom, the remaining substrate is applied to the soil as it supports the growth of beneficial microbes in the soil. During the mushroom cultivation, microbial communities are successively developed in the waste for breaking down organic matter and producing humus. Humus replenishes the soil with its nutritional content and increases the water-retaining ability of soil (Fig. 17.1). Spent mushroom substrate can also be used as substitute for peat (Stoknes et al. 2019).

Many enzymes can also be recovered from spent mushroom substrate. Further, anaerobic digestion of spent mushroom compost provides slurry which can be used as organic amendment for soil and solid residues can be used for animal bedding. Mushroom cultivation by using agricultural waste or food waste not only remediates the waste from the environment but also provides nutrient-rich fruit bodies (Fig. 17.1). Mushroom cultivation technology, if adopted for food or agricultural

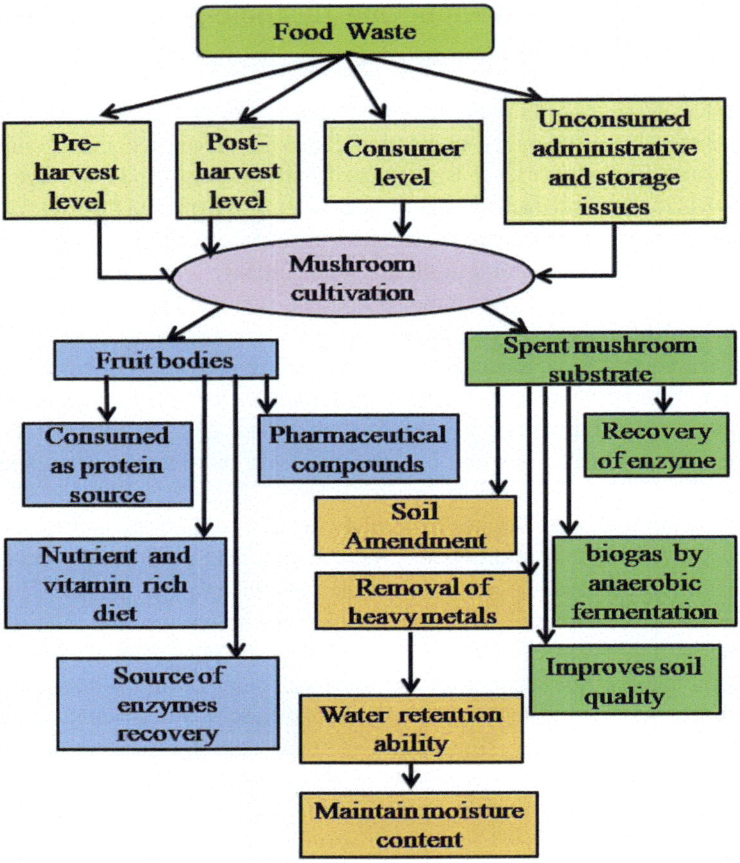

Fig. 17.1 The food waste is produced at different levels, i.e. pre-harvest, post-harvest, consumer level and due to administrative responsibility. In the next step, mushroom cultivation on these wastes is mentioned. Several benefits associated with mushroom cultivation are also mentioned

waste management, will definitely be a boon for our society, environment and country (Grimm and Wösten 2018).

17.3 Vegetable or Food Waste as Substrate for Mushroom Cultivation

The maximum amount of food or vegetable waste is generated at consumer level which can be used for the production of food through mushroom cultivation. The journey of food waste to food is shown stepwise in Fig. 17.2. At first, food waste is segregated from municipal waste and collected by waste collection vehicles and transferred to the nearby compost making fields. This is followed by grinding or shredding of the waste and then drying in sunlight for excessive moisture control.

Fig. 17.2 Processing of food waste into compost and further, its use in mushroom cultivation. Compost is mixed with substrate for mushroom cultivation

Thereafter, food waste is mixed properly for aerobic composting. The prepared compost is mixed with substrate and used for mushroom cultivation. The appeared fruit bodies can be used for consumption (Fig. 17.2). The combinations of date waste, rice straw and corncobs were used by Jwanny et al. (1995) for mushroom cultivation. *Pleurotus ostreatus* NRRL-0366 was cultivated on the date waste alone and its equal combination with rice straw and corncobs which was reported to be the rich source of carbohydrate, protein, essential amino acids like lysine and low RNA content. Similarly, in another experiment, Shashitha et al. (2016) utilized vegetable waste for the production of *Pleurotus ostreatus*. This vegetable waste included the peels of carrot, radish, potato, cucumber and onion was reported to be unsuitable for cultivation of *Pleurotus ostreatus* due to the absence of spawn run, mycelium spread and fruit body formation. However, mixing of this waste with agricultural residues like paddy straw, rice husk, wood shaving and sugarcane bagasse in equal amount provided significant growth with improvement in nutrient quality (Shashitha et al. 2016). When the 20% and 30% of vegetable wastes, i.e. pea pod shell, radish leaves, cauliflower leaves and Brasscia straw, were mixed with 80% and 70% of paddy straw, respectively, for the cultivation of *Pleurotus ostreatus*, a better yield and biological efficiency were achieved as compared to paddy straw alone. A significant increase in protein and amino acid content and decrease in total and reducing sugar content were noticed (Singh and Singh 2014). Likewise, lemon pulp and papaya fruit were mixed with rice straw for the cultivation and fruiting of *Pleurotus ostreatus* NRRL-0366S by solid-state fermentation. The maximum yield was provided by the substrate having equal combination of lemon and rice straw; nevertheless, the maximum activity of amylase was provided by the combination of lemon pulp with papaya and pectinylase by lemon pulp. In contrast to this, cellulase and invertase activities were higher in papaya (Rashad et al. 2009). The peanut waste is produced in the form of hulls and nuts which can also be used for the supplementation of substrate. The substrate with different amount of peanut waste was used for the cultivation of *Pleurotus ostreatus*, and 2% of peanut hulls offered a better yield with 61% biological efficiency and good copper and lower Mn content (Zied et al. 2019). In another study, citrus peel was used in combination with cotton waste for the cultivation of *Pleurotus eryngii*. The combination of cotton waste (96%) and citrus peel (4%) provided a good yield and biological efficiency of *Pleurotus eryngii* as compared to control (Jahangir et al. 2018).

As India is a mega-biodiversity region, it generates huge amount of vegetable waste, plant waste and agricultural wastes which are usually incinerated. In Mexico, wastes from fruits, legumes, pods, leaves, straws and flowers were used for the cultivation of *Pleurotus ostreatus* which grew with 17.65 to 180% biological efficiency. After harvesting, spent mushroom compost was used as an organic amendment for improving the soil quality. The implementation of market and this facility in India will help to obtain nutritious food throughout the year due to availability of one or more of mentioned wastes as it is agriculturally based nation (Aguilar-Rivera et al. 2017).

The method of incorporating vegetable waste in the mushroom substrate was proved to be an efficient way to grow nutrient-rich mushroom. Can these vegetable or fruit wastes be used for biomass production of mushroom? To find out the answer, Behera and Gupta (2015) conducted an experiment by comparing the growth of *Russula, Lentinus* and *Pleurotus* sp. on different chemosynthetic media (Tien and Kirk medium, mushroom complete medium, yeast malt extract medium, glucose yeast extract peptone medium, malt extract broth medium and Sabouraud dextrose broth medium) with different types of vegetable of fruits' wastes like drumstick peel medium, potato peel medium, carrot peel medium, bottle gourd peel medium, litchi peel medium, papaya peel medium, pointed gourd peel medium, chopped grass medium, little gourd peel medium, pumpkin peel medium and rich gourd peel. All three mushroom species were reported to grow well on mushroom complete media as well as papaya peel, drumstick peel, carrot peel and bottle gourd peel medium. Therefore, these vegetable wastes can be used in media for bulk cultivation or biomass production (Behera and Gupta 2015).

Tea is the hot water extract of tea leaves which is discarded as waste after extraction of water-soluble compounds. Tea waste was used by mixing with substrate for cultivation of *Pleurotus* sp. The substrate with 40–60% tea leaves provided a maximum yield. This could be used to prevent the discharges of used tea leaves from hotels, cafes and tea stalls (Yang et al. 2016). Recently, spent coffee grounds, generated from coffee processing industries, were also used for *Pleurotus ostreatus* cultivation in liquid and solid culture. The caffeine was partially degraded to xanthenes via sequential N-demethylation to theophylline (1,3-dimethylxanthine) and 3-methylxanthine, although both paraxanthine and theobromine accumulated in the mushroom substrate. Therefore, consumption of these mushrooms did not provide any health benefit to the consumer due to the presence of partially degraded caffeine constituents (Carrasco-Cabrera et al. 2019).

A large amount of oil palm (*Elaeis guineensis*) wastes is generated in the form of empty fruit bunches which was also used for *Volvariella volvacea* cultivation. The results revealed that cutting of empty fruit bunches into small pieces for *Volvariella* production is nonessential factor. Moreover, the composting is required only for 8 days as further composting reduced the yield. Amendment of this substrate with fertilizer resulted in a higher yield and improved nutritive value of *Volvariella* (Triyono et al. 2019).

Water hyacinth is an edible aquatic plant which is popularly used in soups in Thailand. It is also used to treat water due to its nitrogen and phosphorous absorption ability. This water hyacinth can also be used in the substrate for *Pleurotus sajor-caju* cultivation. The crude protein, ash, cellulose-lignin and reducing sugar content increased; however, crude fibre, lipid carbohydrate, lignin, cellulose and hemicelluloses contents decreased. Enzymatic profiling showed increased activity of carboxy methylcellulases (CMCases), avicelase and amylase, moderate activity of cellobiase and very low activity of pectinase and xylanases enzymes (Shamim et al. 2017). Water hyacinth can also be converted into nutritionally rich food in the form of mushroom. This will be a useful method of disposing water hyacinth waste in the place where it is not used in food and discarded as waste.

17.4 Composted Food Wastes for Mushroom Cultivation

In the aforesaid paragraph, food waste is utilized in its raw form but in different combinations. In many cases, food waste was composted and further used for mushroom cultivation. This is not a new technique and adopted by Block for converting garbage wastes into food. Garbage waste consisted of waste paper and vegetable waste was composted prior to its application for the cultivation of *Agaricus campestris*. This compost did not affect the quality and yield of mushroom fruit bodies and found to be a better option as compared to compost developed by sewage sludge. Jo et al. (2013a) separated the food waste from the municipal waste and used it to prepare compost for the cultivation of *Ganoderma lucidum*. This compost was mixed with rice bran and oak sawdust. 15% of the food waste compost increased the yield of *Ganoderma* fruiting bodies as compared to the control (rice bran +oak dust). The only limitation of this method is high sodium concentration which affects the yield (Jo et al. 2013a). In another experiment, Jo et al. (2013b) composted food waste was used for the cultivation of *Ganoderma lucidum*, *Lentinula edodes*, and *Pholiota adipose*. Results revealed that higher yield of mushroom fruit bodies was obtained by the addition of food waste compost as compared to non-supplemented control. However, calcium, magnesium, sodium and potassium were remaining unaffected by the supplementation of the composted food waste in the substrate. The higher amount of mineral was reported in *Pholiota adipose* as compared to the other mushrooms.

17.5 Anaerobically Digested Food for Mushroom Cultivation

The disposal of unconsumed food and vegetables supports the growth of flies and insects. Moreover, it is not suitable for the agricultural fields. Anaerobic digestion of food waste along with animal manure makes its suitable for mushroom cultivation due to nitrogen-rich digestate. *Pleurotus ostreatus*, when grown on an aerobically digested waste, produced fruit bodies with good biological efficiency. The yield of mushroom on this substrate was equivalent to that of on standard recipes of agricultural waste (O'Brien et al. 2019). When this anaerobically digested nitrogen-rich digestate was mixed with paper, i.e. source of carbon, and used for the cultivation of *Agaricus arvensis*, *A. bitorquis* and *A. subrufescens*, the yield and biological efficiency of mushrooms increased (Jasińska et al. 2014). According to Stoknes et al. (2013), food digestate, obtained by anaerobic digestion, is a good substitute of chicken manure in the straw compost bed for *Agaricus* species. Besides this, biogas generated during anaerobic digestion can be used as fuel to sterilize the mushroom bed. Post-harvest mushroom bed can be used in agricultural fields to improve soil quality.

17.6 Pre- and Post-Harvest Agricultural Waste for Mushroom Cultivation

Agricultural and agro-industrial wastes are generated in massive amount and pose the problem of their disposal. Mushroom consumption is progressively increasing, thereby increasing their demand. Mushroom cultivation is based on the utilization of agricultural and agro-industrial wastes, their supplementation with some nutrients and the combination of different agricultural wastes (Thakur and Lakhanpal 2015). For instance, cotton seed was reported to be a better substrate as compared to the paper waste (Girmay et al. 2016). The detailed literature on the use of agricultural and agro-industrial wastes for the cultivation of different mushroom species is mentioned in Table 17.1. Mushroom cultivation provides many nutritional advantages along with remediating agricultural waste from environment. Cogorni et al. (2014) reported that *Pleurotus sajor-caju* can grow on peach palm leaves and maximum efficiency was achieved on augmenting the substrate with 10% rice bran fraction. The cultivated fruit bodies showed low biological efficiency (4.5%) but good quality. The mixing of harvested *Pleurotus sajor-caju* mushroom powder with wheat flour reduced the content of sugar and fat. 10% of mushroom powder, when supplemented with the wheat flour, augments the fibre, phosphorus, iron, potassium, protein and riboflavin without any change in physicochemical characteristics like moisture, colour and wet gluten (Cogorni et al. 2014). Therefore, mushroom

Table 17.1 The cultivation of mushroom on different agricultural and agro-industrial wastes

S. No	Mushroom	Agricultural or agro-industrial waste	Result	Yield	Biological efficiency	References
1	*Pleurotus florida*	Coffee industry wastes like coffee cherry husk, coffee parchment husk, silver skin, coffee spent wastes, dried leaves with wheat bran	All types of coffee wastes were suitable with wheat bran for *Pleurotus* cultivation	220 g/100 g of substrate in four flushes	Not given	Murthy and Manonmani (2008)
2	*Pleurotus ostreatus*	Solid olive mill wastes from olive oil industry	Wheat straw amendment enhanced the productivity	2 kg/bag	Not given	Mansour-Benamar et al. (2013)
3	*Agrocybe cylindracea, Pleurotus cystidiosus, P. eryngii, P. ostreatus, P. pulmonarius*	Two-phase olive mill waste	Supplementation beyond 60% posed a negative impact		120–135% for *Pleurotus* spp. and 125% for *A. cylindracea*	Zervakis et al. (2013)
4	*Pleurotus sajor-caju*	Peach palm leaves	Good-quality mushroom but low biological efficiency	48.4%	4.5%	Cogorni et al. (2014)
5	*Pleurotus ostreatus*	Date palm wastes with cornmeal and wheat bran	Corn meal was superior over wheat bran as a supplement in all treatments	>140 g	300%	Alananbeh et al. (2014)
6	*Pleurotus ostreatus, Pleurotus cystidiosus*	Sawdust, corncob, sugarcane bagasse	100% of corncob provided the lowest time for the first harvest of both mushrooms	(270.60 g/bag from corncob (100%)	35.7 for *P. ostreatus* and 40.6 for *P. florida*	Hoa et al. (2015)

(continued)

Table 17.1 (continued)

S. No	Mushroom	Agricultural or agro-industrial waste	Result	Yield	Biological efficiency	References
7	*Cordyceps militaris*	Cottonseed shells, corncob particles, Italian poplar saw dusts and substrates spent by *Flammulina velutipes*	Cottonseed shells and corncob particles were the best substrate for growth and production of bioactive compounds like cordycepin, adenosine and d-mannitol	22 and 20 g per bottle	Not given	Lin et al. (2017)
8	*P. eryngii* and *P. ostreatus*	Wheat straw, cardboard, spent coffee ground	Oat grains were the best source for mycelia extension and density levels of *Pleurotus ostreatus*	50% of wheat straw +50% of cardboard combination was best	Varies as per combination	Nguyen and Ranamukhaarachchi (2020)

cultivation is not only an acceptable method to get rid of agricultural waste but also a method to solve the problem of nutritional deficiency.

According to Singh et al. (2007), the extract of agricultural waste can be added to the basal medium to develop a suitable medium for the cultivation of mushroom. For instance, mango leaves extract was supplemented in the basal medium to produce mango leaves extract based medium. The extract containing medium was reported to be better than Sabouraud's dextrose agar. The supplementation of sugarcane bagasse and coconut coir with dried mango leaves extract supported the mycelial growth of *Pleurotus sajor-caju* and *Pleurotus florida*.

17.7 Pre- and Post-Harvest Agricultural Waste for Enzyme Production Through Mushroom Cultivation

Agricultural residues are rich in lignocellulosic material and, therefore, can act as possible substrate for enzymes production. These materials are the source of carbon compounds and induce different enzymes for their biosynthesis and production. The enzymatic profile of the mushroom varies according to the substrate used, culture conditions and different stages of mycelia growth (Nighojkar et al. 2019; Patidar et al. 2018). Sherief et al. (2010) cultivated *Pleurotus ostreatus* on rice straw and sawdust and reported that exoglucanase, endoglucanase, carboxymethyl cellulase and pectinase enzymes produced in higher amount using sawdust than rice straw in both mycelia and fruiting stage. This shows the effect of substrate on enzyme production. In the same experiment and on the same substrate, i.e. sawdust, the production of xyalanase was higher (21.0 U/g) during mycelial growth, while it decreased at the fruiting stage (11.0 U/g). This showed the effect of different stages of mushroom on enzyme production. Therefore, commercial application requires in-depth study about the substrate, conditions and stages of mushroom growth for the production of optimum amount of enzymes. A detailed description of enzyme production and recovery through mushroom substrate is given in Table 17.2.

17.8 Enzymes as by-Product During Mushroom Cultivation on Food Waste

Mushroom cultivation on food waste provides nutrient-rich fruiting bodies. Besides, it produces enzymes for their growth and degradation of substrate for the acquisition of nutrients. In this whole process, the enzymes are accumulated in the substrate that can be recovered after harvesting mushroom fruit bodies. The solid-state fermentation of onion waste by *Pleurotus sajor-caju* provides nutrient-rich fruit bodies as well as pectinase enzyme (Pereira et al. 2017). 29.4 U/g pomace polygalacturonase enzyme was recovered when strawberry pomace was used for the cultivation of *Lentinus edodes*. Further, the polygalacturonase enzyme is produced by same mushroom on apple pomace (20.1 U/g), cranberry pomace (14.0 U) and strawberry pomace (29.4 U/g). In this case, strawberry pomace provided a highest yield, while

Table 17.2 The cultivation of mushroom on different agricultural and agro-industrial wastes for the production of enzymes

S. No.	Mushroom	Agricultural or agro-industrial waste	Enzyme	Yield	References
1	*Lentinus edodes*	Agricultural waste	Polygalacturonase	Produced at relatively lower pH of 5.0	Piccoli Valle et al. (2001)
2	*Pleurotus ostreatus*	Rice straw and sawdust	Exoglucanase, endoglucanase, CMCase and pectinase	Exoglucanase, endoglucanase, CMCase and pectinase produced higher in case of using sawdust than rice straw both in mycelial and fruiting stages.	Sherief et al. (2010)
3	*Pleurotus sajor-caju*	Sugarcane press mud	Cellulase (endo-β-1, 4-glucanase, exo-β-1, 4-glucanase and β-glucosidase)	The highest activities of endo-β-1, 4-glucanase, exo-β-1, 4-glucanase and β-glucosidase were 13.94, 8.84 and 12.29 units (μmole of glucose released/min/g substrate)	Pandit and Maheshwari (2012)
4	*Pleurotus sajor-caju*	Lathyrus sativus (Khesari plant) and sugarcane bagasse	Cellobiase, carboxy methyl cellulase and avicelase	The level of sugar and protein enhanced cellobiase, carboxy methyl cellulase and avicelase activity of crude culture extracts	Karim et al. (2014)
5	*Pleurotus pulmonarius*	Orange waste	Pectinase (hydrolytic enzymes); laccase (oxidative enzyme); amylase, carboxymethyl cellulase and xylanase; manganese peroxidase, β-glucosidase and β-xylosidase	Highest pectinase activity of 9.4 U/mL after 35 days; and 12.2 U/mL of laccase obtained after 20 days of cultivation. Amylase, carboxymethyl cellulase and xylanase were less than 1.5 U/ml. Also produced low amount of manganese peroxidase, β-glucosidase and β-xylosidase	Inácio et al. (2015)

6	*Pleurotus ostreatus* DBUI14 and *P. pulmonarius* DBUI002	Using rice bran, corncob and sawdust	Glucanases and β-glucosidase	*Pleurotus ostreatus* DBUI14 cultivated on corncob had the lowest endoglucanase but highest exoglucanase activity when grown on rice bran	Ekundayo et al. (2017)
7	*Pleurotus sajor-caju*	Water hyacinth, wood chips and paddy	Lignocellulolytic enzymes such as cellulase, xylanase and pectinase	The maximum activity of enzymes was obtained on 28th and 56th day of culture growth	Padhye and Atawane (2017)

cranberry pomace provided least suitable after 40 days of culture (Zheng and Shetty 2000). Onion waste was also used for mushroom cultivation and pectinase enzyme recovery. The yield and biological efficiency of *Pleurotus sajor-caju* on rehydrated onion waste were 45.73 and 4.66%, respectively. Dried onion waste did not support the fruit bodies formation. After harvesting, the pectinase enzymes activity was 4.82 U/ml, which was optimum at pH between 3 and 6 and optimum temperature at 80 °C (Pereira et al. 2017).

According to Economou et al. (2017), laccase enzyme can be recovered through successive cultivation of *Pleurotus ostreatus, Pleurotus pulmonarius, Ganoderma adspersum, Ganoderma resinaceum* and *Lentinula edodes* strains on supplementing spent mushroom substrate with wheat bran and soybean flour. Non-supplemented spent mushroom substrate was an efficient substrate for all mushrooms due to highest growth rate. *G. resinaceum* was the fastest colonizer. Moreover, SMS supplementation did not support the growth, fruiting and laccase production by *Lentinus edodes*. *P. pulmonarius* possessed the highest laccase activity (44,363.22 U/g) (Economou et al. 2017).

17.9 Application of Post-Harvest Mushroom, Mycelium Laiden Agricultural or Food Waste–Beneficial Products and Approaches

In this section, light is shed on the application of mushroom and its spent mushroom substrate derived from food or agriculture waste. Mushroom cultivation technology focuses on the conversion of food waste into protein and other nutrient-rich fruit bodies which can be used for consumption. Again, question arises about the disposal of spent mushroom substrate which is left over after the harvesting of mushroom fruit bodies. As mentioned earlier, spent mushroom substrate is used for recovery of enzymes. Another application is based on their use as organic amendment for improving the soil quality and remediating the heavy metals from soil and wastes.

17.9.1 Recovery of Vitamins and Nutraceuticals from Spent Mushroom Substrate and Mushroom Waste

Mushroom units generate mushroom scrapings as waste. These scrapings include the pieces of mushroom caps and stalk which are discarded due to not meeting the quality standards and customers' or retailers' specification. As mushrooms are rich in many nutraceutical compounds, the discarded mushroom waste can be used for their recovery. Mushrooms are rich in ergosterol which is a precursor of vitamin D_2. This can be converted to the vitamin D_2 after exposure to natural light or UV light (Papoutsis et al. 2020). This recovered vitamin D_2 can be used in food and pharmaceutical industries.

The stalk bases from *Agaricus bisporus* are also discarded as waste from mushroom production units. However, this mushroom is rich not only in vitamin D_2 but also in chitin. Vitamin D_2 was produced by exposing the waste to UV-B light, while chitin was also recovered by N-deacetylation and converted into chitosan (Bilbao-Sáinz et al. 2017). The food waste of mushroom can be recycled back to useful compound which can be useful for food or pharmaceutical industries (Fig. 17.1).

17.9.2 Recovery of Bioethanol

Lentinus edodes is cultivated using wood logs which are discarded as waste after harvesting of mushroom fruit bodies. *L.edodes* inoculation in the wood leads to the degradation of lignin. Lignin is considered as enzymatic hydrolysis inhibitor in sugar production. The yield of total sugar was higher in the *Lentinus edodes* inoculated wood logs as compared to uninoculated wood log. This is the main reason of production of higher amount of alcohol by *L.edodes* inoculated wood logs (12 g/L) as compared to uninoculated wood logs (8 g/L). Therefore, *Lentinus edodes* inoculated wood is considered as a suitable source of fermentable sugars which further used in bioethanol production (Lee et al. 2008).

17.9.3 Improvement of Soil Quality

Spent mushroom substrate, when applied in the soil, can increase the nitrogen content and decrease the carbon content. The reduction in carbon dioxide is the measure of soil functioning and quality. In a study, *Ganoderma* sp. was cultivated on the broad bean stalks, cotton stalks, maize straw, rice straw, sugarcane bagasse and wheat straw supplemented with wheat bran and corn gluten, and post-harvest waste was introduced in the soil (Rashad et al. 2019). These substrates reduced the concentration of carbon dioxide, decreased the carbon and increased the nitrogen content as well as increased the reproduction of reniform nematodes. All these factors together improved the soil quality and plant growth (Rashad et al. 2019).

Spent mushroom substrate was inoculated in the soil for the cultivation of pineapple. The successful cultivation of pineapple in amended soil with high efficiency proved the suitability of spent mushroom compost as organic amendment to improve the soil quality (Adedokun and Orluchukwu 2013).

17.9.4 Mycoremediation of Pollutants

Mushrooms have been used for mycoremediation of pollutants. Mycoremediation is an innovative biotechnological application that uses fungus for in situ and ex situ cleanup and management of contaminated sites (Thomas et al. 2009). Mushroom performs a wide variety of functions in ecosystem and may be a clean, simple and relatively inexpensive method of pollutant remediation, especially if spent mushroom substrate or compost is used. An edible and medicinal mushroom and its spent mushroom substrate also play an important role as natural environment remediator (Kulshreshtha 2018, 2019; Thakur 2019a, 2020).

The toxins (including mercury, PCBs and dioxins, etc.) present in the soil are added to our food chain and become more concentrated at each and every step due to the process of biomagnifications. Mushroom mycelia can destroy these toxins in the soil before they enter our food chain. Mycoremediation through mushroom is thus a biological mechanism to destroy, transform or immobilize environmental contaminants (Adenipekun and Lawal 2012; Kulshreshtha et al. 2014; Thakur 2019b). Hence, it restores and improves the characteristics of depleted and polluted soil. In a study, the spent mushroom substrate improved the quality of soil by nutrient supplementation and by absorbing pollutants or heavy metals from the soil. Hlerema et al. (2017) reported the use of mushroom for cadmium absorption from the pineapple wastes. In this case, cadmium-rich pineapple waste was employed as substrate for the cultivation of mushroom in combination with maize residues. The concentration of cadmium was found to be very high in the fruit bodies obtained from pineapple waste alone, which represented the cadmium adsorption ability of mushroom (Hlerema et al. 2017) reported. Thus, mushrooms and mycelium-coated spent mushroom substrate can be used to remediate cadmium from the soil. The pivotal role of mushroom and their limitations for mycoremediation must be understood in order to utilize them more efficiently.

Now, the question is "how do mushroom and spent mushroom substrate work?" In order to do their work, the extrinsic and intrinsic growth factors viz. right temperature, nutrients and amount of oxygen must be present in the soil along with groundwater. The right combinations of these factors help the mycelium to degrade or adsorb the pollutants. As the process of remediation is over, the mushroom mycelium disappears because there is no more pollutant for them to eat.

Mushrooms can breakdown long-chained toxins into simpler less toxic chemicals. They remove metals from soil by channelling them to mushroom fruiting bodies. They essentially use and digest these toxins as nutrients. Even the enzymes secreted from mycelium can decompose some of the most resistant hazardous toxic materials made by humans or nature. Mushroom possess the biochemical and ecological capacity to degrade environmental organic chemicals and to decrease the risk associated with metals, metalloids and radionuclides, either by chemical modification or by influencing chemical bioavailability. Furthermore, the ability of these mushrooms to form extended mycelial networks, the low specificity of their catabolic enzymes and their independence from using pollutants as a growth substrate make them well suited for bioremediation processes (Hauke et al. 2011). According to Kulshreshtha et al. (2011), mushroom fruit bodies do not absorb pollutants and genotoxicants, if they have the ability of biodegradation.

Mushrooms are among the nature's most powerful decomposers, secreting strong extracellular enzymes due to their aggressive growth and biomass production (Elekes and Busuio 2010). These enzymes include lignin peroxidases (LiP), manganese peroxidase (MnP) and laccase, etc. Thus, carbon sources such as sawdust, straw and corncob can be used to enhance degradation rates by these organisms at polluted sites (Adenipekun and Lawal 2012). *Phanerochaete chrysosporium*, *Agaricus bisporus*, *Trametes versicolor* and *Pleurotus ostreatus* among many mushrooms have been reported in the decontamination of polluted sites (Adenipekun and Lawal 2012). Sesli and Tuzen (1999) reported that mushrooms can be used to evaluate the level of environmental pollution and to remediate the metal-polluted soil. Also, many studies have been carried out to evaluate the possible danger to human health from the ingestion of mushrooms containing heavy metals (Tisma et al. 2010; Ouzouni et al. 2009; Sesli and Tuzen 1999).

Mushrooms have been used for bioremediation of pesticides, degradation of petroleum hydrocarbons and lignocellulolytic wastes in the pulp, paper and cardboard industry (Kulshreshtha et al. 2013). *Lentinus edodes* (Shiitake mushroom) can degrade pentachlorophenol, a broad-spectrum biocide, that is more toxic than DDT in the soil (Adenipekun and Lawal 2012). Chlorobenzoic acids are ubiquitous organic contaminants with different degrees of chlorination in soil. The fungus *L. tigrinus* can mycoremediate CBA from the contaminated soil (Stella et al. 2012). *Lentinus squarrosulus* mushroom has been found to mineralize soil contaminated with various concentrations of crude oil resulting in increased nutrient contents in treated soil. Adenipekun and Fasidi (2005) reported the ability of *L. squarrosulus* to mineralize soil contaminated with various concentrations of crude oil (1–40%). Engine oil-polluted soil can be remediated by *L. squarrosulus* (Adenipekun and Isikhuemhen 2008).

P. ostreatus is able to degrade a variety of polycyclic aromatic hydrocarbons (PAHs) (Sack and Gunther 1993). It has the ability to degrade PAH in non-sterile soil both in the presence and in the absence of cadmium and mercury. It has been reported to catalyze humification of anthracene, benzo(a) pyrene and flora in two PAHs-contaminated soils from a manufactured gas facility and an abandoned electric cooping plant (Bojan et al. 1999). *P. ostreatus* substrate has been used to biotreat Nigerian oil-based drill cuttings containing polycyclic aromatic hydrocarbons (PAHs) under laboratory conditions (Okparanma et al. 2011). *Pleurotus ostreatus* can significantly degrade the toxic chemicals with certain alterations in the biodegradation strategies and can be a potential candidate for the reclamation of sites polluted with polyaromatic hydrocarbon (Baldrian et al. 2000; Eggen and Majcherczyk 1998; Eggen and Sveum 1999; Bhattacharya et al. 2012).

P. tuber-regium has been studied for its ability to ameliorate crude oil-polluted soil. Isikhuemhen et al. (2003) reported that this mushroom had the ability to remediate crude oil-polluted soil and the resulting soil sample supported seed germination and seedling growth of *Vigna unguiculata*. They reported a significant improvement in percentage germination, plant height and root elongation. Adenipekun et al. (2011a) studied bioremediation of contaminated soil by *P. tuber-regium* Singer and reported that there was an improvement in the nutrient status of the soil with an increase in enzyme activity.

Adenipekun et al. (2011b) worked on the management of cement- and battery-polluted soils using *P. pulmonarius*. A general increase was observed in the carbon content, organic matter, phosphorus and potassium and a decrease in the percentage of nitrogen, calcium and pH after 10 weeks of the incubation of mycelium. The lead content was constant in both polluted soils, while a significant decrease was observed in the copper, manganese and nickel contents of the soils.

Aged mycelium from oyster mushrooms (*Pleurotus ostreatus*) mixed in with "compost" made from woodchips and yard waste (50:50 by volume) resulted in far better degradation of hydrocarbons than oyster mushroom mycelium or compost alone (Stamets 2010). *Trametes versicolor* produced three lignolytic enzymes with efficient degradation capacity on lignin, polycyclic aromatic hydrocarbons, polychlorinated biphenyl mixture and a number of synthetic dyes (Tanaka et al. 1999; Novotny et al. 2004). It has also demonstrated the ability to degrade dieldrin in the lab (Morgan et al. 1991; Gadd 2001).

Mushroom *Bjerkandera adusta* has the capacity of PAH degradation. The ligninolytic enzymes involved in the degradation of PAHs are: lignin peroxidase, versatile peroxidase, Mn-peroxidase and laccase (Adenipekun and Lawal 2012). Pozdnyakova (2012) studied the PAH degradation during mycoremediation of PAH-contaminated soils by this fungus.

There are numerous advantages of using mushroom or spent mushroom substrate in the remediation of pollutants. First, it is a natural system and does not introduce any corrosives or harmful chemicals for cleanup. In most cases, researchers use only native species on every site. Mycoremediation basically reduces the amount of wastes to be landfilled, and therefore, it is publicly acceptable technology. Second, the fungal mycelium corrects an imbalance due to a contamination event or situation and restores the natural function and creates a balance in the system. The complete protocol is natural and helps in bringing the contaminants within levels, within the ecosystem that are no longer harmful. The process is environmentally friendly and works on a variety of organic and inorganic compounds. It is safer than most other alternatives of bioremediation. It does not require digging up contaminated products and disposing of it at waste sites. Additionally, this process does not produce any kind of secondary metabolites that require additional cleanup after the initial remediation process. It is a simple technology which requires low maintenance. Moreover, the cost of using mycoremediation is relatively low in comparison to other technologies and treatment methods, as it does not require building of new structures.

The use of mushrooms has been known in the remediation of polluted soil. However, Sasek (2003) reported that the performance of white rot fungus in soil bioremediation depends upon its survival in the soil environment, colonization, relationship and interaction with other soil microflora present in the soil. The nutrient requirement of mushroom has to be completely understood so as to enable it to thrive at a contaminated site.

17.10 Perspective and Conclusion

Waste bioconversion through mushroom cultivation is an advantageous process which can be adopted by industries, and therefore, waste can be minimized in the environment along with financial benefit. This approach can made food or agricultural based industries as "zero waste enterprises". Mushroom cultivation in any food waste industries produces sellable fruit bodies, and spent mushroom substrate can be used for recovery of different enzymes which can be again used in food industries (Fig. 17.3; Part A). Adoption of this technology will give zero discharge from food industries.

- Part A: Food waste generation from food processing industries and its utilization for mushroom cultivation and enzymes recovery for application in food industries.
- Part B: Food waste from agricultural fields can be used in biogas production or for mushroom cultivation, and generated solid by-products can be used for improving the soil quality.
- Part C: Food waste can be directly used as animal feed or processed through mushroom cultivation and then used for feeding animals. Intermediate step of mushroom cultivation provides financial benefit along with nutritionally improved feed.

Agricultural waste can be used for the cultivation of mushroom at the site of origin. Again, produced fruit bodies will provide financial benefit to the farmers, and generated spent mushroom compost can be applied to the agricultural fields. Alternatively, spent mushroom substrate can be digested in anaerobic condition and produced biogas can be used as fuel. Biogas slurry can also be used in agricultural field as an organic amendment. Solid residues can be used as animal bedding. This shows that agricultural waste can be treated onsite with beneficial products with zero discharge (Fig. 17.3; Part B).

Food and agricultural waste can be directly used as animal feed. Alternatively, this waste can be used for mushroom cultivation. Mushroom fruit bodies can be used to generate revenues. Mushroom mycelium-coated waste can be used as nutrient-rich diet for animals. Again, food or agricultural waste can be used as animal feed and feeding human beings by producing mushrooms (Fig. 17.3; Part C). In this case, there is again no waste discharge and, therefore, considered as zero waste enterprises.

Mushroom cultivation can be adopted by food industries for onsite treatment of food and agricultural waste. However, there is a need of awareness among the farmers and owners of food industries. This technology can be implemented by establishing mushroom training centres where any person can get the training to contribute in their own way to reduce food waste.

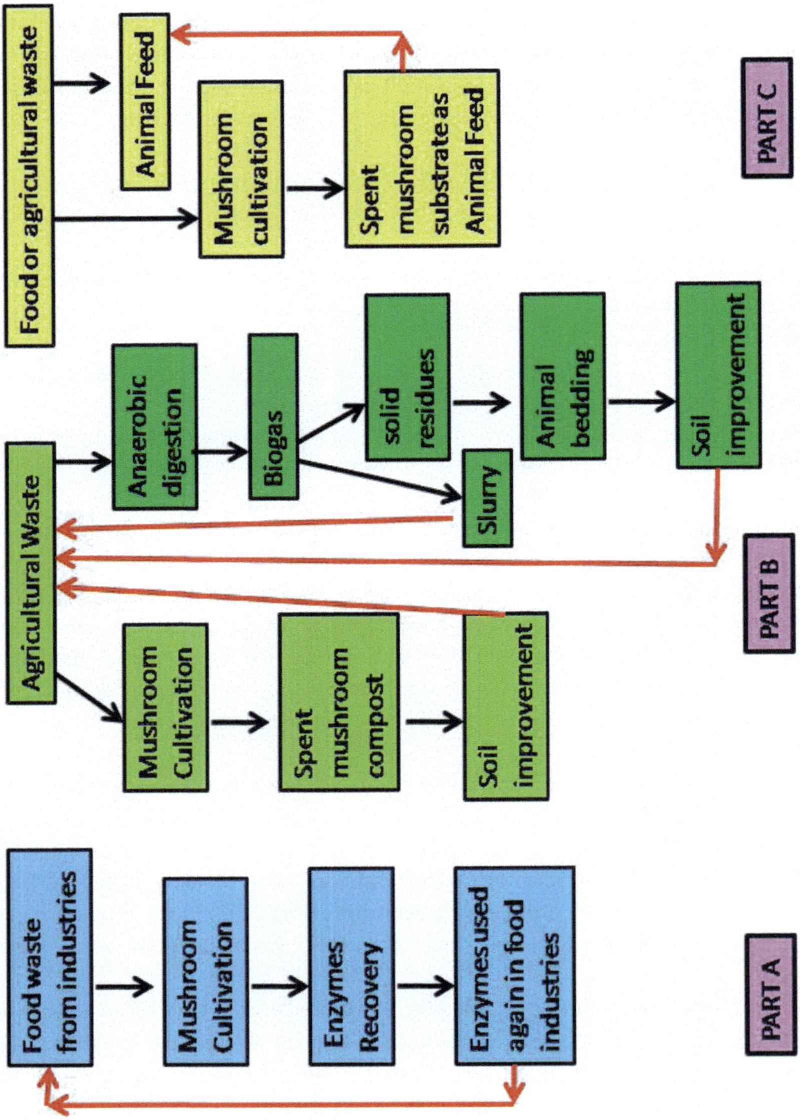

Fig. 17.3 Generation and utilization of food waste at its origin site—A way towards sustainable development

Acknowledgement Authors are thankful to Dr. Nitesh Singh Rajput, Asst. Professor, Amity School of Engineering and Technology, Amity University Rajasthan, Jaipur for his help and valuable suggestions for writing this manuscript.

References

Adedokun O, Orluchukwu JA (2013) Pineapple: organic production on soil amended with spent mushroom substrate. Agric Biol J N Am 4:590–593

Adenipekun CO, Fasidi IO (2005) Bioremediation of oil polluted soil by *Lentinus subnudus*, a Nigerian white rot fungus. Afr J Biotechnol 4:796–798

Adenipekun CO, Isikhuemhen OS (2008) Bioremediation of engine oil polluted soil by the tropical white-rot fungus, *Lentinus squarrosulus* Mont. (Singer). Pak J Biol Sci 11:1634–1637

Adenipekun CO, Lawal R (2012) Uses of mushrooms in bioremediation: a review. Biotechnol Mol Biol Rev 7:62–68

Adenipekun CO, Ejoh EO, Ogunjobi AA (2011a) Bioremediation of cutting fluids contaminated soil by *Pleurotus tuber-regium* singer. Environ Syst Decis 32:11–18

Adenipekun CO, Ogunjobi AA, Ogunseye OA (2011b) Management of polluted soils by a white-rot fungus, *Pleurotus pulmonarius*. AU J Technol 15:57–61

Aguilar-Rivera N, Llarena-Hernández RC, Michel-Cuello C, Gámez-Pastrana MR, Debernardi-Vazquez TJ (2017) Competitive edible mushroom production from nonconventional waste biomass. Intech open

Alananbeh KM, Bouqellah NA, Al Kaff NS (2014) Cultivation of oyster mushroom *Pleurotus ostreatus* on date-palm leaves mixed with other agro-wastes in Saudi Arabia. Saudi J Biol Sci 2014:616–625

Aragie E, Balié J, MoralesOpazo C (2018) Does reducing food losses and wastes in sub-Saharan Africa make economic sense? Waste Manag Res 36(6):483–494

Baldrian P, Der Wiesche C, Gabriel J, Nerud F, Zadrazil F (2000) Influence of cadmium and mercury on activities of ligninolytic enzymes and degradation of polycyclic aromatic hydrocarbons by *Pleurotus ostreatus* in soil. Appl Environ Microbiol 66:2471–2478

Behera S, Gupta N (2015) Utilization of vegetable waste for biomass production of some wild edible mushroom cultures. Tropical Plant Res 2:05–09

Bhattacharya S, Angayarkanni J, Das A, Palaniswamy M (2012) Mycoremediation of Benzo[a] Pyrene by *Pleurotus ostreatus* isolated from Wayanad district in Kerala, India. Int J Pharm Biol Sci 2:84–93

Bilbao-Sáinz C, Bor-Sen C, Williams T, Wood D, Wen-Xian D, Sedej I, Ban Z, Rodov V, Poverrenov E, Vinokur Y, McHugh T (2017) Vitamin D-fortified chitosan films from mushroom waste. Carbohydr Polym 167:97–104

Bojan BW, Lamar RT, Burjus WD, Tien M (1999) Extent of humification of anthrecene, fluoranthene adbenzo (a) pyrene by *Pleurotus ostreatus* during growth in PAH-contaminated soils. Lett Appl Microbiol 28:250–254

Carrasco-Cabrera CP, Bell TL, Kertesz MA (2019) Caffeine metabolism during cultivation of oyster mushroom (*Pleurotus ostreatus*) with spent coffee grounds. Appl Microbiol Biotechnol 103:5831–5841

Cogorni PFBO, Schulz JG, Alves EP, Gern RMM, Furlan SA, Wisbeck E (2014) The production of *Pleurotus sajor-caju* in peach palm leaves (*Bactris gasipaes*) and evaluation of its use to enrich wheat flour. Food Sci Technol (Campinas) 34:267–274

Economou CN, Diamantopoulou PA, Philippoussis AN (2017) Valorization of spent oyster mushroom substrate and laccase recovery through successive solid state cultivation of *Pleurotus, Ganoderma,* and Lentinula strains. Appl Microbiol Biotechnol 101:5213–5222

Eggen T, Majcherczyk A (1998) Removal of polycyclic aromatic hydrocarbons (PAH) in contaminated soil by white-rot fungus *Pleurotus ostreatus*. Int Biodeterior Biodegradation 41:111–117

Eggen T, Sveum P (1999) Decontamination of aged creosote polluted soil: the influence of temperature, white-rot fungus *Pleurotus ostreatus*, and pre-treatment. Int Biodeterior Biodegradation 43:125–133

Ekundayo FO, Ekundayo EA, Ayodele BB (2017) Comparative studies on glucanases and β glucosidase activities of *Pleurotus ostreatus* and *P. pulmonarius* in solid state fermentation. Mycosphere 8:1201–1209

Elekes CC, Busuio G (2010) The mycoremediation of metals polluted soils using wild growing species of mushrooms. In: EDUCATION'10: Proceedings of the 7th WSEAS international conference on engineering education 36–39

Gadd G (2001) Fungi in bioremediation. Cambridge University Press

Girmay Z, Gorems W, Birhanu G, Zewdie S (2016) Growth and yield performance of *Pleurotus ostreatus* (Jacq. Fr.) Kumm (oyster mushroom) on different substrates. AMB Express 6:87

Grimm D, Wösten HAB (2018) Mushroom cultivation in the circular economy. Appl Microbiol Biotechnol 102:7795–7803

Hauke H, Schlosser D, Wick LY (2011) Untapped potential: exploiting fungi in bioremediation of hazardous chemicals. Nat Rev Microbiol 9:177–192

Hlerema IN, Eiasu BK, Koch SH (2017) Pineapple (*Ananas comusus*) plant material as supplement for maize residue-based oyster mushroom substrate and reduction of cadmium soil contamination. HortScience 52:667–671

Hoa HT, Wang C-L, Wang C-H (2015) The effects of different substrates on the growth, yield, and nutritional composition of two Oyster mushrooms (*Pleurotus ostreatus* and *Pleurotus cystidiosus*). Mycobiology 43:423–434

Inácio FD, Ferreira RO, de Araujo CAV, Peralta RM, de Souza CGM (2015) Production of enzymes and biotransformation of orange waste by oyster mushroom, *Pleurotus pulmonarius* (Fr.) Quél. Adv Microbiol 5:1–8

Isikhuemhen OS, Anoliefo G, Oghale O (2003) Bioremediation of crude oil polluted soil by the white-rot fungus, *Pleurotus tuberregium* (Fr.) Sing. Environ Sci Pollut Res 10:108–112

Jahangir MM, Khatana MA, Khan NA, Samin G, Ziaf K, Farooq MU, Iqbal W (2018) Morphological responses of king oyster mushroom against augmentation of cotton waste with citrus peel. Pak J Phytopathol 30:99–107

Jasińska AJ, Wojciechowska E, Krzesiński W, Spiżewski T, Stoknes K, Krajewska K (2014) Cultivation on substrates with addition of anaerobically digested food waste. Acta Hortic 1123:199–206

Jebapriya GR, Gnanasalomi VDV, Gnanadoss JJ (2013) Application of mushroom fungi in solid waste management. Int J Comput Algorithm 2:279–285

Jo E-Y, Cheon J-L, Ahn J-H (2013a) Effect of food waste compost on the antler-type fruiting body yield of *Ganoderma lucidum*. Mycobiology 41:42–46

Jo E-Y, Choi J-Y, Choi J-W, Ahn J-H (2013b) Influence of food waste compost on the yield and mineral content of *Ganoderma lucidum, Lentinula edodes,* and *Pholiota adipose* fruiting bodies. Mycobiology 41:210–213

Johnson LK, Dunning RD, Bloom JD, Gunter CC, Boyette MD, Creamer NG (2018) Estimating on-farm food loss at the field level: a methodology and applied case study on a North Carolina farm. Resour Conserv Recycl 137:243–250

Jwanny EW, Rashad MM, Abdu HM (1995) Solid-state fermentation of agricultural wastes into food through *Pleurotus* cultivation. Appl Biochem Biotechnol 50:71–78

Karim MR, Mahal Z, Iqbal S, Rashid H, Hena MA, Jamal M, Islam MA, Rahman (2014) MM solid state fermentation of Lathyrus sativus and sugarcane bagasse by *Pleurotus sajor-caju*. Int J Agron Agric Res 4(5):1–9

Kulshreshtha S (2018) Mushroom biomass and spent mushroom substrate as adsorbent to remove pollutants. Springer Nature Switzerland AG 2018 281, Crini G, Lichtfouse E (eds) Green Adsorbents for Pollutant Removal, Environmental Chemistry for a Sustainable World 19. https://doi.org/10.1007/978-3-319-92162-4_9

Kulshreshtha S (2019) Removal of pollutants using spent mushrooms substrates. Environ Chem Lett 17:833–847

Kulshreshtha S, Mathur N, Bhatnagar P (2011) Pros and cons of *P. florida* cultivation for managing waste of handmade paper and cardboard industries. IIOAB J spl Issue, 2:45–48

Kulshreshtha S, Mathur N, Bhatnagar P, Kulshreshtha S (2013) Cultivation of *Pleurotus citrinopileatus* on handmade paper and cardboard industrial wastes. Ind Crop Prod 41:340–346

Kulshreshtha S, Mathur N, Bhatnagar P (2014) Mushroom as a product and their role in mycoremediation. AMB Express 4:29

Lee J-W, Bon-Wook K, Joon-Weon C, Don-Ha C, In-Gyu C (2008) Evaluation of waste mushroom logs as a potential biomass resource for the production of bioethanol. Bioresour Technol 99:2736–2741

Lin Q, Long L, Wu L, Zhang F, Wu S, Zhang W, Sun X (2017) Evaluation of different agricultural wastes for the production of fruiting bodies and bioactive compounds by medicinal mushroom *Cordyceps militaris*. J Sci Food Agric Aug 97:3476–3480

Mansour-Benamar M, Jean-Michel S, Chavant L (2013) Valorization of solid olive mill wastes by cultivation of a local strain of edible mushrooms. C R Biol 336:407–415

Morgan P, Lewis ST, Watkinson RJ (1991) Comparison of abilities of white-rot fungus to mineralize selective xenobiotic compounds. Appl Microbiol Biotechnol 34:693–696

Murthy PS, Manonmani HK (2008) Bioconversion of coffee industry wastes with white rot fungus *Pleurotus florida*. Res J Environ Sci 2:145–150

Muth MK, Birney C, Cuéllar A, Finn SM, Freeman M, Galloway JN, Gee I, Gephart J, Jones K, Low L, Meyer E, Read Q, Smith T, Weitz K, Zoubek S (2019) A systems approach to assessing environmental and economic effects of food loss and waste interventions in the United States. Sci Total Environ 685:1240–1254

Nguyen TM, Ranamukhaarachchi SL (2020) Effect of different culture media, grain sources and alternate substrates on the mycelial growth of *Pleurotus eryngii* and *Pleurotus ostreatus*. Pak J Biol Sci 23:223–230

Nighojkar A, Patidar MK, Nighojkar S (2019) Pectinases: production and applications for fruit juice beverages. Process Sustain Bever 2:235–273

Novotny C, Svobodova K, Erbanova P, Cajthaml T, Kasinath A, Lange E, Sasek V (2004) Ligninolytic fungi in bioremediation: extracellular enzyme production and degradation rate. Soil Biol Biochem 36:1545–1551

O'Brien BJ, Milligan E, Carver J, Roy ED (2019) Integrating anaerobic co-digestion of dairy manure and food waste with cultivation of edible mushrooms for nutrient recovery. Bioresour Technol 285:121312

Okparanma RN, Ayotamuno JM, Davis DD, Allagoa M (2011) Mycoremediation of polycyclic aromatic hydrocarbons (PAH) - contaminated oil-based drill-cuttings. Afr J Biotechnol 10:5149–5156

Ouzouni PK, Petridis D, Koller WD, Riganakos KA (2009) Nutritional value and metal content of wild edible mushrooms collected from West Macedonia and Epirus, Greece. Food Chem 115:1575–1580

Padhye R, Atawane S (2017) Cultivation of oyster mushroom and recovery of value added by-products from biodegradable lignocellulosic waste materials by solid state fermentation. Int J Curr Res Aca Rev 5:34–43

Pandit NP, Maheshwari SK (2012) Optimization of cellulase enzyme production from sugarcane pressmud using oyster mushroom - *Pleurotus sajor-caju* by solid state fermentation. J Bioremed Biodegr 3:140

Papoutsis K, Grasso S, Menon A, Bruntona NP, Lyng JG, Jean-Christophe J, Bhuyan DJ (2020) Recovery of ergosterol and vitamin D2 from mushroom waste - potential valorization by food and pharmaceutical industries. Trends Food Sci Technol 99:351–366

Paritosh K, Kushwaha SK, Yadav M, Pareek N, Chawade A, Vivekanand V (2017) Food waste to energy: an overview of sustainable approaches for food waste management and nutrient recycling. Biomed Res Int 2017:2370927

Patidar M, Nighojkar S, Kumar A, Nighojkar A (2018) Pectinolytic enzymes-solid state fermentation, assay methods and applications in fruit juice industries: a review. 3 Biotech 8:199
Pereira GS, Cipriani M, Wisbeck E, Souza O, Strapazzon JD, Gern RM (2017) Onion juice waste for production of *Pleurotus sajor-caju* and pectinases. Food Bioprod Process 106:11–18
Piccoli Valle RH, Passos FML, Passos FJV, Silva DO (2001) Production of pectin lyase by *Penicillium griseoroseum* in bioreactors in the absence of inducer. Braz J Microbiol 32 (2):135–140
Pozdnyakova NN (2012) Involvement of the ligninolytic system of white-rot and litter-decomposing fungi in the degradation of polycyclic aromatic hydrocarbons. Biotechnol Res Int 2012:243217
Prandi B, Faccini A, Lambertini F, Bencivenni M, Jorba M, Droogenbroek BV, Bruggeman G, Schöber J, Petrusan J, Elst K, Sforza S (2019) Food wastes from agrifood industry as possible sources of proteins: a detailed molecular view on the composition of the nitrogen fraction, amino acid profile and racemisation degree of 39 food waste streams. Food Chem 15:567–575
Rashad FM, El Kattan MH, Fathy HM, El-Fattah DAA, El Tohamy M, Farahat AA (2019) Recycling of agro-wastes for *Ganoderma lucidum* mushroom production and *Ganoderma* post mushroom substrate as soil amendment. Waste Manag 88:147–159
Rashad MM, Abdou HM, Mahmoud AE, Nooman MU (2009) Nutritional analysis and enzyme activities of *Pleurotus ostreatus* cultivated on *Citrus limonium and Carica papaya* wastes. Aust J Basic Appl Sci 3(4):3352–3360
Sack U, Gunther T (1993) Metabolism of PAH by fungi and correction with extracellular enzymatic activities. J Basic Microbiol 33:269–277
Sasek V (2003) Why mycoremediations have not yet come into practice. In: The utilization of bioremediation to reduce soil contamination: problems and solution. Kluwer Academic Publishers, The Netherlands, pp 247–266
Sesli E, Tuzen M (1999) Level of trace elements in the fruiting bodies of macrofungi growing in the East black sea region of Turkey. Food Chem 65:453–460
Shamim HM, Abdel-Rahman MA, Hussain MS, Islam MR, Al-Mahin A (2017) Bioconversion of water-hyacinth to nutritionally enriched animal feed by solid state fermentation using *Pleurotus sajor-caju*. J Microbiol Biotechnol Food Sci 6:1165–1169
Shashitha KN, Komal, Shlini P, Singh KG (2016) Vegetable waste-a potent substrate for cultivation of *P. ostreatus*. Int J Res Stud Biosci 4:5–9
Sherief AA, El-Tanash AB, Temraz AM (2010) Lignocellulolytic enzymes and substrate utilization during growth and fruiting of *Pleurotus ostreatus* on some solid wastes. J Environ Sci Technol 3:18–34
Singh RI, Kanojiya A, Sandhu SS (2007) Effect of waste organic substrates supplemented with mango leaf aqueous extract on the mycelial growth of *Pleurotus sajor-caju* and *Pleurotus florida*. J Pure Appl Microbiol 1:307–312
Singh VK, Singh MP (2014) Bioremediation of vegetable and agrowastes by *Pleurotus ostreatus*: a novel strategy to produce edible mushroom with enhanced yield and nutrition. Cell Mol Biol (Noisy-le-Grand) 60:2–6
Stamets PE (2010) Mycoremediation and its applications to oil spills. www.realitysandwich.com/mycoremediation_and_oil_spill. Accessed 15 July 2020
Stella T, Covino S, Křesinová Z, D'Annibale A, Petruccioli M, Cajthaml T (2012) Mycoremidiation of PCBs dead – end metabolites: In vivo and In vitro degradation of chlorobenzoic acids by the white rot fungus *Lentinus tigrinus*. Environ Eng Manag J 11:9
Stoknes K, Beyer DM, Norgaard E (2013) Anaerobically digested food waste in compost for *Agaricus bisporus* and *Agaricus subrufescens* and its effect on mushroom productivity. J Sci Food Agric 93:2188–2200
Stoknes K, Wojciechowska E, Jasinska A, Noble R (2019) Amelioration of composts for greenhouse vegetable plants using pasteurised *Agaricus* mushroom substrate. Sustainability 11:6779
Tanaka H, Itakura S, Enoki A (1999) Hydroxyl radical generation by an extracellular low-molecular–weight substance and phenol oxidase activities during wood degradation by the white–rot basidiomycetes *Trametes versicolor*. J Biotechnol 75:57–70

Thakur M (2019a) Mushroom cultivation: a boon for empowerment of women entrepreneurship. In: Kulshrestha S (ed) Mushroom cultivation development and perspectives. Stadium Press, pp 238–255
Thakur M (2019b) Mushrooms as a biological tool in mycoremediation of polluted soils. In: Jindal T (ed) Emerging issues in ecology and environmental science (Case Studies from India). Springer Nature Switzerland, pp 27–42
Thakur M (2020) Fungi as a biological tool for sustainable agriculture. Springer Nature Switzerland In: Yadav A, Mishra S, Kour D, Yadav N, Kumar A (eds) Agriculturally important fungi for sustainable agriculture. Fungal biology. Springer, Cham, pp 255–273
Thakur M, Lakhanpal TN (2015) Mushroom Cultivation and women Entrepreneurship: prospects & challengers. In: Vasudevan P, Sharma S, Sharma VP, Verma M (eds) Women, technology and development. Narosa Publishers, III Delhi, pp 220–240
Thomas SA, Aston LM, Woodruff DL, Cullinan VI (2009) Field demonstration of mycoremediation for removal of fecal coliform bacteria and nutrients in the dungeness watershed. Pacific Northwest National Laboratory, Washington, Richland
Tisma M, Zelic B, Vasic-Racki D (2010) White-rot fungi in phenols, dyes and other xenobiotics treatment – a brief review. Croatian J Food Sci Technol 2:34–47
Triyono S, Haryanto A, Telaumbanua M, Dermiyati LJ, To F (2019) Cultivation of straw mushroom (*Volvariella volvacea*) on oil palm empty fruit bunch growth medium. Int J Recycl Organic Waste Agric 8:381–392
Yang D, Liang J, Wang Y, Sun F, Tao H, Xu Q, Zhang L, Zhang Z, Chi-Tang H, Wan X (2016) Tea waste: an effective and economic substrate for oyster mushroom cultivation. J Sci Food Agric 96:680–684
Zervakis GI, Koutrotsios G, Katsaris P (2013) Composted versus raw olive mill waste as substrates for the production of medicinal mushrooms: an assessment of selected cultivation and quality parameters. Biomed Res Int 2013:546830
Zheng Z, Shetty K (2000) Solid state production of polygalacturonase by *Lentinus edodes* using fruit processing wastes. Process Biochem 35:825–830
Zied DC, Prado EP, Dias ES, Pardo JE, Pardo-Gimenez A (2019) Use of peanut waste for oyster mushroom substrate supplementation-oyster mushroom and peanut waste. Braz J Microbiol 50:1021–1029

Bioremediation: A Sustainable Biological Tool for Food Waste Management 18

Isha Sai, Vatsala Sharma, Ashmita Singh, and Rukhsaar Sayeed

Abstract

In the past few years, there has been a tremendous increase in the generation of food waste due to rapid industrialization and urbanization. Food waste consists of high levels of sodium and moisture and is usually mixed with other types of waste during its collection. Food waste gathered to have a high level of contaminants which when combines with other components produces many toxic components and has deleterious effects. In order to cope with this food waste production at every level; advanced and effective waste management systems are to be adopted that can overcome the gap between production and management of waste disposal. the microorganisms play a pivotal role in the bioremediation of wastewater generated from various food industries as well. The chapter highlights the role of microbes as a biological tool for a sustainable food waste management system.

Keywords

Food waste · Microbes · Micro-remediation · Waste disposal

18.1 Introduction

Over the last few decades, the waste gathered in the environment is one of the major concerns globally. Food waste is gradually becoming one of the major concerns nowadays and recognized environmental issue globally. On one hand food security

I. Sai (✉)
Department of Biosciences, Sri Sathya Sai Institute of Higher Learning, Anantapur, India
e-mail: ishasai@sssihl.edu.in

V. Sharma · A. Singh · R. Sayeed
Amity Institute of Food Technology, Amity University Uttar Pradesh, Noida, India

M. Thakur et al. (eds.), *Sustainable Food Waste Management*,
https://doi.org/10.1007/978-981-15-8967-6_18

Table 18.1 Characteristics of food industry waste

S. No.	Various section of food industries	References
1.	*Fruit and vegetable processing industry* waste consist of hydrocarbons, proteins, fat, and wastewater have various dissolved solids, herbicides, pesticides, and chemicals used for cleaning purposes	Riggle (1989), Grobe (1994), and Punnagaiarasi et al. (2017)
2.	*Fermentation industries*—the brewing, distilling, and wine manufacture produces liquid waste (with high BODs and CODs) The wastewater has very high concentrations of tannins, phenols, and organic acids	Mayer (1991), Suzuki et al. (1997), and Punnagaiarasi et al. (2017)
3.	*Dairy industries* contain pollutants from the water resources, feed of cattle, and soil	Punnagaiarasi et al. (2017)
4.	*Meat and poultry industry*: Slaughterhouses are the major sources of environmental pollution as wastewater and solid waste	Cournoyer (1996) and Punnagaiarasi et al. (2017)
5.	*Wastewater from food industries*: The concentrations of pollutants (organic matters, fats, oil and grease—FOGS) in food wastewater is increased in wastewater	Punnagaiarasi et al. (2017)

is one of the major challenges, but the impact of food waste on the surrounding environment can no longer be overlooked. As the population is growing day by day and every piece of land is showing urbanization, on one hand, more and more food is being produced and on other side more food is being wasted also. Punnagaiarasi et al. (2017) observed that in urban sectors the food wasted affects environmental health and which in turn has a negative impact upon the environment and human health.

Zafar (2012) also claimed that food industries are the biggest source of the untapped energies, dumped almost in landfills and also releasing greenhouse gases into the atmosphere. Food waste is generally considered as biodegradable but, many times, it is very difficult to treat and recycle food waste due to its composition. The waste generated from the food products has very high levels of moisture and sodium content; which mixes with other types of waste and becomes very toxic. The characteristics of various food industry waste have been shown in Table 18.1.

Mavropoulos (2011) classifies the amount of waste generated into two different types:

- Consumers in the given area
- Consumption pattern of the consumers

Thus, in order to deal with this issue of the large production of food waste; a very effective waste management systems have to be analyzed and adopted so that the gap between production and management of waste disposal shall be fulfilled. A significant fraction of food waste is generally unavoidable, include peels and skins, bones and fats, oils and food mistakenly left to rot. The food waste is one of the major pollution problems which is plaguing the world today. We have to adopt various technologies and processing techniques, so that the same can be tackled at a very faster pace

18.2 Bioremediation

Bioremediation is a naturally occurring process where the microorganisms either immobilize or transform contaminants to the degradable forms in the environment (Thassitou and Arvanitoyannis 2001). NRC (1993) acclaimed the bioremediation process as one of the most widely accepted for the clean-up of the contaminated sited of the environment. Bioremediation is an environment friendly, cost-effective, and innovative approach that utilizes the metabolic potential of microorganisms to clean-up the environmental pollutants by complete mineralization into CO_2, H_2O, N_2, HCl, etc. (Asgher et al. 2008; Haritash and Kaushik 2009; Rhodes 2014).

Bioremediation is still an underutilized technology for food waste management and there is a definite need for these nondestructive types of solutions. Microorganism given encouragement can do most of the reclamation work for us in the environment. Bioremediation is the process by which microbes are used to eliminate contaminants by ingestion and degradation. The process of bioremediation mainly depends on microorganisms which enzymatically attacks the pollutants, but this can be effective only when the extrinsic growth factors are favorable and permit their growth and activity. These microbes are indigenous to the contaminated area and are non-pathogenic. Bacteria, Actinomycetes, fungi, and algae are all commonly used for bioremediation (Table 18.2; Fig. 18.1).

In the current scenario, the climatic conditions are also changing at a faster pace. These changes include a global rise in temperature and atmospheric pollution which adversely affects agricultural productivity. To mitigate the problems arising from the present climatic conditions and to ensure food security, sustainable agricultural practices and minimize the food waste are need of the hour. The pollution generated from the waste gathered in the environment has significant deleterious consequences on the sustainable food system. In the past few, the food waste management system is getting more and more complicated, causing more pollution and remediation of these pollutants again is a very challenging task.

Table 18.2 Biotic community used for bioremediation

Sr. No.	Microorganisms	Examples
1	Bacteria	*Pseudomonas veronii, Burkholderia spp., Kocuria flava, Bacillus cereus*, etc.
2	Fungi	*Aspergillus versicolor, A. fumigatus, Penicillium*, etc.
3	Algae	*Cladophora fascicularis, Spirogyra spp., Spirullina spp.*, etc.
4	Yeast	*Candida utilis, Saccharomyces cerevisiae*, etc.

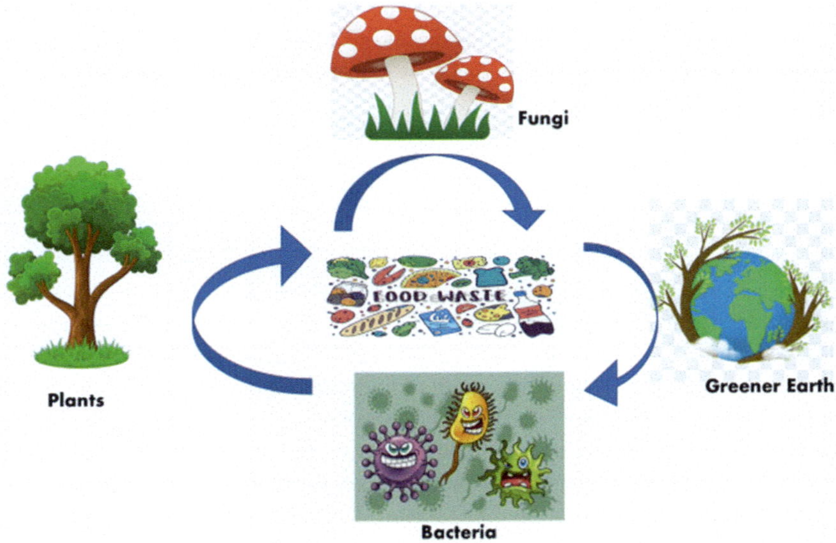

Fig. 18.1 Process of bioremediation of food waste

18.3 Microorganisms: Biological Tools for Sustainable Agriculture Food Waste Management

Microorganism (both bacteria and fungi) plays significant role as a significant role as a biological tool for the establishment of a sustainable agricultural system (Thakur 2014, 2019). Watanabe (2001) reported that naturally occurring microbial consortia including both bacteria and fungi have been utilized in a variety of bioremediation processes. Fungi are more widespread in the environment having rapid adaptability and diverse metabolic activities in agricultural systems. In the soil, the fungal cultures can sustain under diverse environmental conditions in the maintenance and functioning of the ecosystem because of their decomposing behavior. The beneficial fungi play an important role in improving plant growth, increasing plant yield, and involvement in biotic and abiotic stress tolerance, hazardous materials remediation, sustainable crop production, and food safety. Singh et al. (2019) also discussed the use of fungi in mycoremediation and mycocontrol.

Microorganisms also play an important role in attaining the sustainability of the ecosystem by involving itself in the following different ways:

- Fungi as a sustainable biological tool
- Microbial enzymes

18.3.1 Fungi as Sustainable Biological Tool

Finding sustainable solutions for the utilization of food and agricultural wastes is one of the greatest challenges in any ecosystem. Only fungi represent a wide variety of groups like wood-rot fungi, white-rot fungi, brown-rot fungi, leaf decomposing fungi, soil fungi, fungal mycorrhiza, endophytic fungi and aquatic fungi which in association with other microorganism has a greater potential to sustain the ecology of any ecosystem using mechanisms of mycoremediation and myco-control.

The key features of fungi which help in bioremediation are:

- Fungal species grow and symbiotic relationships with other species in the ecosystem and have various positive associations there.
- Fungal hyphae show indeterminate growth, therefore, no limitation of cell division in the hypha as long as resources are available (which they get from the food waste).
- Fungal mycelium form rhizomorphs that grow through vast distance with a robust system of adapting to highly restrictive environmental conditions.
- Fungi are more resistant to high concentrations of toxins in comparison to bacteria.

Mycoremediation is a fungal-based technology that uses the potential of fungi with their high growth rate and high activity of their extracellular enzymes in disintegrating the organic material from different food sources and converting them to simpler compounds (Thakur 2014; Purohit et al. 2018; Pozdnyakova et al. 2018; Thakur 2020). From the few past decades, many researchers have worked on mycoremediation and this technology has been used on oil spills, contaminated and polluted soil, industrial chemicals, contaminated water, and even agricultural farm waste (Alexander 1994; Bennet et al. 2001; Ashoka et al. 2002; Adenipekun and Lawal 2012).

Some specific examples of macro-fungi mycelium especially used for mycoremediation are as (Source: Thakur 2019):

- *Bjerkandera adusta*
- *Irpex lacteus*
- *Lentinus edodes; L. tigrinus; L. squarrosulus*
- *Phanerochaete chrysosporium*
- *Pleurotus ostreatus, P. tuber-regium, P. pulmonarius*
- *Trametes versicolor*

Mycorrhiza Fungi (MF)

Mycorrhiza, a heterogeneous group of fungi is primarily associated with plant roots. Mycorrhizal association assumes different kinds: ectomycorrhiza, endomycorrhiza (VAM or AM), ectendomycorrhiza, ericoid, arbotoid, orchidaceous, and monotropoid mycorrhiza. Fungal mycorrhiza is very helpful for sustainable agriculture and providing resistance to the plants against various pathogens. These associations are very helpful and the plant body gets protection from deadly pathogens, tolerance to drought, and higher pH. Their association helps the plant in improving the nutrient supply also (Basu et al. 2018). They also provide protection to plants against various water and metal toxicity stress by reducing metal translocation and therefore, helping plants to adapt and survive in these contaminated sites containing heavy metals.

18.3.2 Microbial Enzymes

Karigar and Rao (2011) studied the use of various enzymes from microorganisms like—bacteria, fungi, and with the presence of these enzymes, plants help on biodegradation of pollutants:

- Microbial Oxidoreductases
- Microbial Oxygenases
- Monooxygenases
- Microbial Dioxygenases
- Microbial Laccases
- Microbial Peroxidases (*Microbial Lignin Peroxidases*, *Microbial Manganese Peroxidases*, *Microbial Versatile Peroxidases*

18.4 Waste Water from Food Industries

The use of physico-chemical method for wastewater remediation has been replaced by microbial bioremediation method. Utilizing the potential of bacteria, fungi, and even yeasts for remediation purpose is eco-friendly. Microbes have been used for the reduction of wastewater pollutants to acceptable economy level. Microbes synthesize a lot of enzymes that are instrumental in degrading oil pollutants which are major constituent of food wastewater (Table 18.3). Microbes possess some distinctive advantages over other forms of organisms due to ease in handling, mass cultivation ability to withstand various environmental conditions, and high degradable capabilities. Their extremely small size and large surface area relative to their volume make them applicable in many areas of wastewater remediation (Xue et al. 2016). Therefore, the microorganisms play a pivotal role in the bioremediation of waste water generated from various food industries as well.

18.5 Advantages of Bioremediation

Bioremediation process has many advantages over other commercialized technologies available as:

- Natural and environment friendly

Table 18.3 Examples some isolated microorganism that have been identified for food waste water pollutants bioremediation

Sr. No.	Microorganisms	examples
1	Bacteria	*Acinetobacter junii, A. calcoaceticus, A. radioresistens, Achromobacter sp, Alcaligenes sp, Arthrobacter sp, Aeromonas sp, Bacillus subtilis, B. licheniformis, B.amyloliquefaciens, B. laterosporus, B. megatherium , B. cereus, B. Licheniformis , B. laterosporus ,Clostridium sp, Citromonas sp, Cryptococcus sp, Enterococcus sp, Erwinia sp, Enterobacter sp, Escherichia coli, Flavobacterium, Klebsiella sp, Lueconostoc lactis, Mycobacterium, Moraxella lacunata, Nitrosomonas sp. Nitrobacter sp, Pseudomonas aeruginosa, P. fluorescens, , Providences, Proteus vulgaris, Raoultella planticola , Stenotrophomonas sp, staphylococcus aureus, Streptococcus faecalis ,Serratia sp, Zoogloea sp*
2	Fungi	*Aspergillus versicolor, A. fumigatus, Absidia spp., Cunninghamella sp , Fusarium moniliforme, F. oxysporium, Penicillium spp., Rhizopus spp, Thermophilus spp. Alternaria sp, Trichoderma sp* and *Thermoactinomyces sp*
3	Algae	*Spirogyra, Cladophora,* and *Spirulina* species
4	Yeast	*Candida utilis* and *Saccharomyces cerevisiae*

- Safety, simple, and quiet
- Low maintenance and reusable end products
- Economically feasible with very less time consumed for clean-up
- Flexibility and fast
- Directly affects the target and not transferred across different medium
- Better acceptance by the mass

18.6 Limitation of Bioremediation

Some of the challenges faced during the bioremediation process as follows:

- Fungal species has the inability to compete with native microbes in soils. Bacteria could either inhibit the growth of fungi or in combination with fungi, enhance the degradation of pollutants available in the food waste also.
- Nutrient requirement of the microbe has to be completely understood so as to enable it to thrive at a contaminated site.
- Mushroom mycelium should not be used as a starter strain
- Legal issues involved in this process. There are several patents specifically granted for matching fungus against a toxin. This is a major hindrance in preventing wide-scale fungal clean-up of toxins from the food waste polluted site.
- Lack of experienced mushroom cultivators in outdoor trials is a problem in mycoremediation. This lacking has affected the success of several trials.

18.7 Future Scope

Recent advancements with the addition of potential microbial strains to the food waste and the enhancement of the indigenous microbial population have proven to be successful. Whether the fungal mycelium is native or newly introduced to the site, the process of destroying contaminants is important and critical for understanding. A lot of work is focused on the strategic development of the complete process to make it a full-fledged sustainable system. Further, the application of this technology in large scale projects will demand much more work to streamline the methodologies. The use of microorganisms for remediation would allow the commercial concern to offer inexpensive, safe products to their customers. If the underexploited potential of bacterial cultures and fungus mycelium is further exploited, it will go a long way and come as the most efficient biological tool in sustainable food waste management system.

18.8 Conclusion

Over the last few decades, food waste has become an increasingly recognized environmental issue globally. One side food security is one of the major challenges, but the environmental impacts because of food waste can no longer be overlooked. Although the researcher had found the variety of ways by which we can degrade the food waste, but bioremediation also making its leap to tackle the problem associated with different categories of waste with the help of microorganisms. The underexploited potential of bacterial cultures and fungus mycelium is slowly changing the way to tackle one of the most dangerous issues of food waste. More extensive research needs to be carried out on the potential of bacterial and fungal species for bioremediation managing food waste.

References

Adenipekun CO, Lawal R (2012) Uses of mushrooms in bioremediation: a review. Biotec Molecular Biology Rev 7:62–68

Alexander M (1994) Biodegradation and bioremediation, 2nd edn. Academic Press, San Diego

Asgher M, Bhatti HN, Ashraf M, Legge RL (2008) Recent developments in biodegradation of industrial pollutants by white-rot fungi and their enzyme system. Biodegradation 19:771–783

Ashoka G, Geetha MS, Sullia SB (2002) Bioleaching of composite textile dye effluent using bacterial consortia. Asian J Microb Biotech Environ Sci 4:65–68

Basu S, Rabara RC, Negi S (2018) AMF: the future prospect for sustainable agriculture. Physiol Mol Plant Pathol 102:36–45

Bennet JW, Connick WJ, Daigle D, Wunch K (2001) Formulation of fungi for in situ bioremediation. In: Gadd GM (ed) Fungi in bioremediation. Cambridge University Press, Cambridge, pp 97–108

Cournoyer MS (1996) Sanitation and stabilization of slaughterhouse sludges through composting. In: Proceedings of the Canadian meat research institute technology symposium, vol 1, Canadian Meat Research Institute, Ontario, Toronto, pp 1–7

Grobe K (1994) Composter links up with food processor. BioCycle 34(40):42–43
Haritash AK, Kaushik CP (2009) Biodegradation aspects of polycyclic aromatic hydrocarbons (PAHs): a review. J Hazard Mat 169:1–15
Karigar S, Rao SS (2011) Role of microbial enzymes in the bioremediation of pollutants: a review. Enz Res 11:11
Mavropoulos A (2011) Waste management world. http://www.waste-managementworld.com/index/display/article-display/8267238380/articles/waste-management-world/volume-11/issue2/features/wastemanagement_2030.html
Mayer ES (1991) Waste treatment experiments at the Gabriel Sedlmayr Spaten. FranziskanerBraeu K.-G.a.A. Brauwetl. (Ger.) 131:2346
NRC (National Research Council) (1993) In situ bioremediation: National Research Council Report, vol 1. National Academic Press, Washington, DC, pp 2–11
Pozdnyakova NN, Balandina SA, Dubrovskaya EV, Golubev CN, Turkovskaya OV (2018) Ligninolytic basidiomycetes as promising organisms for the mycoremediation of PAH-contaminated Environments. Earth Environ Sci 107:012071
Punnagaiarasi A, Elango A, Rajarajan G, Prakash S (2017) Application of bioremediation on food waste management for cleaner environment. In: Prashanthi M, Sundaram R, Jeyaseelan A, Kaliannan T (eds) Bioremediation and sustainable technologies for cleaner environment. Environmental science and engineering. Springer, Cham. https://doi.org/10.1007/978-3-319-48439-6_5
Purohit J, Chattopadhyay A, Biswas MK, Singh NK (2018) Mycoremediation of agricultural soil: bioprospection for sustainable development. In: Prasad R (ed) Mycoremediation and environmental sustainability. Fungal biology. Springer, Cham
Rhodes C (2014) Mycoremediation (bioremediation with fungi) – growing mushrooms to clean the earth. Chem SpecBioava 26(3):196–198
Riggle D (1989) Revival time for composting food industry wastes. BioCycle 29:35–37
Singh VP, Singh M, Singh SK, Kumar C, Kumar A (2019) Sustainable agricultural practices using beneficial fungi under changing climate scenario. In: Climate change and agricultural systems. Current challenge and adaptation. Woodhead Publishing, Sawston, UK, pp 25–42
Suzuki H, Yoneyama Y, Tanaka T (1997) Acidification during anaerobic treatment of brewery wastewaters. Water Sci Technol 35:265
Thakur M (2014) Mycoremediation – a potential tool to control soil pollution. Asian J Environ Sci 9 (1):24–31
Thakur M (2019) Mushrooms as a biological tool in mycoremediation of polluted soils. In: Emerging issues in ecology and environmental science (case studies from India). Springer, Cham, pp 27–42
Thakur M (2020) Fungi as a biological tool for sustainable agriculture. In: Yadav A, Mishra S, Kour D, Yadav N, Kumar A (eds) Agriculturally important fungi for sustainable agriculture. Fungal biology. Springer, Cham, pp 255–273
Thassitou P, Arvanitoyannis I (2001) Bioremediation: a novel approach to food waste management. Trends Food Sci Technol 12(5–6):185–196
Watanabe K (2001) Microorganisms relevant to bioremediation. Curr Opin Biotechnol 12:237–241
Xue L, Famous E, Jiang J, Shang H, Ma P (2016) Experimental survey on microbial bioremediation of food wastewaters. Intern J Sci Res 6:110–118
Zafar S (2012) BioEnergy consult. http://www.bioenergyconsult.com/trends-in-food-wastemanagement

Recovery of Bioactive Components from Food Processing Waste 19

Chandrakala Ravichandran, Ram Mohan Mutharasu, and Ashutosh Upadhyay

Abstract

With increasing application of bioactives in various fields like pharmaceutical, food and chemical industries, standard method for extraction of bioactives is a major research need. The large amount of food processing waste gets dumped into environment causing pollution and economic loss thereby requiring proper management. Studies have proven that wastes generated from food processing are source of potential bioactive compounds of commercial significance. To humans compounds such as antioxidants and vitamins were shown to exhibit beneficial effect. But finding out appropriate method of recovery which is eco-friendly with higher extraction efficiency is a major challenge. The conventional methods being used for years were replaced by novel green technologies making cost and energy efficient. This chapter will give a brief idea about different extraction techniques which are widely used for extraction of bioactives from food processing waste. It also describes briefly on recent studies in relation to recovery of bioactives from food waste.

Keywords

Food processing waste · Bioactive compounds · Conventional techniques · Innovative non-conventional methods

C. Ravichandran (✉)
Department of Food Processing Technology, Karunya Institute of Technology and Sciences, Coimbatore, India

R. M. Mutharasu · A. Upadhyay
National Institute of Food Technology Entrepreneurship and Management, Sonipat, India

M. Thakur et al. (eds.), *Sustainable Food Waste Management*,
https://doi.org/10.1007/978-981-15-8967-6_19

19.1 Introduction

Food processing industry waste has become one of the most important environmental concern due to its high levels of BOD and COD content (FAO 2013). At present, two methods such as landfill and incineration are widely used to dispose solid wastes almost everywhere. It is of least consideration during waste disposal that landfills result in huge expulsion of methane and carbon dioxide into the environment. Incineration acted as a proven cause releasing pollutants and other residues such as dioxins (environmental pollutant), furan, etc. which also acted as known factor to several environmental threats and health risks. In this regard, there is an urge to utilize abundant food processing waste as a resource for making of value-added products. On the other hand, several works were being carried out to reduce the by-products generation during processing by improving the process efficiency. Advanced processing technologies in several food industries, like in a juice industry, processing apple juice, with major by-products generated including pomace, stem, core, seed, calyx, peel and soft tissues, helped in improving the juice yield. However, these played a less significant role in waste reduction as the wastes generated still stand at 12–20% as pomace (Kosmala et al. 2011).

Sustainable management of food waste was the comprehensively researched and discussed topic in recent years. As per resolution adopted in UN assembly on 25th September, 2015, Sustainable Development goal 12 targets to reduce the losses due to food waste. Environmental, economic and social consequences must be greatly considered to come up with sustainable solutions for food waste management. Wealth out of waste must be the focus for upcoming years which indicates the potential sustainable solution for turning out waste into a highly economic output. Sustainable approach for management of food waste ensures sustainable food production and consumption patterns. Thus, it is a highly sustainable way to divert these wastes into generation of bioactive constituents, to be used in food, pharmaceuticals and other cosmetics industries as well.

The by-products generated during processing acts as a potential source for the recovery of several chemicals and manufacture of metabolites like antioxidants, enzymes, starch, and pigments of natural origin, etc., however, optimization of methodologies for the extraction of bio actives from food by-products are momentous research is last few years. In general, by-products generated are disposed as waste/landfilled that poses environmental risks.

19.2 Need for Recovery

Food waste management in general refers to the gulp of activities performed right from collection of food, through sorting, processing and further conversion into other value-added materials, till the production of new products. It is one of the fruitful ways while addressing environmental issue and assessing manufacturer returns as well. Numerous literatures have reported that by-products are the source

Table 19.1 Bioactive components present in various fruit and vegetable by products

Food commodity	By-product	Waste (%)	Bioactive and phenolic compounds	Reference
Apple	Pomace, peel, and seeds	–	Hydroxycinnamates, glycosides (such as quercetin and phloretin glycosides), catechins, procyanidins	Lu and Foo (1998) and Schieber et al. (2001)
Banana	Peel	35	Carotenoids	Subagio et al. (1996)
Citrus	Peel, and seeds	40–50	Hesperidin, naringin	Coll et al. (1998) and Matharu et al. (2016)
Grapes	Stem, skin and seeds	18–25	Procyanidins, Catechins, anthocyanins, stilbenes, flavonol, Glycosides, epicatechin, epigallocatechin, epicatechin, Gallate	Kallithraka et al. (1995), Saito et al. (1998), Jayaprakasha et al. (2001), and Schieber et al. (2001)
Mango	Peel, stone	40–50	Gallates, gallic acid, ellagic acid	Arogba (2000)
Onion	Outer leaves	–	Quercetin (such as 3,40-*O*-diglucoside and 40-*O*-monoglucoside)	Price and Rhodes (1997)
Papaya	Rind, seeds	10–20	Anthocyanins, β-carotene, Total phenolic content	Sharma and Le Maguer (1996)
Pineapple	Core, skin	33	Anthocyanins, β-carotene, Total phenolic content	
Tomato	Core, skin and seeds	20	Lycopene	Sharma and Le Maguer (1996)
Potato	Peel	15	Chlorogenic acid, gallic acid, protocatechuic acid, chlorogenic acid isomer II	Onyeneho and Hettiarachchy (1993), Rodriguez et al. (1994), and Choi et al. (2016)

of various bioactive compounds that can widen the possibilities to develop new range of products of commercial importance.

In general fruits possess high antioxidants acting as a neutralizer of free radicals. Most of the studies using ferric-reducing activity and trolox equivalent antioxidant capacity enunciated the strong antioxidant power of such by-products. Studies have proven that antioxidant activity was directly proportional to the phenolic compounds present in fruit. The fruit by-products exhibits higher concentration of phenolic content thereby contribute for higher antioxidant activitythan fruit pulps. Thus, fruit residues would be a ready source of bioactive compounds to be used in the food and pharma industries (Deng et al. 2012) (Table 19.1).

19.3 Methods of Bioactive Recovery

There are considerably quite a large number of techniques that have been employed in extraction of bioactives from food wastes. Briefly, in a solvent extraction technique, the extraction power of solvent used, application of heat or pressure, mixing etc. are major factors affecting the rate of extraction. Several of the commonly used extraction methods are discussed below:

Conventional Techniques like

- Soxhlet extraction
- Macreation
- Hydrodistillation

Several new and technological solutions have been brought up which are environmentally friendly for recovery of bioactives those include:

Non-conventional techniques

- Enzyme Assisted Extraction
- Microwave assisted Extraction
- Membrane Filtration
- Pressurized Liquid Extraction
- Supercritical Fluid Extraction
- Other techniques

19.3.1 Conventional Techniques

19.3.1.1 Soxhlet Extraction

Scientist named Soxhlet from Germany in 1879 who first proposed Soxhlet extractor, which stands as a comparative test for any new extraction technique being developed, used a finely ground dry sample kept in a porous bag of filter paper or cellulose. The thimble placed in a distillation flask containing selective solvent was heated in the round bottom flask which vaporizes into sample thimble and drips back after getting condensed in the condenser. After reaching an overflow level, the solution of thimble is aspirated by siphon which refills the solution back to flask. The process is carried out until the extraction is complete. But this method had a shortfall of more extraction time and large amount of solvent used (Azmir et al. 2013a; Azwanida 2015)

19.3.1.2 Maceration, Infusion, Percolation

Maceration was being used since years for homemade preparation of tonic. This method involves soaking of sample (finely or coarsely grounded) with a solvent in a stoppered bottle for predetermined period of time followed by pressing or grinding. The mixture is pressed and strained by filtration. During grinding, the sample surface area will be increased thereby aiding proper mixture of solvents. Agitation cum grinding can provide improved extraction by increasing the rate of diffusion and in removing the extracted portion from surface of the sample. Infusion and maceration are almost same but varying only with the time of soaking (Azwanida 2015). Decoction is used for extraction of heat stable compounds, specifically oil soluble compounds. The main disadvantage of this method is large amount of solvents being used.

19.3.1.3 Hydrodistillation

The Common and widely employed methods for extraction of plant essential oils include hydrodistillation involving three main physicochemical process: namely, hydrodiffusion, hydrolysis and decomposition by heat, which consumes around 6–8 h and does not require solvent. It takes place in a single step, extracting and separating both organic volatile and non-volatile compounds. However, this method is time consuming and consumes high energy (Soquetta et al. 2018).

19.3.2 Non-Conventional Techniques

19.3.2.1 Enzyme Assisted Extraction (EAE)

Enzymes act as catalysts that increase the rate of product formation from a substrate under mild conditions such as temperature, pH, oxygen concentration, etc., without involving in the reaction. Enzymes are derived from wide sources such as bacteria, fungi, vegetable extracts and fruits and even animal organs. Its classification includes these categories: hydrolyzing enzymes, oxidation-reduction enzymes, ligases, group transfer enzymes, desmolases, isomerizing enzymes, and carboxylation enzymes. Based on their catalytic property, they catalyse definite reactions, by acting on a specific part of a substrate known as active site.

Enzymes are an ideal catalyst which has the ability to assist in extraction of complex bioactive compounds by degrading the cell wall matrix and membranes covering the cell wall. Therefore, it increases cell wall permeability and thereby achieving higher extraction. Hence, enzymatic pre-treatment is found to be an effective method to release compounds bound within cell structures and increase overall yield as the phytochemicals are dispersed in cytoplasm in plant matrices and compounds present are retained in polysaccharide-lignin network through bondings such as hydrogen/hydrophobic which are not possible using solvents (Rosenthal et al. 1996). Hydrolyzing the structural polysaccharides and lipid bodies using enzymes during extraction improves recovery (Rosenthal et al. 1996; Singh et al. 1999). Enzymes like cellulases, β-glucosidases, xylanase, β-gluconase and pectinases are mostly used as they help to break the cell wall structure and to break the polymers in cell wall polysaccharides, thereby facilitating the release of bound components (Moore et al. 2006)

There are two major approaches followed generally in enzyme assisted extraction as suggested by Latif and Anwar (2009)

- Enzyme Assisted Aqueous Extraction (EAAE)

Table 19.2 Recovery of bioactive compounds from different fruit by-products

Fruit by-products	Bio-active compound	Amount of recovery (mg GAE/g)	Reference
Mango peel	Total phenolic compound	22.95	Deng et al. (2012)
Chinese olive peel		13.16	
Grape seed		22.95	
Avocado seed		7.54	

Table 19.3 Yield of bio-actives from various food processing wastes using enzyme assisted extraction

Product	Source	Enzyme	Yield %	References
Oligosaccharide	Rice bran	Cellulase	39.9	Patindol et al. (2007)
Lignan	Flax		0.407	Renouard et al. (2010)
Anthocyanin	Saffron		0.67	Lotfi et al.(2015)
Quercetin	Onion	Pectinase	75	Choi et al. (2015)

- Enzyme Assisted Cold Pressing (EACP)

EAAE methods developed majorly for extraction of seed oils showed a higher recovery of Total Phenolic Contents (TPC) using celluzyme MX enzyme from five citrus peels such as Yen Ben lemon, Meyer lemon, grapefruit, mandarin and orange (Hanmoungjai et al. 2001; Rosenthal et al. 1996; Puri et al. 2012; Li et al. 2006).

EACP method was employed to hydrolyze the polysaccharide cell walls of seeds and to liberate bioactive compounds which is used in combination with a filtration step (Concha et al. 2004). It is an excellent variant for the extraction of bioactive components from oilseeds as compared to other methods, on account of its non-toxic and non-inflammable properties (Bhattacharjee et al. 2006). Better free fatty acids and phosphorus contents retention can be achieved here than any of the traditional hexane extraction (Dominguez et al. 1995). EAE is also an eco-friendly alternative for the extraction of bioactive compounds and oil as no organic chemicals are used rather water is used as the solvent (Puri et al. 2012).

In grape pomace, a combination of pectinolytic and cellulolytic enzymes was used in a ratio of 2:1 to extract bioactive compounds such as phenolic acids, anthocyanins and non-anthocyanin flavonoids with obtained higher yields compared to sulphite-assisted extraction (Maier et al. 2008). The antioxidant composition also was reported to be higher in black carrot juice through the use of EAE and recently EAE is also used to obtain vegetable oils (Khandare et al. 2010) (Tables 19.2, 19.3 and 19.4).

Celluclast, novoferm and pectinex are commonly employed to extract phenolic compounds from grape wastes, with Novoferm greatly showing stronger effect on the release of phenolic compounds from grape wastes (Gomez-García et al. 2012) (Fig. 19.1).

Table 19.4 Enzyme and their operating conditions employed for bioactive extraction from various food by-products

Enzymes used	Conditions used for extraction	Material tested	Bioactive compounds	Reference
Pectinase, cellulose, and beta-glucosidase	30–35 °C/18 h, pH 3.5–4	Pigeon pea leaves	Legume Flavones: Luteolin and Apigenin	
Alpha-amylase and glucoamylase	2% w/w enzyme, 60 min. pH 4.5	Turmeric (*Curcuma longa L.*)	Oleoresin	Fu et al. (2008)
Cellulase	1.5 % w/w enzyme, 50 °C/3 h	Citrus peels, grapefruit, mandarin, and orange	Total phenolics	Kurmudle et al. (2010)
Glucanase and xylanase	0.05 % and 0.1% w/w, 20 min/70 °C and 20 min/50 °C	Rapeseed	Phenolics, phospholipids and tocopherols	Li et al. (2006)
Protamex, koizyme, neutrase, flavourzyme, and alcalase	12 h	Ecklonia cava	Phenolic compounds	Heo et al. (2003)

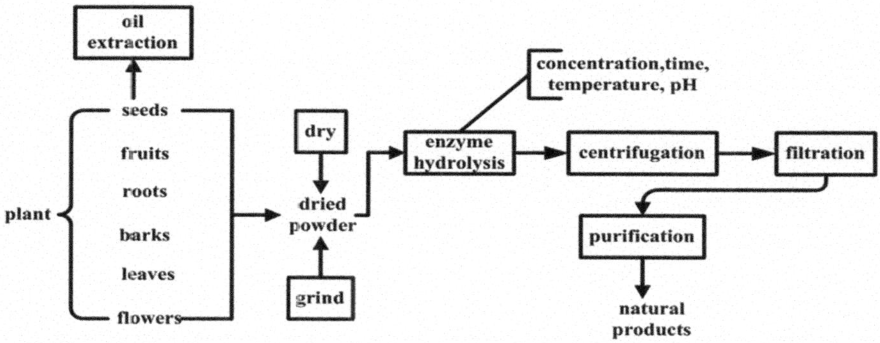

Fig. 19.1 Enzyme assisted extraction method for plant products (reproduced from Xian Cheng et al. 2015)

Commercial Applications

EAE is recommended for extraction of oil and phytochemicals bound within the cell structures. Gomez-García et al. (2012) illustrated EAE as an best approach to advantageously extract bioactive compounds from agro-industrial by-products.

Advantages
1. Eco-friendly
2. Suitable to extract any bound compounds
3. High extraction rate

Limitations
1. High enzyme costs
2. Not feasible at industrial levels due to different enzyme nature and behaviour

Adequate differences in extraction yield was observed with moisture content of the sample, hydrolysis time, particle size, composition and concentration of the enzyme.

19.3.2.2 Microwave Assisted Extraction (MAE)

In 1986, microwave energy as used and described firstly by Gedye et al. (1986) and Giguerre et al. (1986) for simple extraction was then used for extraction of biological samples and for analysis of organic compounds Ganzler et al. (1986). Microwaves are electromagnetic radiation with a wavelength between 0.0010 and 1 m in which it can be transmitted. When it passes through a medium, microwave energy is absorbed and converted into thermal energy. Microwave assisted method of extraction is used as a high-efficiency method which uses microwave radiation to heat solvents when it comes in contact with the sample. In chitin/chitosan extraction from fish and fishery products such as crustacean wastes, etc generally it involves several steps such as derivatization which further makes changes such as chain elongation, substitution of compounds, depolymerisation of which microwave assisted extraction is physical method of depolymerisation.

As it is said that MAE is a physical method which cleaves glycosidic bonds without the use of chemical agents. A research conducted by Makuuchi (2010) states that chitosan gets degraded by hydrogen peroxide, which is enhanced by UV-radiation, gamma radiation and microwaves. Though this kind of microwave assisted extraction is a better method, the extraction with a percentage recovery of chitosan was very low (20 min/98 °C/650 W) produced 55% reduction in molecular weight (Tishchenko et al. 2011). This could be overcome by the use of Ionic Liquid (IL) solution along with microwave and is considered as an alternative method to improve the chitosan hydrolysis (Chen et al. 2012)' as 1L solution interacts efficiently with microwaves through the action of ionic conduction mechanism. It also rapidly heats at rates in excess of 10 °C/s easily without consequently increasing the pressure.

Polar materials are immediately affected by the microwaves. Dipolar rotation and ionic interactions are involved in transcending microwaves into thermal energy and the resultant heat produced due to the resistance of medium during action of ionic flow or conduction of ions while aligning themselves toward the direction of field and changing randomly.

Heating effect of the microwaves causes the moisture present inside the cells to get evaporated which in turn produces an increased pressure build up and modifies the physical properties of the biological tissues. This improves the porosity of sample matrices (Zhang et al. 2011). As a result of this a better diffusion property into the extracting solvent through the matrix can be obtained which improves the yield of desired components.

As suggested by Alupului et al. (2012), MAE technique is of three steps:

1. Solute molecules are split and separated from the sample matrix (as a result of increase in temperature and pressure)
2. Solvent diffusion into the sample matrix.
3. Release of required solute component into the solvent from the sample matrix.

Classification

Through the advancements emerged in microwave extraction techniques, there are two classes of practices widely used for extraction.

1. Microwave Assisted Solvent Extraction (MASE)
2. Microwave Solvent-Free Extraction (MSFE)

Microwave Assisted Solvent Extraction is done by treating organic solvent with the sample using microwave energy. The analytes separate from the sample and diffuse into the organic solvents varying with the temperature conditions and polarity of the solvent.

Based on the dielectric properties of the solvent and the sample matrix, two conditions need to be considered:

1. The solvent (ethanol, methanol, water, etc.) in general imbibes the microwaves, followed by rapid conversion of microwave energy into thermal energy driven by two basic principles discussed above. This heats the sample upto the boiling point of the solvent, which leads to alteration of extraction environment, which sees an increase in the diffusion of the solvent into the sample matrix with the sample matrix responding by the increased solubility of the analytes in the solvent. Note: For sample of solid matrices heat transfer can be accomplished by conduction from the solvent.
2. Direct heat treatment to sample of wet matrix can directly absorb microwaves: as target compound tends to move from matrix through transparent non polar solvents such as hexane, toluene, etc.

Microwave energy is generally applied to samples by using two methods such as

1. Closed extract vessels (controlled pressure/temperature): This method is widely advised for extractions to be carried out at extreme conditions such as high temperature. It is advantageous to operate at higher pressure so that the solvent used for extraction can be made to achieve a higher temperature usually higher than its corresponding boiling point under atmospheric pressure, thereby exhibiting a higher control over accelerated extraction of the target compounds by rapid mass transfer from the sample.
2. Open vessels under atmospheric pressure: In this technique, as the extraction is to be carried out at atmospheric pressure, the maximum temperature that can be attained ultimately depends on the boiling point of the solvent.

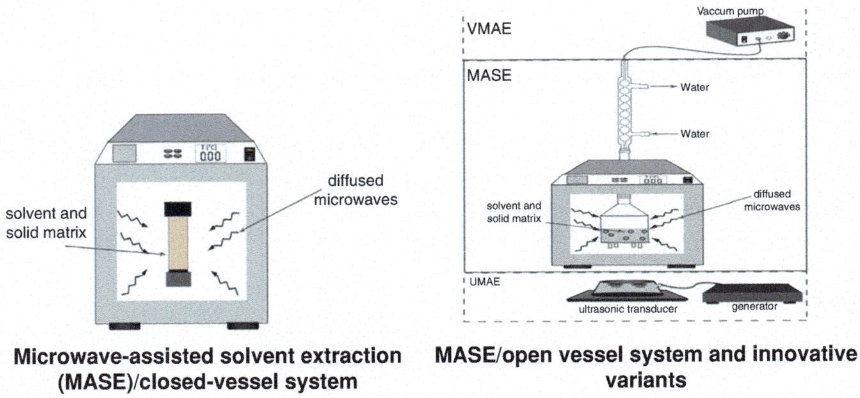

Fig. 19.2 Schematic diagram of microwave assisted extraction system (reproduced from Li et al. 2013)

Penetration power of microwaves into sample matrices of biological nature and interaction with polar molecules (eg: water) generates heat leading to an increase in the temperature which in turn leads to the improved efficiency of the extraction than traditional techniques. Hayat et al. (2010) studied the effect of treatment time in extraction of phenolic compounds from citrus peel stated that an increase in microwave energy along with time duration, substantially increased the free phenolic content in the extracts and decreasing the bound-phenolic content.

This also stipulates that the effect of microwaves over the tissue matrices led to the cleavage and liberation of phenolics thereby allowing them in freely available form rather than in a bound form. They also stated that longer treatment to microwave degrades flavonol compounds (Fig. 19.2).

Extraction of natural bioactive compounds through microwave assisted method of extraction can be affected through factors like temperature, pressure used for extraction, frequency of extraction, processing time, % moisture content of the sample, particulate size, power applied, ratio of solvent used for solid sample and so on. Among all these factors listed, solvent used for the process of extraction plays a key and crucial role as its solubility, dissipation factor and dielectric factor have to be considered. The dissipation factor is the efficiency of the solvent with which it gets heated up under microwave energy.

Polar solvents with higher dielectric constant as that of water are preferred over non-polar counterparts. Dissipation factor ultimately determines the process efficiency in terms of rapid heating of moisture from the cell matrix eventually leading to the generation of higher pressure and rupturing of cell structures thereby facilitating the rapid extraction of valuable bioactive compounds. Solvents such as ethanol/methyl alcohol with higher dissipation factor exhibited higher recovery rate for phenolic compounds Wang and Weller (2006) (Table 19.5).

Table 19.5 Studies on MW assisted extraction of bio-actives and their operating conditions

Sample	Extracted compound	Extraction	Conditions maintained	Advantage	Reference
Soybeans	Iso flavoids	Microwave assisted extraction	Atmospheric pressure, 600 W, acetonitrile/water (80:20v/v), 1 min MAE	Excellent efficiency, low sample, solvent and time consumption	Careri et al. (2007)
Soybeans	Iso flavones	Microwave assisted Extraction	Atmospheric pressure, 500 W, $T = 50$ °C	High reproducibility without degradation.	Rostagno et al. (2007)
Onion (*Allium cepa*L.)	Flavonol	Microwave Hydrodiffusion and gravity assisted	Atmospheric pressure, 500 W, 23 min	Shorter extraction time, no solvent or water used	Zill-e-Huma et al. (2008)
Olive leaves	Bio-phenols	Microwave assisted Extraction	Atmospheric pressure, 200 W, 8 min	–	Japon-Lujan et al. (2006)
Grape skin and seeds	Phenolic compounds	Microwave assisted extraction	Atmospheric pressure, 500 °W, $T = 65$–140 °C, 20 min	–	Liazid et al. (2007)
Grape seed	Polyphenols	Microwave assisted extraction	66 °C, 200 s and 30 W	Total phenolic content (115%)	Pan et al. (2008)
Grape skins	Anthocyanins	Microwave assisted extraction	40% methanol, 100 °C, 5 min, 500 W	Total anthocyanins (118%)	Liazid et al. (2007)
Green tea leaves	Polyphenols, caffeine	Microwave assisted extraction	Atmospheric pressure, 700 W, $T = 65$–140 °C, 4 min	Short time	Pan et al. (2008)
Mandarin pomace	Phenolic acids	Microwave assisted extraction	125 W, 5 min, 2.45 GHz	Total phenolic acids (99%)	Hayat et al. (2010)
Curcuma longa L., turmeric	Curcumin	Microwave assisted extraction	Atmospheric pressure, 140 W, acetone (20:1), 4 min	Efficient extraction	Mandal et al. (2008)
Red raspberries	Anthocyanins	Microwave assisted extraction	Atmospheric pressure, 366 W, ethanol (15:85), 12 min	High yield up to 98.33%	Sun et al. (2007)

(continued)

Table 19.5 (continued)

Sample	Extracted compound	Extraction	Conditions maintained	Advantage	Reference
Paprika	Carotenoids	Microwave assisted extraction	Atmospheric pressure, 50 W, $T = 60\ °C$, 2 min	Extraction efficiency increases with organic solvents	Csiktusnádi Kiss et al. (2000)
Grape skins	Anthocyanins	Microwave assisted extraction	Atmospheric pressure 500 W, $T = 100\ °C$, 5 min	–	Liazid et al. (2007)
Peanut skins by-product	Phenolic compounds	Microwave assisted extraction	30% ethanol, 30 s, 855 W	Pharmaceutical applications of health-promoting compounds including cancer prevention	Ballard et al. (2010)

Commercial Applications

Rapid extraction of bioactive compounds (especially polyphenols) can be possible. It has been employed in the extraction of compounds such as colours and antioxidants, over a range of matrices including plants, flowers, algae, etc. Extraction of polyphenols during drying of wood at an industrial scale has also been done using this technique. In industry recent area of research is to recover non-polar compounds as it hinders the development of this technology and also to prevent the modification of structure of the compounds so that the possibility in bio-activity of some target compounds can be obtained.

Advantages

1. Better quality
2. Compact equipment
3. High yield
4. Short extraction time
5. Cost-effective
6. No chemical is required

Limitations

1. Expensive apparatus and equipment
2. Operation is difficult
3. Poor recovery rate for non-polar components
4. Heat sensitive components are difficult to extract

19.3.2.3 Membrane Filtration

Membrane filtration technique is gaining wide importance in food processing and recovery industries as it is operational at room temperature, pressure driven and advantageous over low energy consumption when compared to conventional techniques. It is a method of separation/fractionation. Membrane filtration techniques such as microfiltration, nanofiltration, reverse osmosis and ultrafiltration are widely employed in many areas such as desalination, demineralisation, concentration, etc.

Microfiltration is particularly used in industries for removal of finer particles, removal of bacteria, to reduce turbidity and so on. It has a pore size of 0.2–2 mm which can filter particles with molecular weight more than 200 kDa. Driving force is through pressure gradient. Minimum of 0.1 bar and Maximum of 2 bar pressure are required for microfiltration. Mass transfer mechanism is through convection process and its application are widely used in cold sterilization and concentration.

Ultrafiltration is applied in general for fractionating macromolecules such as proteins. It has a pore size of 0.01 mm and molecular weight ranging between 1 and 300 kDa. Driving force is through pressure gradient. Minimum of 0.1 bar and maximum of 7 bar pressure are required for ultrafiltration. Mass transfer mechanism is through convection process.

Nanofiltration, reverse osmosis mainly used in water treatment plants such as in the process of desalination (removal of salt), demineralization (removal of minerals) also for concentration of extracts and juices. It has a pore size of 0.5–1 nm and it has molecular cut-off range between 100 and 1000 Da. Driving force is through pressure gradient. Minimum of 3 bar and Maximum of 25 bar pressure are required for ultrafiltration. Mass transfer mechanism is through diffusion/convection process and its application is widely used in concentration and purification of smaller organic compounds and salt separation.

For reverse osmosis, mass transfer mechanism is through solubilisation/diffusion process. RO uses pressure 5–10 times higher than those used in ultrafiltration.

The pressure applied between the membrane and flow/velocity with which it is allowed to pass through the membrane plays a key role in extraction, concentration and so on.

Physical barrier is applied through which materials can pass or get retained due to the driving force which can be of change in pressure (pressure difference), temperature and concentration across the membrane. The materials which pass through is called permeate and the one which is retained in the membrane is known as retentate (Fig. 19.3).

The performance of the membrane is highly characterized by its separation capability influenced by characteristics which are inherent to the membrane such as membrane and feed composition, and process variables such as temperature, pressure and feed flow. In addition to this molecular interaction between the components of the feed and the surface of the membrane is to be considered as well.

Castro-Muñoz et al. (2015) suggested that UF can be used in fractionating and concentrating both macromolecules to micromolecules with lower molecular weight such as dietary fibre, proteins, phenolic compounds and antioxidant components from various food wastes. Lower molecular weight bioactives can be recovered efficiently using nanofiltration.

Galanakis et al. (2013) did a process modelling for ultrafiltration process to recover and fractionate phenolic compounds from wine sludge from red grapes. Three membranes each with different molecular weight cut-off as 1 kDa, 20 kDa and 100 kDa and showed difference in compounds being recovered. 20 kDa membrane had the highest retention capacity for phenolic compounds and sugars up to 60%, while100 kDa membrane showed separation of polar solutes from Pectin. Hydroxy cinnamic acid, obtained by the conversion of anthocyanins and flavonoids, can be selectively separated by1 kDa membrane from diluted as well as concentrated extracts.

Nawaz et al. (2006) described the efficiency of ultrafiltration by concentrating polyphenolic extracts from grape seeds using alcohol-water mixture. Mixture of ethanol and water increased the extraction rates and also showed higher selectivity in a shorter duration of extraction and also it reduces the labour required to carry out the process. Maximum of 11.4% of polyphenols were recovered by using small pore size of 0.22 mm.

Figueroa et al. (2011) carried out a study to recover antioxidant from orange press liquors. In this method process parameters of ultrafiltration membrane were optimised in order to achieve higher polyphenol retention upto 28.5% and

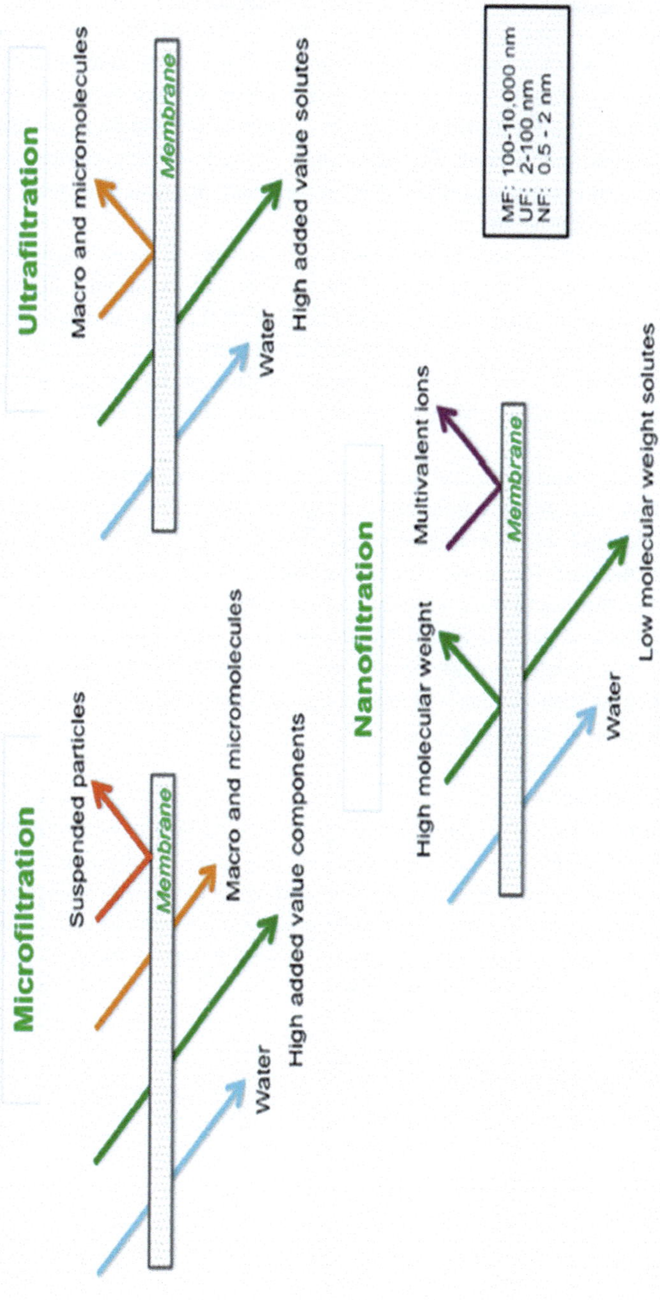

Fig. 19.3 Description of micro, ultra and nanofiltration (reproduced from Castro-Muñoz et al. 2016)

antioxidants upto 32.28 mM Trolox at 0.02 MPa at 20 °C, with feed flow being 254 l per hour.

Nyamien et al. (2017) performed a study on extraction and concentration of bioactive compounds from Cola nitida using membrane filtration process in series of combination for better separation and extraction. It has been stated that the crude extract obtained by extraction using ethanol and water mixture was clarified using microfiltration and ultrafiltration to remove macromolecules and suspended solids. Then, permeate obtained through ultrafiltration is carried through nanofiltration unit for concentration and permeate obtained from each step are tested for total phenolic content, caffeine, proteins, etc.

Microfiltration employs use of membranes of pore size 0.2 μm, ultrafiltration with filters having molecular weight cut-off of 5 kDa, 15 kDa and 50 kDa and Nanofiltration with molecular weight cut-off of 200–400 Da were used. The results obtained showed that by using microfiltration there were about 6.82% loss of total phenols, 2.10% loss of caffeine in addition to that it has good retention of macromolecules and it shows high purity of caffeine and total phenols (El-Rayess et al. 2011).

Nanofiltration showed higher retention for total phenols and proteins as a result of their greater molecular weight and catechin, epicatechin and caffeine were reduced 13.89%, 16.07% and 27.42% respectively. Gosch et al. (2013) used cross-flow microfiltration method for improvement in isolation of bio-active components from bovine colostrum and the results have been tabulated in Table 19.6.

Membrane filtration application for the treatment of fruit residues using pressure-driven process has greater impact on extraction of polyphenols, proanthocyanins, flavonoids, etc. Many researches had been made and the conditions, membrane used have been tabulated in Table 19.7

Commercial Applications

Membrane filtration though it has wide range of application it is of importance in concentration and desalination process. But in this case of bio-active compound extraction it is widely used in extraction from fruit residues as it shows greater purity and recovery %. It has been widely used in purification and concentration from fruit processing wastes and it has been successfully implemented in food colourants, supplements, pharma and cosmetic industries.

Advantages

1. High purity products
2. Low energy requirements
3. Easy scale up
4. Can be operated in room temperature
5. Higher efficiency

Limitations

1. High cost of operation
2. Pressure-driven process and hence control of pressure is critical step
3. Maintenance cost

Table 19.6 Studies on membrane filtration of various bioactive compounds and their recovery

Filtration	Material	Recovery	TS (%)	Protein (%)	Casein (%)	Whey protein (%)	Lactoferrin (μg/ml)	IgG (mg/ml)	Reference
1.4 μm microfiltration	Skimmed raw milk	% component	9.2	3.6	2.8	0.7	160	0.3	Gosch et al. (2013)
		Permeate	8.4	3.3	2.6	0.7	170	0.2	
		% recovery	91	93	92	100	>100	68	
	Skimmed colostrum	% component	16.5	11.5	5.1	6.4	220	34.1	
		Permeate	14.1	9.9	4.3	5.6	120	29.8	
		% recovery	85	86	85	87	55	87	
0.8 μm microfiltration	Skimmed raw milk	% component	9.2	3.6	2.8	0.7	160	0.3	
		Permeate	7.6	2.8	2.1	0.6	150	0.2	
		% recovery	83	78	74	91	94	79	
	Skimmed colostrum	% component	17.3	12	5.4	6.6	190	37	
		Permeate	12.6	8.3	2.9	5.4	110	28.6	
		% recovery	73	69	53	81	58	77	

Table 19.7 Studies on membrane filtration of bioactive and their operating conditions

Fruit residue	Membrane filtration used	Condition used for extraction	Permeate	Reference
Almond skin	Ultrafiltration	50 kDa	Proanthocyanidines	Prodanov et al. (2008)
Grape pomace	Ultrafiltration/ nanofiltration	250–100 Da	Polyphenols	Díaz-Reinoso et al. (2009)
Grape seed	Microfiltration	0.22 and 0.45 μm solvent/ethanol (50:50)	Polyphenols	Nawaz et al. (2006)
Grape juice	Ultrafiltration	Polyvinylidene fluoride flat sheet 10–100 kDa	Monomeric anthocyanins	Kalbasi and Cisneros-Zevallos (2007)
Olive mill waste water	Ultrafiltration/ nanofiltration	Polyvinylidene fluoride 1 kDa	Tyrosol	Cassano et al. (2013)
Pigmented orange peel	Nanofiltration	–	Total anthocyanins, flavonoids	Conidi et al. (2012)
Orange press liquor	Ultrafiltration/ nanofiltration	PES 1 kDa	Flavonoids	Cassano et al. (2014)
Soybean/ peanut oil residues	Ultrafiltration	20–40 °C 3–5 MPa	Tocopherols	Subramanian et al. (1998)

19.3.2.4 Pressurized Liquid Extraction (PLE)

To replace solvent extraction techniques which is commonly used, some newer extraction techniques have been proposed such as pressurized liquid extraction (PLE). This has gained increased attention in recent years and being employed widely in extraction of bioactive compounds for the production of nutraceuticals, food and drugs from natural sources.

PLE is known popularly as accelerated solvent extraction and pressurized solvent extraction because of the organic solvents used here at higher temperatures of 50–200 °C with pressure conditions kept to about 1450–2175 psi to facilitate shorter extraction time (Dunford et al. 2010). With an increase in the temperature, dielectric constant of the solvent decreases which accordingly lowers the polarity of the solvent. Hence the polarities of solvent and the target compound need to be matched, since temperature is the only way that could be used apparently adjusted to achieve the desired rate of recovery.

As, the name itself claims that it is pressurized process and hence higher the pressure applied in the cell helps the sample to get filled faster and also it forces the liquid to the inner most part of the solid matrix. Thereby, by using less amount of solvents, extraction rate is higher and shorter and hence higher yield is obtained compared to that of traditional method of solvent extraction. This technique is widely employed in extracting food-grade extracts, as this technique allows the food grade extracts to get attained in the solvents such as water or other GRAS solvents like ethanol, etc. (Plaza et al. 2010a).

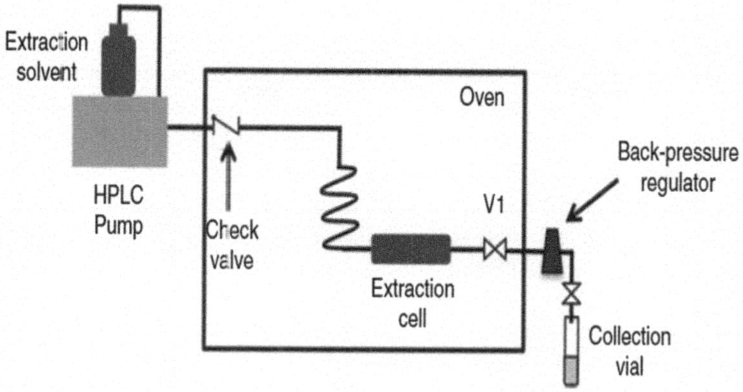

Fig. 19.4 Schematic diagram of PLE (reproduced from Hernandez-Ledesma and Herrero 2014)

In PLE, the application of pressure through which the solvents can be heated above their boiling point facilitates the extraction to be carried out efficiently. This also provide an added advantage that extraction above boiling point or higher temperature could change their mass transfer and surface equilibrium. This will increase mass transfer and extraction rates leading to an increase in the solvent capacity to solubilize the solutes present in the sample. An increase in the rate of diffusion, causes the bonds between solute and matrix of the cells to disrupt and decreases the viscosity and surface tension of the solvent (Ramos 2002).

Temperature is the most crucial factor that needs optimization for performing extraction using PLE. It is considered that temperature increase is directly proportional to the solubility of the analyte. Therefore, selection of optimal temperature of extraction is vital for the recovery of an analyte to aid in increased selectivity.

When 100% water is used as a solvent in extraction then PLE extraction can be called as superheated water extraction, pressurized low polarity water extraction, subcritical water extraction or pressurized hot water extraction (Pronyk and Mazza 2009). For the recovery of bioactives from any by-product of food industries, food grade solvents such as aqueous ethanol or water need to be used. For polar compounds extraction, it is considered as a superior technique to supercritical fluid extraction.

Howard and Pandjaitan (2008) reported that flavonoids extraction from spinach using PLE with ethanol and water (70:30) as a solvent at 50–150 °C increased the rate of extraction than using water solvent alone at 50–130 °C. Mroczek and Mazurek (2009) suggested that PLE method was more effective than solvent extraction, microwave and ultrasound assisted extraction for extracting lycorine and galanthamine (alkaloids) from *Narcissus jonquilla*. Erdogan et al. (2011) extracted individual phenolic compounds such as gallocatechin, catechin, epicatechin gallate, etc. from various parts of *Anatolia propolis* using PLE at a temperature of 40 °C, and a pressure of about 1500 psi for 15 min (Fig. 19.4).

Paula milani et al. (2015) did a comparison of Supercritical Fluid Extraction (SFE) assisted with ultrasound extraction (US) cum pressurized solvent extraction (PLE) for the recovery of phenolic compounds from blackberry residues. In this SFE-US three variables optimized at different levels were as follows: temperature, pressure and ultrasound. Temperature at the range of 40, 50 and 60 °C, pressure at 15 MPa, 20 MPa and 25 MPa, ultrasound power of 0 W, 200 W and 400 W and extraction duration of 120 min are kept constant. Yield obtained is 6.35% for SFE-US. In PLE, four different solvents including water, ethanol, acidified water at pH 2.5, ethanol water (50:50) at different temperatures 60, 80 and 100 °C and pressures of 7.5 MPa, feed ratio being 18 kg of solvent to 18.0 kg residue, and extraction time 30 min were kept constant. Yield obtained is 6.33% (Table 19.8)

The optimized conditions were as follows:

- For SFE-US 60 °C, 15 MPa, and 200 W
- For PLE 100 °C, ethanol water (50:50).

Commercial Applications

Pressurized liquid extraction of flavourings from plant materials comes as an alternative to steam distillation. It can also be employed in the extraction and removal of organic pollutants from environment that are stable at higher temperatures. Extraction of bioactives from marine sponges has also been facilitated using technique.

Advantages

1. Suited for solid samples
2. Better for polar compounds
3. Short time for extraction
4. less solvent required

Limitations

1. Not suitable for thermolabile compounds due to high temperature.
2. Higher equipment costs
3. Unsuitable for matrices with very low concentration of target analyte because of less weight of sample say 5 to 10 taken for extraction.

19.3.2.5 Supercritical Fluid Extraction

Supercritical fluids (SF), having higher solvation capacity, were reported a century ago. Supercritical fluids exihibiting properties of both liquids and gases are capable of diffusing into matter like gases and dissolving like liquids. The most commonly used solvent in SFE is carbon dioxide for its great versatility and more than 90% of analytical extraction is performed using supercritical carbon dioxide (SC-CO_2). This technique has attracted a wide away of sectors like environmental, food, pharmaceutical, polymer applications, etc. When a substance is exposed to a characteristic temperature and pressure beyond its critical point they attain a supercritical state, where its properties get altered. In this supercritical fluid state the substance possesses the gas like properties of diffusion, viscosity and surface tension, liquid

Table 19.8 Studies on extraction of bioactives by pressurized liquid method and their operating conditions

Sample	Extracted compound	Conditions maintained	Extraction	Reference
Algae: *Himanthalia elongata* and *Synechocystis* sp.	Volatiles, fatty acids, and carotenoids	Hexane, ethanol, and water, 50–200 °C, 20 min	•7.59% yield with hexane •36.91% with ethanol •46.43% with water	Plaza et al. (2010b)
Wheat straw, germ, and bran	Policosanols	n-hexane, ethanol, petroleum ether, 80–125 °C, 3 × 15 min, 1500 psi	•1026 ± 18 mg/100g with n-hexane, •27.6 ± 0.04 mg/100g with ethanol, •43.2 ± 3.7 mg/100 g with petroleum ether	Dunford et al. (2010)
Oregano, tarragon, and wild thyme	Phenolics	Ethanol and water 25:75, 50:50, 75:25, 1500 psi, 20 min, 50–200 °C	184.9 mg GAE/g in oregano	Miron et al. (2010)
Red grape pomace	Procyanidins	6.8 MPa, 140 °C, 50% ethanol	Total procyanidins (115%)	Monrad et al. (2010a)
	Anthocyanins	10 MPa, 100 °C, 5 min, 0.1% HCl in water	Total anthocyanins (100%)	Ju and Howard (2003)
		6.8 MPa, 100 °C, 50% ethanol	Total anthocyanins (112%)	Monrad et al. (2010b)
Spirulina	PUFA	Hexane and chloroform/ methanol (2:1), 10.3 MPa,100 °C, $T = 8$ min (two extraction cycles)	–	Zheng et al. (2012)
Pomegranate peel	Polyphenols	10.2 MPa, 40 °C, 5 min, 100% water	Total phenolic content (100%)	Çam and HisIl (2010)
Onion skin	Flavonols	9–13.1 MPa, 160 °C, 15 min, 100% water	Quercetin (92%)	Ko et al. (2011)
Blackberry pomace	Phenolic compounds	100 °C, 7.5MPa, ethanolwater (50:50), 30 min	Total phenolic content (6.33%)	Paula milani et al. (2015)

(continued)

Table 19.8 (continued)

Sample	Extracted compound	Conditions maintained	Extraction	Reference
Micro algae: *Haematococcuspluvialis*	Carotenoids	Ethanol, 100 °C, 10.3 MPa, $T = 20$ min	–	Jaime et al. (2010)

like density and solvation power making it suitable for extraction (Azmir et al. 2013b). At the supercritical temperature and pressure, these properties of CO_2 could be modified for desired improvements.

CO_2 forms the major solvent utilized for the extraction of non-polar molecules, in many cases, whereas subcritical water system was used for polar compounds (Henry and Yonker 2006). As the critical point of water is very high (374 °C, 22.064 MPa) as compared to CO_2 (31.1 °C, 7.4 MPa), superheated water cannot be used to extract heat sensitive compounds (Lang and wei 2011). Moreover, CO_2 is safe to use and food grade. Their wide availablity at lost cost and higher purity broadens their applicability.

SFE due to its high suitability to extract of non-polar substances has lower capacity to recover bioactive components from wastes rich in water. In such cases, co-solvents such as water and ethanol are added. The addition can expand the range of polar compounds and intermediate polar compounds that can be extracted. In most of the studies, processing conditions like temperature, particle size, pressure, solvent flow rate and co-solvents were optimized to achieve higher yield and desired physical and chemical properties of product. As wide array of parameters are need to be optimized in this technique, it becomes more complex. Among these parameters extraction pressure, temperature of fluid and extraction time are most important ones directly affecting the extraction.

Extraction of phytochemicals from fruits and vegetables using SFE is of great significance due to the higher levels purity of final products and also their value in international market (Singh and Saldaña 2011). In case of bioactives extraction, temperature forms a critical factor affecting their efficiency in SC-CO2 system. High temperature may aid in extraction; However, heat degradation may occur to sensitive components. Optimum temperature ranges between 55 and 59 °C generally for for flavour extracts of β-carotene. Extraction pressure between 30 & 40 MPa possesses best recovery of lycopene and β-carotene. Modification of extraction pressure can eventually change the solvent density to affect the solvation power of SC-$CO_{2.}$

The SFE system consists majorly of a solvent tank, a gas pressure pump, co-solvent tank and pump, oven containing extraction vessel and a controller for maintaining higher pressure inside the system and a trapping vessel. The raw materials for extraction are to be placed in an extractor vessel under controlled temperature pressure conditions. The fluid is then pressurized and moved into the vessel by pump. After extraction, the fluid and dissolved components are collected in the collector where separation takes place. Finally, the fluid is recollected or recycled

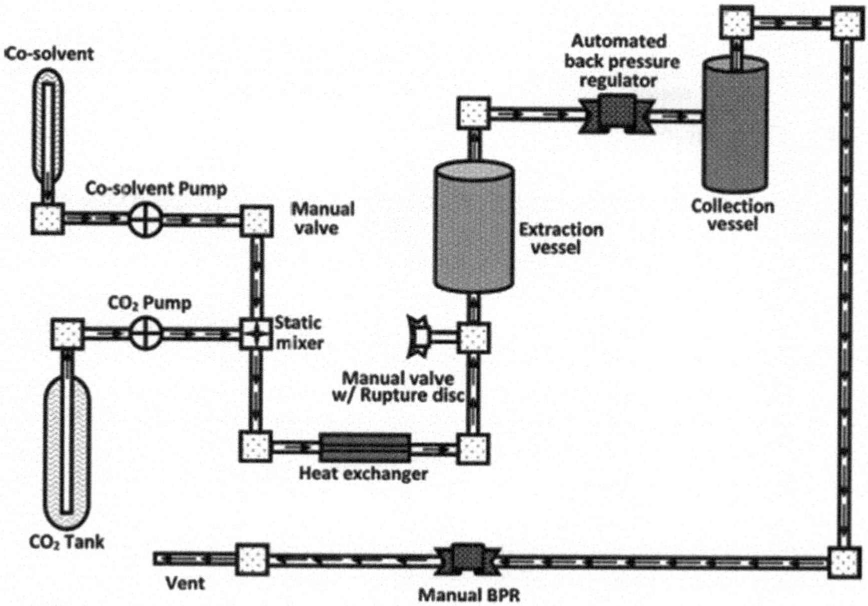

Fig. 19.5 Schematic diagram of supercritical fluid extraction system

or can be released to the environment. The typical schematic set up of SFE system is shown in Fig 19.5.

Advantages

1. SFE causes no degradation of heat sensitive components.
2. It does not leave toxic solvent residues in solute after process.
3. It can be operated at room temperature
4. Small amount of sample can be extracted thereby reducing extraction time.

Disadvantages

1. SC-CO2 is not very suitable for the extraction of polar analytes
2. The SFE process is largely influenced by particulate size, and the consumption of SC-CO_2 depends on grinding efficiency.

19.3.2.6 Other Techniques

Several other techniques include pulsed electric field (PEF) extraction whereby the electric potential induces the electroporation of cell membrane, aiding extraction yield. Similarly, ultrasound assisted extraction (UAE) uses ultrasound of 20–2000 KHz resulting in acoustic cavitation. The cavitation increases the permeability of cell walls facilitating release of compounds and mass transfer of solvents. However, this technique may result in formation of free radicals during extraction.

These techniques can be used as a preliminary extraction step in extracting bioactives

19.4 Conclusion

The advancements in analytical process and increasing environmental concern are the major reasons for the development of novel non-conventional methods. All these extraction methods result in crude extract which again requires extensive purification process making them complex and time consuming. Even though the international market for bioactives has been increasing in recent years, the extraction as well as purification methods must be well optimized prior to commercialization in large scale levels. Among the optimization studies, the most influential parameter is solvent type and strength. Apart from these other parameters like temperature pressure, solvents, agitation speed make tedious in choosing the suitable method. As far as extraction of bioactives is concerned, there is no ideal method is suitable for all by-products and each extraction procedure is unique to each type of by-product. Replacement of conventional techniques with newer techniques will result in eventual improvement of extraction yields apart from reduction of processing times and environmental damages due to the release of toxic solvents. Hence, a combination of newer techniques discussed can possibly improve these processes.

References

Alupului A, Calinescu I, Lavric V (2012) Microwave extraction of active principles from medicinal plants. UPB Sci Bull Series B 74:129–142

Arogba SS (2000) Mango (*Mangifera indica*) kernel: chromatographic analysis of the tannin, and stability study of the associated polyphenol oxidase activity. J Food Compos Anal 13:149–156

Azmir J, Zaidul ISM, Rahman MM (2013a) Techniques for extraction of bioactive compounds from plant materials: A review. J Food Eng 117(2013):426–436

Azmir J, Zaidul ISM, Rahman MM, Sharif KM, Mohamed A, Sahena F, Jahurul MHA, Ghafoor K, Norulaini NAN, Omar AKM (2013b) Techniques to extract bioactive compounds from food by-products of plant origin. Food Res Int 46:505–513

Azwanida NN (2015) A Review on the Extraction Methods Use in Medicinal Plants, Principle, Strength and Limitation. Med Aromat Plants 4(3):1–6

Ballard TS, Mallik arjunan P, Zhou K, O'Keefe S (2010) Microwave-assisted extraction of phenolic antioxidant compounds from peanut skins. Food Chem 120:1185–1192

Bhattacharjee P, Singhal RS, Tiwari SR (2006) Supercritical carbon dioxide extraction of cottonseed oil. J Food Eng 79(3):892–989

Çam M, HisIl Y (2010) Pressurised water extraction of polyphenols from pomegranate peels. Food Chem 123:878–885

Careri M, Corradini C, Elviri L, Mangia A (2007) Optimization of a rapid microwave assisted extraction method for the liquid chromatography-electrospray-tandem mass spectrometry determination of iso flavonoid aglycones in soybeans. J Chromatogr A 1152:274–279

Cassano A, Conidi C, Giorno L, Drioli E (2013) Fractionation of olive mill wastewaters by membrane separation techniques. J Hazard Mater 248–249:185–193

Cassano A, Conidi C, Ruby-Figueroa R (2014) Recovery of flavonoids from orange press liquor by an integrated membrane process. Membranes 4:509–524

Castro-Muñoz R, Yáñez-Fernández J, Álvareza R (2015) Recovery of bioactive compounds from food processing waste waters by ultra and nanofiltration: A review. Adv Biores 6:152–158
Castro-Muñoz R, Yáñez-Fernández J, Fíla V (2016) Phenolic compounds recovered from agro-food by-products using membrane technologies: An overview. Food Chem 213:753–762
Chen Y, Xie M, Li W, Zhang H, Nie S, Wang Y, Li C (2012) An effective method for deproteinization of bioactive polysaccharides extracted from lingzhi (Ganoderma atrum). Food Sci Biotechnol 21(1):191–198
Cheng X, Liangwu Zhendong Z, Chen Y (2015) Advances in enzyme assisted extraction of natural products. 3rd International conference on material, mechanical and manufacturing engineering (IC3ME 2015)
Choi IS, Cho EJ, Moon MH, Bae HJ (2015) Onion skin waste as a valorization resource for the by-products quercetin and biosugar. Food Chem 188:537–542
Choi SH, Kozukue N, Kim HJ, Friedman M (2016) Analysis of protein amino acids, non-protein amino acids and metabolites, dietary protein, glucose, fructose, sucrose, phenolic, and flavonoid content and anti-oxidative properties of potato tubers, peels, and cortexes (pulps). J Food Compos Anal 50:77–87
Coll MD, Coll L, Laencina J, Tomas-Barberan FA (1998) Recovery of flavanones from wastes of industrially processed lemons. Zeitsf'urLebensmund-Forsch A 206:404–407
Concha J, Soto C, Chamy R, Zuniga ME (2004) Enzymatic pretreatment on Rose- Hip oil extraction: hydrolysis and pressing conditions. J Am Oil Chem Soc 81(6):549–552
Conidi C, Rodriguez-Lopez AD, Garcia-Castello E, Cassano A (2012) Valorization of artichokewastewaters by integrated membrane process. Water Res 48:363–374
Csiktusnádi Kiss GA, Forgács E, Cserháti T, Mota T, Morais H, Ramos A (2000) Optimisation of microwave-assisted extraction of pigments from paprika (Capsicum annum L.) powders. J Chromatogr A 889:41–49
Deng GF, Shen C, Xu XR, Kuang RD, Guo YJ, Zeng LS, Gao LL, Lin X, Xie JF, Xia EQ, Li S, Wu S, Chen F, Ling WH, Li HB (2012) Potential of fruit wastes as natural resources of bioactive compounds. Int J Mol Sci 13(7):8308–8323
Díaz-Reinoso B, Moure A, Domínguez H, Carlos Paroja J (2009) Ultra- and nanofiltration of aqueous extracts from distilled fermented grape pomace. J Food Eng 91:587–593
Dominguez H, Ntiiiez MJ, Lema JM (1995) Enzyme-assisted hexane extraction of soybean oil. Food Chem 54(2):223–231
Dunford N, Irmak S, Jonnala R (2010) Pressurised solvent extraction of policosanol from wheat straw, germ and bran. Food Chem 119(3):1246–1249
El Rayess Y, Albasi C, Bacchin P, Taillandier P, Devatine A (2011) Cross-flow microfiltration of wine: effect of colloids on critical fouling conditions. J Membr Sci 385-386:177–186
Erdogan S, Ates B, Durmaz G, Yilmaz I, Seckin T (2011) Pressurized liquid extraction of phenolic compounds from Anatolia propolis and their radical scavenging capacities. Food Chem Toxicol 49(7):1592–1597
FAO (2013) Food wastage footprint on line accessed on December 2015. http://www.fao.org/docrep/018/i3347e/i3347e.pdf
Figueroa RAR, Cassano A, Drioli E (2011) Ultrafiltration of orange press liquor: Optimization for permeate flux and fouling index by response surface methodology. Sep Purif Technol 80:1–10
Fu YJ, Liu W, Zu YG, Tong MH, Li SM, Yan MM, Efferth T, Luo H (2008) Enzyme-assisted extraction of luteolin and apigenin from pigeonpea [Cajanuscajan (L.) Millsp.] leaves. Food Chem 111(2):508–512
Galanakis CM, Markouli E, Gekas V (2013) Recovery and fractionation of different phenolic classes from winery sludge using ultrafiltration. Sep Purif Technol 107:245–251
Ganzler K, Salgo A, Valko K (1986) Microwave extraction. A novel sample preparation method for chromatography. J Chromatogr 371:299–306
Gedye RN, Smith FE, Westaway KC, Ali H, Baldisera L, Laberge L, Roussel J (1986) The use of microwave ovens for rapid organic synthesis. Tetrahedron Lett 27:279–282

Giguerre RJ, Bray TL, Duncan SM, Majetich G (1986) Application of commercial microwave ovens to organic synthesis. Tetrahedron Lett 27:4945–4948

Gomez-García R, Martínez-Ávila GCG, Aguilar CN (2012) Enzyme-assisted extraction of antioxidative phenolics from grape (Vitis vinifera L.) residues. 3Biotech 2(4):297–300

Gosch T, Apprich S, Kneifel W, Novalin S (2013) Improved isolation of bioactive components of bovine colostrum using cross-flow microfiltration. Int J Dairy Technol. 66(2):175–181

Hanmoungjai P, Pyle DL, Niranjan K (2001) Enzymatic process for extracting oil and protein from rice bran. J Am Oil Chem Soc 78:817–821

Hayat K, Zhang X, Chen H, Xia S, Jia C, Zhong F (2010) Liberation and separation of phenolic compounds from citrus mandarin peels by microwave heating and its effect on antioxidant activity. Sep Purif Technol 73(3):371–376

Heo SJ, Lee GW, Song CB, Jeon YJ (2003) Antioxidant activity of enzymatic extracts from brown seaweeds. Algae 18(1):71–81

Hernández-Ledesma B, Herrero H (2014) J Aquat Food Prod Technol 23(3):313–317. https://doi.org/10.1080/10498850.2014.893407

Henry MC, Yonker CR (2006) Supercritical fluid chromatography, pressurized liquid extraction, and supercritical fluid extraction. Anal Chem 78(12):3909–3916. https://doi.org/10.1021/ac0605703. PMID: 16771531

Howard L, Pandjaitan N (2008) Pressurized liquid extraction of flavonoids from spinach. J Food Sci 73(3):151–157

Jaime L, Rodríguez-Meizoso I, Cifuentes A (2010) Pressurized liquids as an alternative process to antioxidant carotenoid extraction from Haematococcus pluvialis microalga. LWT—Food Sci Technol 43:105–112

Japon-Lujan R, Luque-Rodriguez JM, Luque de Castro MD (2006) Multivariate optimization of the microwave-assisted extraction of oleuropein and related biophenols from olive leaves. Anal Bioanal Chem 385:753–759

Jayaprakasha GK, Singh RP, Sakariah KK (2001) Antioxidant activity of grape seed (Vitis vinifera) extracts on peroxidation models in vitro. Food Chem 73:285–290

Ju ZY, Howard LR (2003) Effects of solvent and temperature on pressurized liquid extraction of anthocyanins and total phenolics from dried red grape skin. J Agric Food Chem 51:5207–5213

Kalbasi A, Cisneros-Zevallos L (2007) Fractionation of monomeric and polymeric anthocyanins from Concord grape (Vitis labrusca L.) juice by membrane ultrafiltration. J Agric Food Chem 55:7036–7042

Kallithraka S, Garcia-Viguera C, Bridle P, Bakker J (1995) Survey of solvents for the extraction of grape seed phenolics. Phytochem Anal 6:265–267

Khandare V, Walia S, Singh M, Kaur C (2010) Black carrot (Daucus carota ssp. sativus) juice: processing effects on antioxidant composition and color. Food Bioprod Process 89(4):482–486

Ko MJ, Cheigh CI, Cho SW, Chung MS (2011) Subcritical water extraction of flavonol quercetin from onion skin. J Food Eng 102:327–333

Kosmala M, Kołodziejczyk K, Zdunczyk Z, Juskiewicz J, Boros D (2011) Chemical composition of natural and polyphenol-free apple pomace and the effect of this dietary ingredient on intestinal fermentation and serum lipid parameters in rats. J Agric Food Chem 59(17):9177–9185

Kurmudle NN, Bankar SB, Bajaj IB, Bule MV, Singhal RS (2010) Enzyme-assisted three phase partitioning: a novel approach for extraction of turmeric oleoresin. Process Biochem 46 (1):423–426

Latif S, Anwar F (2009) Physicochemical studies of hemp (Cannabis sativa) seed oil using enzyme-assisted cold-pressing. Eur J Lipid Sci Tech 111:1042–1048

Li Y, Guo C, Yang J, Wei J, Xu J, Cheng S (2006) Evaluation of antioxidant properties of pomegranate peel extract in comparison with pomegranate pulp extract. Food Chem 96:254–260

Li Y, Fabiano-Tixier A-S, Abert-Vian M, Chemat F (2013) Microwave-assisted extraction of antioxidants and food colors. In: Chemat F, Cravotto G (eds) Microwave-assisted Extraction for Bioactive Compounds: Theory and Practice, Food Engineering Series 4. Springer, Boston, MA

Liazid A, Palma M, Brigui J, Barroso CG (2007) Investigation on phenolic compounds stability during microwave-assisted extraction. J Chromatogr A 1140:29–34

Lotfi L, Kalbasi-Ashtari A, Hamedi M, Ghorbani F (2015) Effects of enzymatic extraction on anthocyanins yield of saffron tepals (*Crocos sativus*) along with its color properties and structural stability. J. Food Drug Anal 23:210–218

Lu Y, Foo LY (1998) Constitution of some chemical components of apple seed. Food Chem 61:29–33

Maier T, Göppert A, Kammerer DR, Schieber A, Carle R (2008) Optimization of a process for enzyme-assisted pigment extraction from grape (Vitis vinifera L.) pomace. Eur Food Res Technol 227(1):267–275

Makuuchi K (2010) Critical review of radiation processing of hydrogel and polysaccharide. Radiat Phys Chem 79:267–271

Mandal V, Mohan Y, Hemalatha S (2008) Microwave assisted extraction of curcumin by sample-solvent dual heating mechanism using Taguchi L9 orthogonal design. J Pharm Biomed Anal 46:322–327. https://doi.org/10.1016/j.jpba.2007.10.020

Matharu AS, de Melo EM, Houghton JA (2016) Opportunity for high value-added chemicals from food supply chain wastes. Bioresour Technol 215:123–130

Miron TL, Plaza M, Bahrim G, Ibáñez E, Herrero M (2010) Chemical composition of bioactive pressurized extracts of Romanian aromatic plants. J Chromatogr A 1218(30):4918–4927

Monrad JK, Howard LR, King JW, Srinivas K (2010a) Subcritical solvent extraction of procyanidins from dried red grape pomace. J Agric Food Chem 58:4014–4021

Monrad JK, Howard LR, King JW, Srinivas K (2010b) Subcritical solvent extraction of anthocyanins from dried red grape pomace. J Agric Food Chem 58:2862–2868

Moore J, Cheng Z, Su L, Yu L (2006) Effects of solid-state enzymatic treatments on the antioxidant properties of wheat bran. J Agric Food Chem 54(24):9032–9045

Mroczek T, Mazurek J (2009) Pressurized liquid extraction and anticholinesterase activity-based thin-layer chromatography with bioautography of Amaryllidaceae alkaloids. Anal Chim Acta 633(2):188–196

Nawaz H, Shi J, Mittal GS, Kakuda Y (2006) Extraction of polyphenols from grape seeds and concentration by ultrafiltration. Sep Purif Technol 48:176–181

Nyamien Y, Belleville MP, Coulibaly A, Amisa AA, Henri Marieus GB (2017) Extraction and concentration of bioactive compounds of cola nitida using membrane processes: analysis of operating parameters and membrane fouling. Eur J Nutr Food Saf 7(1):84–100

Onyeneho SN, Hettiarachchy NS (1993) Antioxidant activity, fatty acids and phenolic acids compositions of potato peels. J Sci Food Agric 62:345–350

Pan Y, Wang K, Huang S, Wang H, Mu X, He C, Ji X, Zhang J, Huang F (2008) Antioxidant activity of microwave-assisted extract of longan (Dimocarpus longan Lour.) peel. Food Chem 106:1264–1270

Patindol J, Wang L, Wang YJ (2007) Cellulase-assisted extraction of oligosaccharides from defatted rice bran. J Food Sci 72(9):516–521

Paula Milani G, Pratob A, Antonio Monteiroc RG, Mariane Maiorala F, Livia Marchia B, Sílvio da Costa C (2015) Assessment of operating parameters, membrane fouling and juice quality during Acerola ultrafiltration. Chem Eng Trans 44:325–330

Plaza M, Amigo-Benavent M, del Castillo MD, Ibanez E (2010a) Neo formation of antioxidants in glycation model systems treated under subcritical water extraction conditions. Food Res Int 43 (4):1123–1129

Plaza M, Amigo-Benavent M, del Castillo MD, Ibanez E (2010b) Screening for bioactive compounds from algae. J Pharm Biomed Anal 51(2):450–455

Price KR, Rhodes MJC (1997) Analysis of the major flavonol glycosides present in four varieties of onion (Allium cepa) and changes in composition resulting from autolysis. J Sci Food Agric 74:331–339

Prodanov M, Garrido I, Vacas V, Lebrón-Aguilar R, Duenas M (2008) Ultrafiltration as alternative purification procedure for the characterization of low and high molecular-mass phenolics from almond skins. Anal Chim Acta 609:241–251

Pronyk C, Mazza G (2009) Design and scale-up of pressurized fluid extractors for food and bioproducts. J Food Eng 95:215–226

Puri M, Sharma D, Barrow CJ (2012) Enzyme-assisted extraction of bioactives from plants. Trends Biotechnol 30(1):37–44

Ramos L (2002) Current use of pressurised liquid extraction and subcritical water extraction in environmental analysis. J Chromatogr A 975:3–29

Renouard S, Hano C, Corbin C, Fliniaux O, Lopez T, Montguillon J, Barakzoy E, Mesnard F, Lamblin F, Laine E (2010) Cellulase-assisted release of secoisolariciresinol from extracts of flax (*Linum usitatissimum*) hulls and whole seeds. Food Chem 122(3):679–687

Rodriguez SD, Hadley M, Holm ET (1994) Phenolics in aqueous potato peel extract: extraction, identification and degradation. J Food Sci 59:649–651

Rosenthal A, Pyle DL, Niranjan K (1996) Aqueous and enzymatic processes for edible oil extraction. Enzyme Microb Technol 19(6):402–420

Rostagno MA, Palma M, Barroso CG (2007) Microwave-assisted extraction of soy iso flavones. Anal Chim Acta 588:274–282

Saito M, Hosoyama H, Ariga T, Kataoka S, Yamaji N (1998) Anti-ulcer activity of grape seed extract and procyanidins. J Agric Food Chem 46:1460–1464

Schieber A, Stintzing FC, Carle R (2001) By-products of plant food processing as a source of functional compounds - recent developments. Trends Food Sci Technol 12:401–413

Sharma SK, Le Maguer M (1996) Lycopene in tomatoes and tomato pulp fractions. Italian J Food Sci 8:107–113

Singh PP, Saldaña MDA (2011) Subcritical water extraction of phenolic compounds from potato peel. Food Res Int 44:2452–2458

Singh RK, Sarker BC, Kumbhar BK, Agrawal YC, Kulshreshtha MK (1999) Response surface analysis of enzyme-assisted oil extraction factors for sesame, groundnut, and sunflower seeds. J Food Sci Technol 36(6):511–514

Soquetta MB, Marsillac Terra LD, Bastos CP (2018) Green technologies for the extraction of bioactive compounds in fruits and vegetables. Cyta-J Food 16(2):400–412

Subagio A, Morita N, Sawada S (1996) Carotenoids and their fatty-acid estersin banana peel. J Nutr Sci Vitaminol 42:553–566

Subramanian R, Nakajima M, Kawakatsu T (1998) Processing of vegetable oils using polymeric composite membranes. J Food Eng 38:41–56

Sun Y, Liao X, Wang Z, Hu X, Chen F (2007) Optimization of microwave-assisted extraction of anthocyanins in red raspberries and identification of anthocyanin of extracts using high performance liquid chromatography-mass spectrometry. Eur Food Res Technol 225:511–523

Tishchenko G, Simunek J, Brus J, Netopilik M, Pekarek M, Walterova Z, Koppova I, Lenfeld J (2011) Low-molecular-weight chitosans: preparation and characterization. Carbohydr Polym 86:1077–1081

Wang L, Weller CL (2006) Recent advances in extraction of nutraceuticals from plants. Trends Food Sci Technol 17(6):300–312

Zheng G, Li C, Guo L, Ruo W, Wang S (2012) Purification of extracted fatty acids from microalgae Spirulina. J Am Oil Chem Soc 89:561–566

Zhang H, Tao Y, Guo J, Hu Y, Su Z (2011) Hypolipidemic effects of chitosan nanoparticles in hyperlipidemia rats induced by high fat diet. Int J Immunopharmacol 11:457–461

Zill-e-Huma, Abert Vian M, Maingonnat JF, Chemat F (2008) Clean recovery of antioxidant flavonoids from onions: optimising solvent-free microwave extraction method. J Chromatogr A 1216(45):7700–7707

20 Food Processing Waste to Biofuel: A Sustainable Approach

Divya Agarwal and Dipti Sharma

Abstract

The present paper is reviewing utilization of FPW in generating biofuels with respect to technologies on the global perspective. The sustainable food management, substantial food wastage occur at consumer and supply chain levels in various steps of storage, packaging, transportation and delivery. Damage to crop and food products occur during natural and technological disasters. Climate change is responsible for accelerating natural disasters, that is again a trap for crop failures. Hunger, malnutrition due to unequitable distribution of resources and natural resource crisis due to population explosion are the major problems faced today. Thus role of FPW based biofuels, a step to combat climate change is introspected as an important tool towards circular bio-economy, i.e. an overall sustainable framework ensuring reduction in global greenhouse gaseous emission, reduction in poverty, economic upliftment, wasteland utilization and utilization of food processing waste.

Keywords

Food processing waste (FPW) · Biofuel · Sustainability

D. Agarwal
Environmental Science, Jesus and Mary College, University of Delhi, New Delhi, India

D. Sharma (✉)
Department of Food Technology, Shyama Prasad Mukherji College, University of Delhi, Delhi, India

M. Thakur et al. (eds.), *Sustainable Food Waste Management*,
https://doi.org/10.1007/978-981-15-8967-6_20

20.1 Introduction

Rate of population increase is exponential putting excessive pressure on natural resources to fulfil the expanding needs of food as well as fuel, which again accelerates the cost of food production, indicating a linkage of fuel production process and food production technologies. During the process of food processing by various food technology means, there are enough generation of food processing waste (FPW) as the last by-product, which needs to addressable in waste management. Traditionally looking towards a closed ecosystem approach, there exist a sustainable food management hierarchy explaining the importance of refuse, reduce, reuse, recycle, recover, composting, vermicomposting and landfill disposal. There is a substantial increase in generation of food processing waste every year due to increase in demand of food production (Qasim and Mane 2013). Utilization of the same in production of biofuels can be a significant step in self-reliance of a country in ecofriendly food production, reducing the demand of fuels from conventional energy sources. Thus, it is a vital step towards sustainability. FPWs being rich in carbohydrates, oils, fats, proteins and organic acids can be sustainably utilized for the production of various important biofuels, biochemicals, enzymes, etc. (Girotto et al. 2015). The McKinsey consulting reports if the wastage at consumer level is curtailed by 30%, it would conserve approximately 0.100 billion acres of cropland ecosystems by 2030 (Dobbs et al. 2011). In the same way use of the food processing waste towards biofuel production reduces the pressure on cropland reducing toxic gaseous emissions due to eco-friendly combustion of biofuels.

20.2 Sustainable Framework of Food Processing Waste Based Biofuels

Accounting economic value of food processing waste, it costs approx. 310 billion dollars for the developing countries and US$ 680 billion for the industrialized countries. It has been estimated that almost 30% of the cereals; 40–50% root crops, fruits and vegetables; 20% of the oil seeds, meat and dairy products and 35% of the fish get wasted every year. 95% of the food waste is reported to be dumped at landfill sites where it gets degraded naturally releasing various greenhouse gases by anaerobic digestion, revealed in various studies (Melikoglu et al. 2013). The approx. annual cost of food processing waste is 2600 billion USD, which is equivalent to the gross domestic product (GDP) of France or the UK (FAO 2014).

Waste is presently a major concern worldwide, more importantly in the developing countries (China, India, etc.) and in Europe. Wastes can be classified into industrial, sanitary, agricultural and solid urban residues based on their origin. Wastes generated by food processing companies are a good example of a pre-consumer type of waste, produced on a large scale globally (Kiran et al. 2014). Food processing and agriculture are the two vital components for the well-being of society. The production of fruits, vegetables, cattle, grains, fish, etc.; the transportation and storage of the farm products to processing plants; and the production of food in ready-to-use forms improve and sustain the quality of human life. Potato-processing industry has the highest rate of losses. The lowest proportion of food losses is in the cereals and baking industry (21,000 tonnes) (Hegde et al. 2018).

Food losses are globally occurring with at different levels of consumption. An interesting observation is recorded that industrialized countries are encountering similar amount of food losses as in developing countries. Former facing more than 40% food loses at retail and consumer levels and latter facing the same at post-harvest and processing levels (Dar et al. 2019). The European Union contributes most of the food waste from households (47 million tonnes) and the processing sector (17 million tonnes). In Switzerland, there is almost 61% of food waste generated in the agricultural sector, 22% in the processing industry, 13% in the catering industry and 4% in the large supermarket chains.

20.2.1 Energy Context with Respect to Indian Scenario

India is an agriculture-based country. Agriculture and dairy farming produce large amount of biomass, *e.g.* agricultural waste (crop residues) and cow dung (gobar). The major crop residues were paddy straw, wheat straw, jowar straw, sugarcane trash, groundnut haulms, maize stalks and bajra straw, accounted for almost 88% of the total residues in the country. India processes only 2% of its produce although it being the second largest producer of cereals and fruits, third in marine production and also one of the leading countries in terms of livestock in the world. As per United Nations Development Programme, up to 40% of the food produced in India is wasted. The agriculture ministry, Govt. of India, records food loss of Rs. 50,000 crore worth every year in the country. The Indian government's initiative of self-reliance approach should also take forward the green growth opportunities where coal reservoirs are to be exploited along with the potential of food processing waste based biofuel production technology. Research is exploring Best Practicable Environmental Option, with cheaper technologies along with social and environmental integrity.

Biomass is burnt in 'chulhas' in villages is an example of directly using biomass as fuel. The burning of cow dung destroys essential nutrients of soil, *viz.* N (Nitrogen) and P (Phosphorous). It is, therefore, more useful to convert biomass into biogas and biofuels. Methanogenic bacteria undergo anaerobic decomposition of organic waste at high temperature (38–45 °C) (thermophilic bacteria) under the ground known as digestion well or anaerobic digestor (AD). The semi-solid waste of biogas production is sludge which is highly nutrient rich and used as manure. Biogas is an eco-friendly clean fuel. Without storage tank, it can be supplied directly to the homes from plant. Pathogens and parasites cannot come in contact of faecal material as the digestion of waste takes place in closed chamber. Waste of biogas (sludge) is used as highly nutrient rich manure. It can be supplied only to few kilometres of production of biogas (Mittal et al. 2019).

20.2.2 Sustainability of Energy Plantations vs FPW Based Biofuels

Large-scale plantation of trees for production of energy from biomass is known as energy plantation. Latex of trees of the family Euphorbiaceae, e.g. *Jatropha* and *Euphorbia* species are used for energy plantations. Biodiesel is produced from the oil of Jatropha. Plantations on rural wastelands also provide employment to local people. Diversion of agricultural croplands to bioenergy plantations will intensify the pressure on land and forest resource which will indirectly contribute towards large releases of CO_2 in the atmosphere, also lead to increase in global food costs (Fargione et al. 2008; Searchinger et al. 2008). In a study by Spawn (2019) because of legal binding of usage of transportation ethanol in USA, usage of approximately 40% of domestic corn gets shifted to ethanol which further led to the conversion of 1.5 million hectares of grasslands into croplands releasing huge amount of CO_2, in the period of 2008 and 2012. However, utilizing FPW puts no extra pressure on agricultural land, helps in reduction of greenhouse gaseous emissions, effective management of organic waste with lots of employment opportunities. Thus, approaches of utilizing food processing waste into biofuels are the need in today's global scenario towards sustainable food processing waste management (Zhang et al. 2013) (Fig. 20.1).

20.2.3 Food Processing Waste Based Biofuels vs Fossil Fuels

Due to high water content of food processing waste, the pathogenic microorganisms grow on it, which makes it unsuitable for conversion to biofuels. The meat industry waste with high-fat content is prone to oxidation and the spoilage due to the enzymatic actions (Fargione et al. 2008). Researchers are exploring the practically

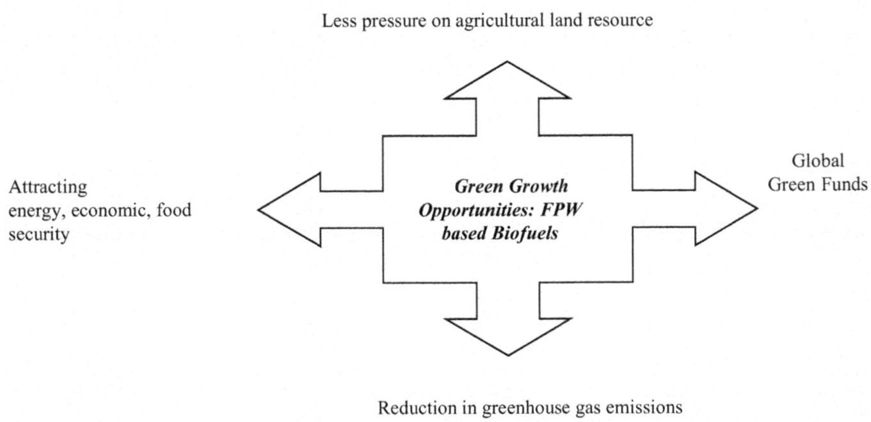

Fig. 20.1 Some important green growth opportunities of food processing waste based biofuels

Table 20.1 List of food processing industries and respective waste used in biofuel production

Food processing industries	Food processing waste	Potential characteristics of FPW utilized for biofuel production
Meat industry (Butchering or Slaughtering)	The liquid and solid wastes (grease, oils and fats, cooking waste, hair, feathers and the animal weight remains as by-products, from butchering or slaughtering sites) (Sahu 2016; Pap et al. 2016; Prazeres et al. 2012)	The poultry and meat processing industries generate effluents rich in both fat and proteins
Seafood wastes	The examples of solid by-products are heads, rejected fish, offal, skin, bones and tail (Helkar and Patil 2016)	The seafood and meat processing waste streams are rich in nitrogen. Phosphorous is added due to the use of detergents
Dairy sources	Damaged or out-dated products, solids, cheese, whey, curd and milk sludge including lactose, proteins, fats, etc. (FAO 2012)	Dairy waste is rich in dissolved protein, lactose and suspended fat
Fruit and vegetable processing	Stems, stalks and rotten fruits and pulp, seeds, peels and pomace	The wastewater of fruit and vegetable processing plants contains sugars, starches, pectin, vitamins and other components of the cell wall. They have higher BOD levels and are rich sources of many nutrients like fibres, minerals, vitamins, etc. (Mabrouk and El-Ahwany 2008)
Pulses and oil seed industry	A number of waste streams are generated during the processing of oil seeds. Emission in the air also takes place like grain dust, hexane solvent loss, odour during the meal drying and deodorization. The wastes from this industry are generated in the major processes like milling, extraction, deodorizing, caustic refining, acidulating, packaging, tank car washing, margarine production, bleaching winterizing, mayonnaise production and salad dressing (Tiwari et al. 2011)	The processing of legumes creates an effluent rich in proteins. Effluents from oilseed processing contain fats as suspended matter
Miscellaneous sources	Wastewater from juice, soda or fruit bottling; breweries, bakeries, distilleries and sugar processing units release carbohydrate-rich effluent	

viable ways of extraction of polysaccharides like cellulose and pectins, processing of dietary fibre, and vegetable and fruit processing wastes towards efficient production of FWP (Joonsson et al. 2013; Jitputti et al. 2006; Kim and Day 2011) (Table 20.1)

20.3 Conventional (Non-Renewable) Energy Resources

Coal is the most abundantly found fossil fuel in the world. It contains carbon, water, sulphur and nitrogen. Coal meets 70% of the total energy needs of the world and 87.4% of all commercial energy. In India about 58% of commercial energy is obtained from coal and 38% from petroleum along with natural gas. Coal is used for cooking, heating, in industries and thermal power plants. Petroleum is useful for transportation, agricultural equipment and some industries. Natural gas is used both in cooking and in industries. Burning of coal produces SO_2, *i.e.* cause air pollution. Release of CO_2 and SO_2 gas in the atmosphere causes greenhouse effect and global warming. In thermal power plants, burning of coal also generates large amount of fly-ash. Fly-ash is a toxic waste, contains toxic heavy metals. Workers in the coal mines suffer from the following lung diseases: Black-lung disease, asthma, bronchitis, lung cancer. The gaseous fuels are basically derived from petroleum, formed due to decomposition of micro plankton deposited upon the sea beds, lakes and rivers for millions of years. The decomposition takes place by the action of bacteria, under lack of oxygen and also by catalytic cracking. It is also referred as crude oil. Liquid fuel (Petroleum) is easy to transport. Liquid fuel (Petroleum) is comparatively cleaner. Environmental impacts of petroleum include emission of toxic and greenhouse gases (NO_2, SO_2, CO, NO_2, CO_2, etc.) as combustion by-products, contamination in the water in case of the leakage. Level of awareness among general public about the adverse health impacts of environmental pollution is very low. We are hopeful that people will adapt sustainable lifestyle in the present scenario of Covid-19 pandemic (Kaushik et al. 2020). Coal and petroleum also contribute to acid rain and urban pollution. One-third of the global food production goes waste (FAO) (Lin et al. 2013). Food processing industry generates substantial high organic wastes along with high energy uses. The recovery of food processing wastes as renewable energy sources represents a sustainable option for the substitution of fossil energy, contributing to the transition of food sector towards a low-carbon economy.

FPWs are defined as residual wastes which stay after processing a primary product. Food industries release large proportions of solid and liquid wastes due to faulty processes of preparation, production and consumption of food; losses in processing steps; inappropriate transport, storage and packaging (Sahu 2016). Food processing wastes are rich in biodegradable components including vegetable and fruit waste, pits, seeds, blood, bone, process water, dairy waste, wastewater treatment sludge. They show high chemical oxygen demand (COD) and biological oxygen demand (BOD) contents. If FPWs are left untreated and unmanaged, their uncontrolled decomposition will pollute the environment due to the release of toxic materials and methane.

20.4 Innovative Management Approaches of Biofuel Production Technology

Food processing industry must be converted to an enclosed circular economy, wherein FPWs need to be used as raw materials for other industries, as a value added by-product or waste treatment substrate, integrating industrial processes involving biochemical reactions, thermochemical reactions, biotechnology, management, business continuity.

20.4.1 Basic Chemistry of Biodiesel Production

Technologies converting food-waste-to-energy include biological, e.g. anaerobic digestion and fermentation, thermal and thermochemical technologies (e.g. pyrolysis, gasification, incineration, liquefaction and hydrothermal carbonization) (Schieber et al. 2001; McDougall and Hruska 2000). Thermochemical technologies reduce the amount of waste generation drastically.

Incineration
As per the principle of cogeneration, the heat generated by incineration of wastes can be utilized for electricity generation or could be used to operate machinery. Food wastes have lower oxygen content but higher nitrogen, ash and energy contents than wood (Digman and Kim 2008).

Pyrolysis
The conversion of biomass into syngas at very higher temperatures (around 1000 °C) in the absence of oxygen is known as pyrolysis. It can also be defined as the thermal decomposition of the organic matrix to produce solid, liquid and gaseous products (Canabarr et al. 2013; Balat et al. 2009).

Gasification
Syngas is a mixture of combustible gases, viz. carbon monoxide (CO), hydrogen (H_2), carbon dioxide (CO_2), methane (CH_4), nitrogen (N_2) and water vapours (H_2O) manufactured by burning of biomass at moderately high temperatures by the simultaneous oxidation and pyrolysis under inadequate oxygen supply (Barman et al. 2012; Brown and Brown 2003).

Liquefaction
Biomass is thermochemically converted into a liquid product operating at high pressure and low temperature in the presence of a catalyst. It is a costlier process because of expensive reactors. Complex fuel feeding systems also make it difficult to operate (Demirbas 2011).

Hydrothermal Carbonization
A process of conversion of food wastes having high moisture content (80–90%), into a valuable, energy resource under endogenous pressures and moderately low temperature (180–350 °C) is known as hydrothermal carbonization (Hoekman et al. 2011; Liu et al. 2013)

20.4.2 Biochemical/Biological Conversion Processes

The processes categorized under this category are anaerobic digestion, alcohol fermentation and biodiesel formation.

Anaerobic Digestion

This technology seems very suitable for the food processing wastes, *viz.* dairy waste, distillery effluent digestion (Zarkadas et al. 2015), fruit and vegetable processing wastes (Krylova et al. 2008; Noomtim and Cheirsilp 2011; Sompong et al. 2012; Scano et al. 2014), slaughterhouse wastes (Ware and Power 2016). Anaerobic digestion (AD) is a process where microorganisms break down (hydrolysis, fermentation, acetogenesis and finally methanogenesis) biodegradable materials, such as food wastes, manure and sewage sludge, in the absence of oxygen, can be carried out at both mesophilic (25–45 °C) and thermophilic (55–70 °C) temperatures, to produce biogas (Chandra et al. 2012). The semi-solid and a solid waste are the by-products of anaerobic digestion, which is used as manure and prevents soil erosion (Naik et al. 2010). Co-digestion is another process related to anaerobic digestion.

Alcohol Fermentation

Fermentation is the age-old process familiar to humans. It is used to produce foods, feeds and other well-known compounds by the utilization of microorganisms. The food processing wastes can be used to produce bioethanol and biobutanol (the liquid biofuels). These biofuels are the potential alternative sources of energy to replace petrol and diesel (conventional liquid fuels). Alcoholic fermentation producing liquid biofuels is advantageous as it can significantly reduce the greenhouse gas emissions (Macedo et al. 2008).

Biodiesel Formation

Biodiesel, which is a potential alternative biofuel to conventional fuels can be generated from food processing wastes either through chemical reaction of transesterification (by using acid or alkaline catalysts) or by microbial transesterification. Commercial generation of biodiesel is by transesterification of alcohol in the presence of a catalyst. It consists of the conversion of triglycerides (oil) to methyl esters (biodiesel) and a by-product (glycerol) (Ramadhas et al. 2005). Most of the biodiesel produced today use a base catalysis reaction (Meher et al. 2006) which provides advantages of high conversion yields at low temperatures and pressures and the minimal side reactions as well as reaction times (Baskar et al. 2019).

20.5 Biofuels from Food Processing Wastes

Mode of generation, uses and environmental impact have been discussed in Table 20.2

Table 20.2 General mode of generation, uses and environmental impacts of biofuel

Biofuel	Generation	Uses	Environmental impacts
Bioalcohols (bioethanol)	Action of microorganisms and enzymes through the fermentation of sugars (wheat, corn, sugarbeets, sugarcane, molasses) or starches (potato and fruit waste) or carbohydrates (grain processing waste). *Saccharomyces cerevisiae* (Adaganti et al. 2014). The hydrolysis of starch to sugars (glucose) is enhanced by enzymes like α-amylase and glucoamylase. Sugar juice obtained from food wastes like fruit pomace can be directly used for ethanol production	Bioethanol is used as fuel for vehicles (USA and Brazil)	A gasoline additive to increase octane and improve vehicle emissions
Biodiesel (monoalkyl esters of long chain fatty acids)	Biodiesel is produced from animal fat, waste cooking oils, restaurant grease and vegetable oils by transesterification (Canakci 2007). Food-grade vegetable oils like soybean oil and rapeseed are popular for the production of biodiesel in the USA and Europe. The production from vegetable oils is not economically viable	Due to similarity with diesel, both are easily blended without any need for modifications	Less air pollution from diesel-powered vehicles. Pure biodiesel (B100) is the lowest emission diesel fuel (Canakci and Gerpen 1999)
Green diesel or renewable diesel	It is produced from oils of Canola, Algae, Jatropha and Salicornia by traditional fractional distillation process of oils, popularly used in Ireland	Used as fuels along with diesel	Less pollution
Fuel ethers or oxygenated fuels	When *petrol* is formulated with *fuel ethers*, they raise its *oxygen* content, leading to more complete	Cost-effective compounds that act as octane rating enhancers	Significantly reducing engine wear and tear and toxic exhaust emissions including level of ozone

(continued)

Table 20.2 (continued)

Biofuel	Generation	Uses	Environmental impacts
	combustion. This means that all *fuel* is burned inside the engine rather than expelled through the exhaust system into our air, allowing for cleaner-running engines		
Syngas	Mixture of carbon monoxide, hydrogen and other hydrocarbons, produced by partial combustion of biomass	May be burned directly in internal combustion engines, turbines or high-temperature fuel cells	Less toxic emissions
Biobutanol	It can be easily produced from food processing wastes and cellulosic substrates with high water content. Food processing waste is reported to be better substrate for butanol production being economic, providing higher yield with less residual sugars and no dependency on hydrolytic enzymes. All the above-mentioned merits make FPWs potential feedstocks for n-butanol production at commercial scale (Huang 2015)	Unlike ethanol, it can be blended with gasoline (Huang et al. 2015)	
Biogas	Anaerobic digestion is recommended for biogas (methane) production from food processing wastes (Chandra et al. 2012). Complex food processing wastes consisting of animal fats, protein and lignocellulosic biomass are to be undergone hydrolysis to form simpler units (sugars, fatty acids, amino acids). Afterwards converted to volatile fatty acids which are	Heat or electricity generation	

(continued)

Table 20.2 (continued)

Biofuel	Generation	Uses	Environmental impacts
	later converted to acetate, to be acted upon by methanogenic bacteria to produce methane (Chandra et al. 2012)		
Biohydrogen	It is carbon neutral and has a tendency to replace the conventional fuels, thus saving the depletion of oil reserves. Fermentative H_2 generation via light-dependent and dark fermentation processes is advantageous. Light-dependent processes like photolysis and photo-fermentation are aerobic in nature while as the dark fermentation occurs under anaerobic conditions. The combined photo-dark fermentations for food processing wastes have been economical. The dark fermentation among all these processes is comparatively favourable due to higher yield and lesser production cost (Sangyoka et al. 2016)	The high calorific value, easy availability and environment-friendly nature make biohydrogen a prospective biofuel (Sangyoka et al. 2016)	It releases clean by-product after combustion (Kotay and Das 2008)

Biodiesel is an eco-friendly alternative of fossil fuel and its production from food processing waste is one of the sustainable options depending upon availability of economically viable technology. Combustion of biodiesel generates less air pollutants as compared to fossil fuels. Biodiesel synthesis is a transesterification reaction of lipids (present in animal fats and vegetable oils) in the presence of alcohol, base, acid, enzyme or solid catalyst. (Sarkar et al. 2020)

Waste eggshell was found as highly active and reusable catalyst in triglyceride transesterification reaction making it economically viable and environmental friendly option to reuse the eggshell waste (Wei et al. 2009). Another study to optimize biodiesel production from the waste oil of pork grilling process in the food factory in Udon Thani, Thailand used it as a raw material. KOH was reported as an effective catalyst in the transesterification reaction (Singhasiri and Tantemsapya 2016). Biodiesel production is also successfully reported from *Stauntonia chinensis* seed oil waste. It showed an ideal fatty acid composition. SC biodiesel was found with better fuel properties as compared to various feedstock oils (Wang et al. 2014).

20.6 Biodiesel Production from Various Food Processing Wastes

Food-grade vegetable oils are inexpensive raw material for biodiesel production but still prove to be more expensive than diesel fuel. In a study to analyse the free fatty acids and moisture in the waste material, it was found that former content varies from 0.7% to 41.8% and latter varies from 0.01% to 55.38%, reducing the efficiency of transesterification in converting these feedstocks into biodiesel. (Canakci and Gerpen 1999)

An integrated approach of co-digestion of sewage sludge (SS) and food waste (FW) is analysed in a study in China. The results showed that a biggest synergistic effect happened in the co-digestion of SS and FW in volatile solid (VS) mixing ratio of 1:1 with the highest methane yield of 415.3 mL/g VS (Wang et al. 2000).

Genetic Engineering Strategies

Use of genetic engineering strategies to produce biodiesel from yeast, plant and algae is one of the latest approaches. The oil content of the plant can be increased by genetically modifying the lipid biosynthetic pathway. It has been observed that an increase in the activity of acetyl-CoA carboxylase increases the lipid synthesis by enhancing the utilization of malonyl-CoA in the biosynthetic pathway. Also, changing the composition of the fatty acid of plant seed oil improves the biodiesel production. In an attempt to enhance the biodiesel production from *Arabidopsis*, it was observed that overexpression of WRINKLED1 (WRI1) mRNA is directly associated with the increase in seed oil. WRI1 regulates the seed storage metabolism and its expression under the CMV 35S promoter increases the seed oil content and also accumulates triacylglycerols in developing seedlings.

Various metabolic engineering methods have been used for constructing lactose-consuming *Saccharomyces cerevisiae* strains, using the lactose genes of the yeast *Kluyveromyces lactis*, *Escherichia coli* and *Aspergillus niger*.

Hybrid Processes for Improving Biofuels Production

In the past few years, studies have been carried out to produce hydrogen by integrating dark fermentation process (first stage) and photo fermentative process (second stage). The volatile fatty acids (major deterrent for the hydrogen production) formed in the first stage of the fermentation can be used as a substrate in the second stage. A two-stage process for improved hydrogen production was developed where glucoamylase was first used to hydrolyze the food waste. The hydrolysate was then further utilized as a substrate for batch fermentation and continuous fermentation processes. The highest cumulative production of hydrogen was observed with a yield of 245.7 mL H_2/g glucose in the batch system (Tenca et al. 2011).

Similarly, studies have been carried out to produce hydrogen and methane from food waste in a two-stage anaerobic digester. Both the fermentations were carried out in continuously stirred tank fermentors and the effect of continuous circulation on the efficiency and stability of processes was also investigated. Maximum H_2 production achieved was 3 L hydrogen per litre per day and maximum methane was 2.9 L methane per litre per day and about 70% degradation of volatile fatty acids was also observed (Silva et al. 2018).

The suitable materials for biohydrogen production should be rich in carbohydrate but be nitrogen deficient. Among the various substrates assessed, food processing wastes being comparatively cheaper and have been an ideal biodegradable organic matter for biohydrogen production through dark fermentation (FOEN 2018). The type of cultures whether pure, co-culture or mixed also influences the biohydrogen yield. Giannis studied the H_2 production from various types of cheese under different pH values (6.5–7.5) and reported the highest yield of 170 ml H_2/kg of total organic content (TOC).

20.7 Conclusion

Food processing wastes are produced at an alarming rate. The management of these wastes is of utmost importance. Biofuel production from these wastes is the most feasible option as it helps in alleviating the dependence on conventional fuels and also reduces greenhouse emissions. The various approaches of biofuel production (like anaerobic digestion, alcoholic fermentation and thermochemical conversion methods) from FPWs are promising and well-consolidated methods. There is still the need to explore and research the various aspects of valorization of FPWs, the dissemination of knowledge to common masses regarding the efficient utilization of food processing wastes. Traditional knowledge with respect to utilization of food processing waste as biofuels should also be given priority over technological options because that can be the quickest, low-resource demanding innovative method. Technological inputs to such traditional innovations can be worth magical solutions. The government and industries should join hands to give research wings from laboratories to commercial scale.

References

Adaganti SY, Yaliwal VS, Kulkarni BM, Desai GP, Banapurmath NR (2014) Factors affecting bioethanol production from lignocellulosic biomass (*Calliandra calothyrsus*). Waste Biomass Valoriz 5:963–971. https://doi.org/10.1007/s12649-014-9305-8

Balat M, Balat M, Kirtay E, Balat H (2009) Main routes for the thermo-conversion of biomass into fuels and chemicals. Part 1: Pyrolysis systems. Energy Convers Manag 50:3147–3157. https://doi.org/10.1016/j.enconman.2009.08.014

Barman NS, Ghosh S, De S (2012) Gasification of biomass in a fixed bed downdraft gasifier - a realistic model including tar. Bioresour Technol 107:505–511. https://doi.org/10.1016/j.biortech.2011.12.124

Baskar G, Kalavathy G, Aiswarya R, Selvakumari I (2019) Advances in bio-oil extraction from nonedible oil seeds and algal biomass. In: Advances in eco-fuels for a sustainable environment. Woodhead Publishing series in energy, vol 1. Elsevier, Amsterdam, pp 187–210. https://doi.org/10.1016/B978-0-08-102728-8.00007-3

Brown RC, Brown TR (2003) Biorenewable Resources Engineering: New Products from Agriculture, second ed. Wiley, Hoboken, NJ

Canabarr N, Soares JF, Anchieta CG, Kelling CS, Mazutti MA (2013) Thermochemical processes for biofuels production from biomass. Sustain Chem Process 1:1–10. https://doi.org/10.1186/2043-7129-1-22

Canakci M (2007) The potential of restaurant waste lipids as biodiesel feedstocks. Bioresour Technol 98:183–190

Canakci M, Gerpen J (1999) Biodiesel prouction via acid catalysis. Trans ASAE 42(5):1203–1210. https://doi.org/10.13031/2013.13285

Chandra R, Vijay VK, MV SP, Khura TK (2012) Production of methane from anaerobic digestion of jatropha and pongamia oil cakes. Appl Energy 93:148–159

Dar RA, Yaqoob M, Parmar M, Phutela UG (2019) Biofuels from food processing wastes, microbial fuel cells: materials and applications. Materials Research Forum LLC, Millersville, PA. https://doi.org/10.21741/9781644900116-10

Demirbas A (2011) Waste management, waste resource facilities and waste conversion processes. Energy Convers Manag 52:1280–1287. https://doi.org/10.1016/j.enconman.2010.09.025

Digman B, Kim DS (2008) Review: alternative energy from food processing wastes. Environ Prog 27:524–537. https://doi.org/10.1002/ep.10312

Dobbs R, Oppenheim J, Thompson F, Brinkman M, Zomes M (2011) Resource revolution: Meeting the world's energy, materials, food, and water needs. McKinsey Global Institute. https://www.mckinsey.com/business-functions/sustainability-and-resource-productivity/our-insights/resource revolution

FAO (2012) Key facts on food loss and waste you should know! save food: global initiative on food loss and waste reduction, Food and Agriculture Organization of the United Nations. http://www.fao.org/save-food/resources/keyfindings/en/ Accessed on 25 Apr 2020

FAO (2014) Food wastage footprint: full cost-accounting. www.fao.org/publications/card/en/c/5e7c4154-2b97-4ea5-83a7-be9604925a24/

Fargione J, Hill J, Tilman D, Polasky S, Hawthorne P (2008) Land clearing and the biofuel carbon debt. Science 319:1235–1238

Federal Office for the Environment (FOEN) (2018) Food waste. https://www.bafu.admin.ch/bafu/en/home/topics/waste/guide-to-waste-az/biodegradable-waste/types-of-waste/lebensmittelabfaelle.html. Accessed on 29 Apr 2020

Girotto F, Alibardi L, Cossu R (2015) Food waste generation and industrial uses: a review. Waste Manag 45:32–41

Hegde S, Lodge JS, Trabold TA (2018) Characteristics of food processing wastes and their use in sustainable alcohol production. Renew Sustain Energy Rev 81:510–523. https://doi.org/10.1016/j.rser.2017.07.012

Helkar PB, Patil SN (2016) Review: food industry by-products used as a functional food ingredients. Int J Waste Resour 6:1–6

Hoekman SK, Broch A, Robbins C (2011) Hydrothermal carbonization (HTC) of lignocellulosic biomass. Energy Fuels 25:1802–1810

Huang H, Singh V, Qureshi N (2015) Butanol production from food waste: a novel process for producing sustainable energy and reducing environmental pollution. Biotechnol Biofuels 8:147

Jitputti J, Kitiyanan B, Rangsunvigit P, Bunyakiat K, Attanatho L, Jenvanitpanjakul P (2006) Transesterification of crude palm kernel oil and crude coconut oil by different solid catalysts. Chem Eng J 116:61–66

Joonsson LJ, Alriksson B, Nilvebrant NO (2013) Bioconversion of lignocellulose: inhibitors and detoxification. Biotechnol Biofuels 6:16
Kaushik M, Agarwal D, Gupta AK (2020) Cross-sectional study on the role of public awareness in preventing the spread of COVID-19 outbreak in India. Postgraduate Medical Journal Published Online First. https://doi.org/10.1136/postgradmedj-2020-138349
Kim M, Day DF (2011) Composition of sugar cane, energy cane, and sweet sorghum suitable for ethanol production at Louisiana sugar mills. J Ind Microbiol Biotechnol 38:803–807
Kiran EU, Trzcinski AP, Ng WJ, Liu Y (2014) Bioconversion of food waste to energy: A review. Fuel 134:389–399
Kotay SM, Das D (2008) Biohydrogen as a renewable energy resource prospects and potentials. Int J Hydrogen Energy 33(1):258–263
Krylova AY, Kozyukov EA, Lapidus AL (2008) Ethanol and diesel fuel from plant raw materials: a review. Solid Fuel Chem 42(6):358–364
Lin CSK, Pfaltzgraff LA, Herrero-Davila L, Mubofu EB, Abderrahim S, Clark JH, Koutinas AA, Kopsahelis N, Stamatelatou K, Dickson F, Thankappan S, Mohamed Z, Brocklesby R, Luque R (2013) Food waste as a valuable resource for the production of chemicals, materials and fuels. Current situation and global perspective. Energy Environ Sci 6:426–464. https://doi.org/10.1039/c2ee23440h
Liu Z, Quek A, Kent Hoekman S, Balasubramanian R (2013) Production of solid biochar fuel from waste biomass by hydrothermal carbonization. Fuel 103:943–949. https://doi.org/10.1016/j.fuel.2012.07.069
Mabrouk MEM, El-Ahwany AMD (2008) Production of β-mannanase by Bacillus amylolequifaciens 10A1 cultured on potato peels. Afr J Biotechnol 7:1123–1128. https://doi.org/10.5897/AJB08.047
Macedo IC, Jea S, Jear S (2008) Green house gases emissions in the production and use of ethanol from sugarcane in Brazil: the 2005/2006 averages and a prediction for 2020. Biomass Bioenergy 32:582–595. https://doi.org/10.1016/j.biombioe.2007.12.006
McDougall FR, Hruska JP (2000) Report: the use of life cycle inventory tools to support an integrated approach to solid waste management. Waste Manag Res 18:590–594. https://doi.org/10.1177/0734242X0001800610
Meher LC, Dharamagadda VSS, Naik SN (2006) Optimization of alkali-catalyzed transesterification of Pongamia pinnata oil for production of biodiesel. Bioresour Technol 97:1392–1397
Melikoglu M, CSK L, Webb C (2013) Analysing global food waste problem: pinpointing the facts and estimating the energy content. Open Eng 3(2):157–164
Mittal S, Ahlgren EO, Shukla PR (2019) Future biogas resource potential in India: a bottom-up analysis. Renewable Energy 141:379–389. https://doi.org/10.1016/j.renene.2019.03.133
Naik SN, Goud VV, Rout PK, Dalai AK (2010) Production of first and second gen-eration biofuels: a comprehensive review. Renew Sustain Energy Rev 14(2):578–597
Noomtim P, Cheirsilp B (2011) Production of butanol from palm empty fruit bunches hydrolyzate by Clostridium acetobutylicum. Energy Procedia 9:140–146
Pap N, Pongrácz E, Myllykoski L, Keiski R (2016) Waste minimization and utilization in the food industry. In: Introduction to advanced food processing engineering, 1st edn. CRC Press, Taylor & Francis Group, Boca Raton, FL, pp 595–630. https://doi.org/10.1201/b16696-23
Prazeres AR, Carvalho F, Rivas J (2012) Cheese whey management: A review. J Environ Manage 110:48–68. https://doi.org/10.1016/j.jenvman.2012.05.018
Qasim W, Mane AV (2013) Characterization and treatment of selected food industrial effluents by coagulation and adsorption techniques. Water Resour Ind 4:1–12. https://doi.org/10.1016/j.wri.2013.09.005
Ramadhas AS, Jayaraj S, Muraleedharan C (2005) Biodiesel production from high FFA rubberseed oil. Fuel 84(4):335–340
Sahu JK (2016) Introduction to advanced food processing engineering. CRC Press, Taylor & Francis Group, Boca Raton, FL

Sangyoka S, Reungsang A, Lin CY (2016) Optimization of biohydrogen production from sugarcane bagasse by mixed cultures using a statistical method. Sustain Environ Res 26(5):235–242

Sarkar N, Jeon B-H, Kumar P, Ganguly A (2020) Food waste, a good option for biodiesel production. Bioresour Utilization Bioprocess 1:267–273

Scano EA, Asquer C, Pistis A, Ortu L, Demontis V, Cocco D (2014) Biogas from anaerobic digestion of fruit and vegetable wastes: experimental results on pilot-scale and preliminary performance evaluation of a full-scale power plant. Energy Convers Manage 77:22–30

Schieber A, Stintzing F, Carle R (2001) By-products of plant food processing as a source of functional compounds—recent developments. Trends Food Sci Technol 12:401–413. https://doi.org/10.1016/S0924-2244(02)00012-2

Searchinger T, Heimlich R, Houghton RA, Dong F, Elobeid A, Fabiosa J, Tokgoz S, Hayes D, Yu TH (2008) Use of US croplands for biofuels increases greenhouse gases through emissions from land-use change. Science 319(5867):1238–1240

Silva F, Mahler MS, Claudio F, Oliveira LB, Bassin João P (2018) Hydrogen and methane production in a two-stage anaerobic digestion system by co-digestion of food waste, sewage sludge and glycerol. Waste Manag 76:339–349

Singhasiri T, Tantemsapya N (2016) Production of biodiesel from food processing waste using response surface methodology. J Energy Sources 38(19):2799–2808. https://doi.org/10.1080/15567036.2015.1117543

Sompong O, Boe K, Angelidaki I (2012) Thermophilic anaerobic co-digestion of oil palm empty fruit bunches with palm oil mill effluent for efficient biogas production. Appl Energy 93:648–654

Spawn SA, Lark TJ, Gibbs HK (2019) Carbon emissions from cropland expansion in the United States. Environ Res Lett 14(4):45009

Tenca A, Schievano A, Perazzolo F, Adani F, Oberti R (2011) Biohydrogen from thermophilic co-fermentation of swine manure with fruit and vegetable waste: maximizing stable production without pH control. Bioresour Technol 102(18):8582–8588

Tiwari BK, Gowen A, McKenna BM (2011) Pulse foods: processing, quality and nutraceutical applications, 1st edn. Academic Press, Cambridge, MA

Wang N, Zheng T, Yingqun M (2000) New insights into the co-locating concept on synergistic co-digestion of sewage sludge and food waste towards energy self-sufficient in future WWTPs Bioresource Technology Reports

Wang R, Sun L, Xie X, Ma L, Liu Z, Liu X, Ji N, Xie G (2014) Industrial Crops and Products 62:8–13

Ware A, Power N (2016) Biogas from cattle slaughterhouse waste: energy recovery towards an energy self-sufficient industry in Ireland. Renewable Energy 97:541–549

Wei Z, Chunli X, Baoxin L (2009) Application of waste eggshell as low-cost solid catalyst for biodiesel production. Bioresour Technol 100(11):2883–2885. https://doi.org/10.1016/j.biortech.2008.12.039

Zarkadas IS, Sofikiti AS, Voudrias EA, Pilidis GA (2015) Thermophilic anaerobic digestion of pasteurised food wastes and dairy cattle manure in batch and large volume laboratory digesters: focussing on mixing ratios. Renew Energy 80:432–440. https://doi.org/10.1016/j.renene.2015.02.015

Zhang C, Xiao G, Peng L, Su H, Tan T (2013) The anaerobic co-digestion of food waste and cattle manure. Bioresour Technol 129:170–176

Utilization of Fly Ash as a Sustainable Waste Management Technique

21

Ayushi Varshney, Sumedha Mohan, and Praveen Dahiya

Abstract

Fly ash is an important combustion residue and the most familiar industrial by-product. It is a source of various micronutrients and macronutrients which are essential for plant growth. The striking variations towards the sensitivity of fly ash are observed in the plant growth performance as fly ash comprises heavy metals which may alter the physical, biochemical, and biological properties of the plant thereby alter the plant growth. The growth and metabolism of different plant species were studied by various researchers in different soil–fly ash amendments. Certain observations were made for cotton yield and grain yield of wheat which revealed that 20% fly ash application is significant for increased growth and rice yield or wheat crops. An improved yield and growth in potato, barley and rye was reported at optimum rate of application of fly ash. Ten percent fly ash doses with soil increased the growth of tomato, petunia, and Boston fern as it increased the height of seedling and plant, increased the girth, enhanced the number of leaves, area of leaves, length of spike, and the plant dry weight. Plant growth and yield, carotenoids and chlorophyll contents of leaves were significantly increased at high fly ash rates in chilly, eggplant, and tomato. Changes in the soil characteristics and the cropping system due to the addition of fly ash were also studied and investigated by various workers that show an enhanced production of food crops. This study is an attempt to know the impact of fly ash on morphological and biochemical parameters of different plant species and also to determine the possible utilization in various agricultural practices. At high concentration of fly ash–soil doses, the experiment shows deleterious effect, the maximum being 100% fly ash with no amendment of soil. This shows that at high concentration of

A. Varshney · S. Mohan (✉) · P. Dahiya
Amity Institute of Biotechnology, Amity University, Noida, Uttar Pradesh, India
e-mail: smehta1@amity.edu

M. Thakur et al. (eds.), *Sustainable Food Waste Management*,
https://doi.org/10.1007/978-981-15-8967-6_21

heavy metals generated from fly ash can cause genetic damage to plants due to their toxic concentration. At low concentration of fly ash–soil doses, heavy metals are better for the plant growth performance as well as plant productivity. Large-scale use of fly ash in agriculture holds the prospective to increase on an average 10–15% of yield and growth performance of the plants. This agro based technique can also be used with the concept of sustainable waste management by using fly ash–soil combination as a suitable growing medium for growing different plant species.

Keywords

Fly ash · Heavy metals · Plant growth · Biochemical parameters · Morphological parameters · Sustainable waste management

21.1 Introduction

In the present time, agricultural production is mainly dependent upon the different conventional farming techniques and methods of making use of chemical fertilizers, pesticides, and herbicides to enhance essential soil nutrients such as potassium, magnesium, phosphorus, etc., for the better plant growth. These techniques although play a significant role in crop production, improve growth performance and crop yield but its indiscriminate and long-term application can degrade soil fertility, cause environmental pollution, depletion of non-renewable resources and thereby adversely affect the plant productivity. There are certain fertilizers which are known to be carcinogenic in nature which can affect human health and livestock.

These concerns have made many researchers to use fly ash in soil for enhancing physical and chemical properties of fly ash–soil mixture. As management of fly ash is becoming extremely difficult because of human and environmental concern its utilization in different industrial, construction, and agricultural sectors needs to be investigated. According to the data provided by the Government of India 169 million tons of fly ash is produced in the year 2016–2017. After almost three decades with many technical advancements and research, even then we have not reached 50% of its utilization as is same is readily available in any thermal power plant. Fly ash is mainly utilized in making of bricks, construction of dams, road embankments, in making buildings, etc., but its utilization in agriculture is only 1% (Fig. 21.1).

Fly ash is an inorganic waste produced after combustion from thermal power plant every year. At present it is utilized as a building material in construction areas because of various advantages—environment friendly, cost effective, high ultimate strength, lower permeability. Fly ash may show acidity or alkalinity depending upon coal type and combustion source which may directly utilize with soil to buffer the soil pH. In general, Fly ash contains various major and micronutrients like Zn, B, F, Al, K, Mg, Mn, Si, Ca, Cu, Fe, and Na which are essential for plant growth. There are various attempts made by different scientists for the fly ash usage as a mixture in soil to improve WHC, soil texture, crop yield, soil productivity which thereby

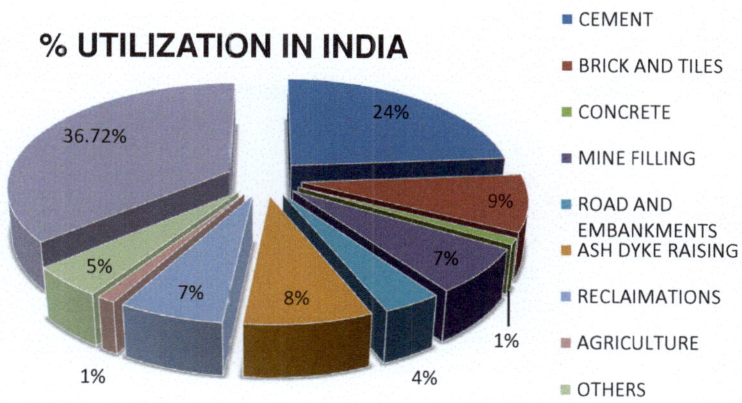

Fig. 21.1 Fly ash utilization in different sectors (2017–2018)

improves plant growth. Fly ash addition in soil generally resulted in favorable impact on nutrient uptake and plant productivity.

21.2 Fly Ash: Physical and Chemical Properties

Physical and chemical properties of fly ash generally depend upon coal type being used in combustion process, combustion procedure, and storage and handling techniques in thermal power plant. Therefore, different types of coal—anthracite, bituminous, and lignite have different composition of ash generated. Chemically synthesized major elements present in majority of fly ash are composed of Si, Ca, Na, Mg, Fe, K, and Al (Yue et al. 2019). Types of fly ash classified as: Class C and Class F generally differ in content of lime, silica, alumina, and iron (II, III) oxide content. However, it is generally rich in certain element which includes Mo, S, Se, As, B, Sr, and Ca (Page et al. 1979) (Table 21.1).

21.3 Fly Ash Effect on Soil Fertility

Consequence of fly ash application on the engineering properties of soil and hence its utilization as a soil ameliorate for effective cultivation of plants has been studied by various scientists (Inam 2007). The changes produced on the soil physico-chemical and biological properties due to addition of fly ash are illustrated in Table 21.2.

Table 21.1 Physical and chemical properties of fly ash

Properties	Values
Physical properties	
Color	Gray to black
Shape	Spherical
pH	4.5–12
Bulk density (g/cm^3)	1–1.8
Texture	Sandy silt or silty loam
Surface area (m^2/kg)	500–5000
Specific gravity (g/cm^3)	1.9–2.5
Plasticity	Non plastic
Moisture content (%)	18–38
Maximum dry density (gm/cc)	0.9–1.6
Lime reactivity (MPa)	4.57–7.20
Porosity (%)	47.10
Cohesion (kg/m^2)	Negligible
Clay (%)	1–10
Silt (%)	8–85
Sand (%)	7–90
Gravel (%)	0–10
Chemical properties	
Organic carbon (%)	0.114
Loss on ignition (%)	1.42
Total N (%)	0.86
Total P (%)	0.97
Total K (%)	1.42
Aluminum (ppm)	11,500–144,000
Boron (ppm)	10–3000
Calcium (ppm)	5400–177,000
Chlorine (ppm)	13–25,000
Cobalt (ppm)	6–1500
Copper (ppm)	30–3020
Manganese (ppm)	31–4400
Phosphorus (ppm)	600–2500
Potassium (ppm)	1534–34,000
Sulfur (ppm)	0.11–0.25
Iron (ppm)	7800–289,000
Sodium (ppm)	1180–20,300
Nickel (ppm)	11.8–8000

Source: (Adriano et al. 1980; Carlson and Adriano 1993; Dhindsa et al. 2016; Thomas et al. 2015)

Table 21.2 Fly ash impact on physical, chemical, and biological properties of soil

Physical properties	Chemical properties	Biological properties
Improves soil texture, increases porosity, and improves the WHC of soil (Page et al. 1979)	Fly ash addition to the soil enhances its nutrient grade (Gupta et al. 2010)	Fly ash application to soil alters the respiration of microorganism and nitrification process activities in the soil (Chou et al. 2005)
Fly ash application rates reduces bulk density, improves the soil organization, thereby increases porosity, workability, root penetration, and moisture retention capacity of soil (Kene et al. 1991)	With the upsurge in the chemical properties—like soil pH, there is a significant change observed in the soil moisture content, crop growth (Mittra et al. 2003)	Due to the presence of heavy metals at high application rate of fly ash, total bacteria, actinomycetes, fungi as well as enzymatic activities such as soil phosphates, sulfatase, dehydrogenase, and invertase in soil decrease (Pitchel and Hayes 1990)
Fly ash application rate decreases the density of the soil because of its low specific gravity (Prabakar et al. 2004)	Nitrogen content is usually absent in fly ash and little availability of Phosphorus in plant (Jala and Goyal 2006). At high Fly ash concentration, it causes P deficiency to the soil	Conditions of soil may also be enhanced with amendment of fly ash–sewage sludge as it increases cation exchange capacity, which thereby result in the immobilization of certain toxic metals and further increase in the availability of Mg, K, and Ca (Pitchel and Hayes 1990)
Increases carbon content in soil but limits the water availability to the plant (Eisenberg et al. 1986)	Depending upon the rate of application of fly ash to soil resulted in the improved concentrations of Ca, Mo, Se, S, Cd, Ba, B, and Cd (Basu et al. 2009)	Fly ash, in general, acts as a conditioning agent to block soil erosion (Katiyar et al. 2012)
At high application rate, the soil pH increases because of effective discharge of Ca, Na, Al, and OH^- from fly ash (Wong and Wong 1986)	Fly ash provides essential nutrients to nutrient deficient soil to promote plant growth and it has already been reported to correct deficiencies of B, Mo, Mg, Zn, and S (Singh et al. 2000)	Low rate of fly ash application to soil acts as a good substrate by making use of certain trace elements for the growth of microorganism which can act as a soil ameliorate (Hodgson, and Holliday 1966)
In soil electrical conductivity increases by the consequence of fly ash application as fly ash comprises soluble major and minor inorganic ingredients (Adriano and Weber 2001)	Lime which is readily present in fly ash combines with the acidic components of the soil and releases some of the nutrients like—S, B, Mo in some minimum amount beneficial for the crop plant (Rautaray et al. 2003)	

21.4 Effect of Fly Ash on the Performance of Plants

Fly ash when mixed with air, water, and soil drastically pollutes the surroundings. Fly ash disposal has adverse impacts on terrestrial and aquatic ecosystems due to leaching of heavy metals from the ash into soil and groundwater which results in reduction of plant growth (Adriano et al. 1982; Wong and Wong 1990). Thus, fly ash poses a serious threat to both animals and plants of the biome. Phytoremediation technology is being used to overcome this problem by various scientists to clean and revegetate fly ash from wastelands by suitable plantation of trees and shrubs to develop bio-esthetic surroundings for native inhabitants and to stopover fly ash from expanding into the environment (Rai et al. 2000). But on the contrary the rate of application of fly ash for changing the soil fertility which is affiliated with the performance of plant and its changed productivity varies according to the dosage of fly ash added to the degraded soils.

Various researches have been conducted on utilization of fly ash in growing suitable food crops and it was observed that application of fly ash causes an enhanced growth rate and productivity of cereal crops, plantation crops, and vegetable crops. Besides increasing the yield of food crops, it also has the potential of improving the uptake of nutrients and soil fertility status (Yu et al. 2019). There is hence a great scope to utilize fly ash in agriculture with the aim to improve the soil physical, chemical, and biological properties thereby enhancing the world food production with a cost-effective approach to achieve environmental sustainability.

21.5 Impact of Fly Ash on Morphological Parameters of Plant

Fly ash utilization in the recent years has gained interest in the areas of crop production and crop yield especially for redeveloping wasteland in agriculture. These soils of wasteland are predominantly poor in fertility and require high quantities of nutrients for their restoration. Thus, Fly ash has been utilized in mixing with soil as it provides minerals needed for plant growth. It was found that potential benefits of using Illinois coal fly ash to amend growth media for tomato plants, turfgrass, and chrysanthemums (Chou et al. 2005). Both turfgrass and tomato plants were grown in sandy soil/compost/fly ash mixed media in different combinations. The results indicated that Illinois coal fly ash might be beneficial for typical plant growth if it is applied at an adequate proportion with soil. The effects of fly ash at various doses with soil on growth performance and plant biomass were observed in *Solanum nigrum* grown in pot culture conditions (Robab et al. 2010). An upsurge in the amount of fly ash doses up to 30% showed an increased availability of elements in the soil and thereby changes in the physico-chemical characteristics of soil. There was presence of most of the essential elements in the soils ameliorated with fly ash; however, both the nitrogen and available phosphorus content showed no significant increase due to fly ash addition. It was studied that the soils amended with low doses of fly ash (up to 30%) boosted the growth performance and biomass production in the plant (Rizwan et al. 2016). Similarly, the effect of different concentrations of fly ash (0%, 5%, 10%, 20%, 30%, 40%, and 50%) added to the soil was studied for the growth rate and productivity patterns of eggplant. It was reported that there was a significant increase (5–30%) in the growth and productivity of eggplant grown in soil amended with low concentrations of fly ash (5%–30%). However, at higher rate of application of fly ash in soil (40 and 50%) plant growth and productivity parameters reduced significantly. Hence as per the conclusion drawn from this study, it was reported that 20% dose of fly ash given to soil was the most suited base material for the cultivation of eggplant in terms of growth and productivity (Rizvi and Khan 2009).

Number of leaves and branches, root and shoot weight, and height of *Eucalyptus globulus*, *Syzygium cumini*, *Azadirachta indica*, and *Annona squamosa* were affected significantly with different combinations of fly ash and soil combination. Sunflower (*Helianthus annuus L.*) plants treated with fly ash showed enhanced growth (Pani et al. 2015). Leaf area of the treated plants also improved. The reduced growth parameters at high application rate occur due to the existence of heavy metals in the fly ash. Heavy metals present in plants causes the overproduction of reactive oxygen species (ROS) and causes oxidative destruction thereby promotes physiological constrains that decrease plant strength and reduces plant growth (Saba et al. 2000). It was detected that tomato plant grown in fly ash–soil mixture presented flourishing growth with better and greener leaves. At high application rate of fly ash, yields were significantly reduced. The best fly ash level for incorporation in soil was 40%, which enhanced the yield and nutritious value of tomato fruit (Punjwani et al. 2011).

In *Brassica parachinensis* and *B. chinensis,* increase in the leaf length at 3 and 6% of fly ash and soil treatment was recorded, whereas at high application rate, i.e. 12, 30, and 50% reduction in the length of the leaves and all the growth parameters were observed (Wong and Wong 1986). Fly ash contains many trace elements like Cd, Ni, and other metals including Co, Fe, Cu, Mn, Zn, Ni in high amount. Generally, the growth rate of plant is a degree of plant productivity and expression of the plant efficacy to produce natural plant product (Hortensteiner 1999). Relative growth including all the morphological parameters increases at low concentration of fly ash shows that fly ash increases the growth of the plant at low doses. Many chemical constituents present in the fly ash at low rate of application might be favorable for the plant growth performance and could increase the engineering properties of soil as at low doses the nutrient availability in fly ash is in the form essential for the plant growth (Pandey et al. 1994).

As fly ash is deficient in nitrogen, its application at high doses lead to very little or no nitrogen in the soil resulted in the major deficiency of nitrogen in the soil and in the plant tissues which is one of the main causes of suppressed growth and plant yield (Khan and Khan 1996). The toxic chemical effects of fly ash at high doses also eliminate the microbial respiration in the soil (Wong and Wong 1986).

Some workers supported out a study on influence of fly ash on the tomato plant. They identified the suitable dose of fly ash–soil mixture for the plant. As per their results, in treatments which contained 20–80% of fly ash, the shoot and root growth were increased by 40–90%. The most optimal growth was of 50–60% of fly ash amended soil for tomato plant (Khan and Khan 1996). Similar observations were made on conducting the field experiment up to 50 t/ha to study its effect on soil properties and wheat yield (*Triticum aestivum* L.), mustard (*Brassica juncea* L.), rice (*Oryza sativa* L.), and maize (*Zea mays* L.) (Kalra et al. 2003). The yield of wheat expands for 20 t/ha fly ash amended soil, whereas mustard and rice were observed to increase growth with 10 t/ha of fly ash amended soil, their results showed that all the three crops showed improved growth as compared to control.

21.6 Impact of Fly Ash on Plant Biochemical Parameters

Fly ash application with sewage sludge and soil in different proportions is gaining importance and has become a substitute to chemical fertilizers in different countries. Different ratios of sewage sludge and fly ash were made in soil to study metal availability and yield and their uptake by *Brassica campestris* (Pillai and Chaturvedi 2012). Increase in content of oil of *B. campestris* was recorded in lower amendments of soil with mixture; however, total chlorophyll content improved significantly, and carotenoid content showed non-significant increase in all the doses with mixtures A and B as compared to control. It has been reported that significant impact of fly ash on chlorophyll a and b content as well as significantly increased phenolic component in fly ash containing soil for the two plant species, *Coleus forskohlii* and *Andrographis paniculata* (Sinha et al. 2013). Similarly, reports on increase in chlorophyll content and antioxidant potential in *Coleus forskohlii* and *Andrographis paniculata* (Tripathi et al. 2014).

The decrease in the chlorophyll content in different plant species in response to the heavy metals was utilized as a marker to determine the heavy metal toxicity in the plant tissue (Hamp et al. 1976). The decline in the pigment content through the heavy metal stress suggested that somewhere these metals interfere with the pigment metabolism in plant leaves which thereby hinder with the synthesis of proteins and structural components of the chloroplast (Bhattacharya and Choudhari 1994). The heavy metal stress in plants blocks the enzyme proteins accountable for chlorophyll biogenesis. The reduction in the photosynthetic area and inhibition of photosynthetic rate are responsible for the decrease in the chlorophyll content (Setia et al. 1993).

It has been reported that in *Capsicum annuum*, the photosynthetic pigment was found to be increased at 5% of fly ash–soil amendment on 40th day of the growing period further followed by the declining trend (Dash et al. 2015). Pot experiment was conducted on *Tagetes erecta* (Marigold species) with different fly ash-soil combinations to study the changes in photosynthetic pigments-chlorophyll a, chlorophyll b and carotenoids. It was found that photosynthetic pigments were found to be maximum at 10% fly ash–soil doses on 40th day of growth status followed by a declining trend toward higher doses (Pradhan et al. 2015). 10–15% of fly ash application with soil showed the beneficial results in terms of morphological and pigment content in leaves of *Withania somnifera* (L.) Dunal (Kumar et al. 2017).

There have been reports on the increase in the nitrate reductase activity in leaves of *Phaseolus vulgaris* at low fly ash doses in soil but reduced as the doses of fly ash in soil increased (Kamenova et al. 1995). Similar findings were also reported in sourgrass (Sajwan et al. 1996). Nitrogen is the major component in the plant for increasing the metabolic activity of nitrate reductase with increase in the nitrogenous metabolites (Pandey et al. 1994). With exposure of heavy metal stress in plants at high application rate of fly ash may resulted into the vacuolar accumulation of heavy metals (Rauser and Dumbroff 1981). Hence, the heavy metal accumulation in plant leads to the decrease in the leaf nitrate reductase activity which further leads to the blockage of translocation of the metal to the enzyme active site in the leaves (Bharti et al. 1999). A field trial was conducted in growing *Beta vulgaris* in different combination of fly ash–soil mixtures at different concentrations and the results revealed that low quantity of fly ash application favored plant growth. Sugar content increases at low doses of fly ash amended soil which signifies that low doses of fly ash improves the possible use of fly ash in crop improvement and plant productivity whereas at high application rates, fly ash shows adverse effect due to the presence of the toxic metals (Singh et al. 1994). Similar findings were studied on the plant *Datura innoxia* where low doses increase the sugar content of leaves and the antioxidative enzymes (Satyanarayana et al. 1988).

Fly ash addition in soil modifies the texture of soil from sandy, clayey soil to loamy soil. Fly ash effect on soil mainly depends upon the properties of coal and the soil used. With the addition of Fly ash into soil increases the electrical conductivity of the soil mixture as the major and minor inorganic constituents are enhanced with the addition of fly ash in the mixture. High concentration of Fe and Al in Fly ash translates soluble P into insoluble P compounds which are not freely available to plants. It was found that soil when added to the fly ash enhanced the WHC of the mixture (Mishra et al. 2017). No organic content is present in fly ash and is also rich in electrolytes and hence it enhances the hydraulic activity of soil but only when used in lower rates which thereby increases the pH of the soil and affects the plant growth (Yeledhalli et al. 2007). Some strains of plant growth promoting bacteria NBRI K24 and NBRI K3 from fly ash contaminated soil decrease the toxicity of Cr and Ni in *Brassica juncea* (Indian mustard) (Gonzalez and Navia 2009). On 10-hectare area where fly ash dump is present, a field experiment was directed by the use of bio fertilizer with nitrogen fixing bacteria strains. When *Azotobacter* and *Bradyrhizobium* species were used the nitrogen content improved by 4.5 times and the toxicity of different heavy metals in soil was decreased in soil due to the bioaccumulation (Juwarkar and Jambhulkar 2007).

Some researchers had conducted an experiment to find the increase of heavy metals in *Cajanus cajan L.* at different fly ash amended soil proportions. Fly ash incorporation at low application rates up to 25% demonstrates positive response in many studied plant growth and yield parameters as compared with the control. Mean concentration of certain trace heavy metals Zn, Cu, Cr, and Cd in edible parts of the plant was found below the respective critical value (Pandey et al. 1994). Researchers have revealed that accumulation of heavy metal in plant leaves may range between 5 and 10% on a dry weight basis. A study was conducted by Kumari and Prasad 2014 in which the utilization of fly ash in different amendments of soil was taken for the analysis of heavy metals in the ornamental plant (*Calendula officinalis*). In this study, total heavy metal status found in the fly ash amended soil was found in the order—Mn > Pb > Cu > Cd. It was observed that fly ash is beneficial to the plant growth at 40% fly ash treatment (Page et al. 1979).

Fly ash incorporation at low dose usually triggered into the positive impression on plant biomass production and nutrient acceptance. The related study was accompanied by Sale et al. 1996 in overseas. Related discoveries in India were conducted by diverse workforces in the identical field (Mishra et al. 2005; Gupta et al. 2006; Gupta et al. 2007). Similar findings on low concentration of fly ash in soil amplified chlorophyll a and b and plant progress performance (Mishra et al. 2007). Diverse doses of fly ash–soil affect the yield of seed in several crop plants in different manner. Outcome of fly ash on seed physiology was studied overseas by Jaramillo-Lopez 2011 and in India, related findings were performed by Kalra et al. 2003; Mishra et al. 2007; Shrivastava et al. 2008; Pradhan et al. 2013). Outcome of fly ash on photosynthetic pigments like carotenoids, chlorophyll a, chlorophyll b, and total chlorophyll was studied by many scientists (Rai et al. 2004; Mishra et al. 2005; Sinha and Gupta 2005; Dwivedi et al. 2007; Gupta et al. 2007; Mishra et al. 2007). Many studies have been performed to govern the effect of fly ash integration into soil and its responses on various biochemical parameters like pigments, protein, proline, sugar, nitrate, and nitrate reductase activity on crop plants, viz. wheat, corn, soybean, tomato, rye, maize, rice, gram under pot culture experiment, green house, and ground study by overseas scientists (Clien et al. 2000; Jaramillo-Lopez et al. 2011) and Indian workers (Sharma et al. 2001; Kumar and Singh 2003; Gupta et al. 2006; Yeledhalli et al. 2008; Mishra et al. 2007; Raj and Mohan 2018) (Figs. 21.2 and 21.3).

21.7 Conclusion

Fly ash can be used as a nutrient supplement for degraded agricultural soil and improved crop improvement which helps in solving the problem of solid waste management to some extent. Amendment of agriculture soil with Fly Ash can affect the soil by improving soil texture, water holding capacity, and the overall chemical composition of soil. However, care is needed while using Fly Ash as a source of nutrients, since overapplication could result in phytotoxic level of few elements leading to food chain risks. The future poses challenge to the scientist, technologists, and engineers towards various novel and judicious management of Fly ash disposal and utilization techniques. Identification of crops capable of growing well in presence of Fly Ash particulates has become necessary for getting sustained agricultural yields under the changed scenario.

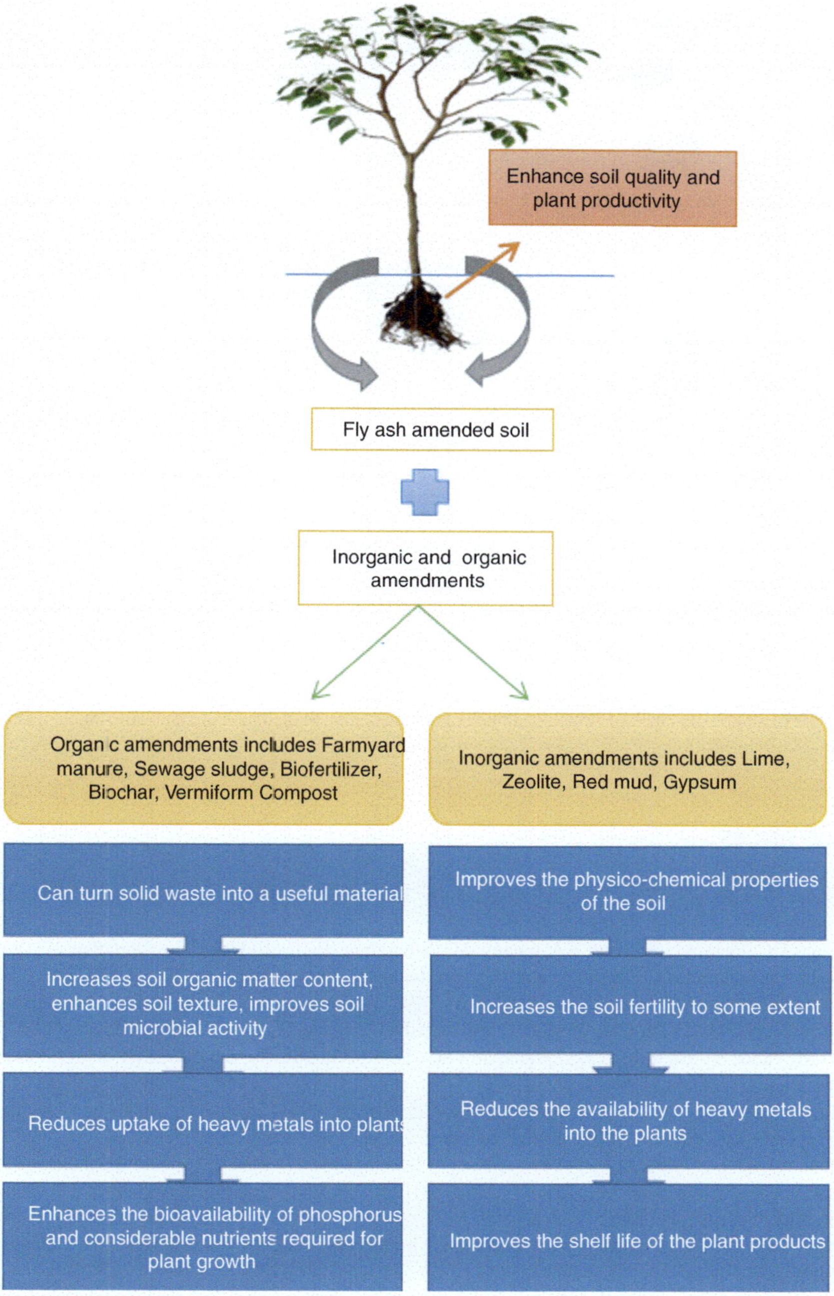

Fig. 21.2 Benefits of fly ash application with other amendments

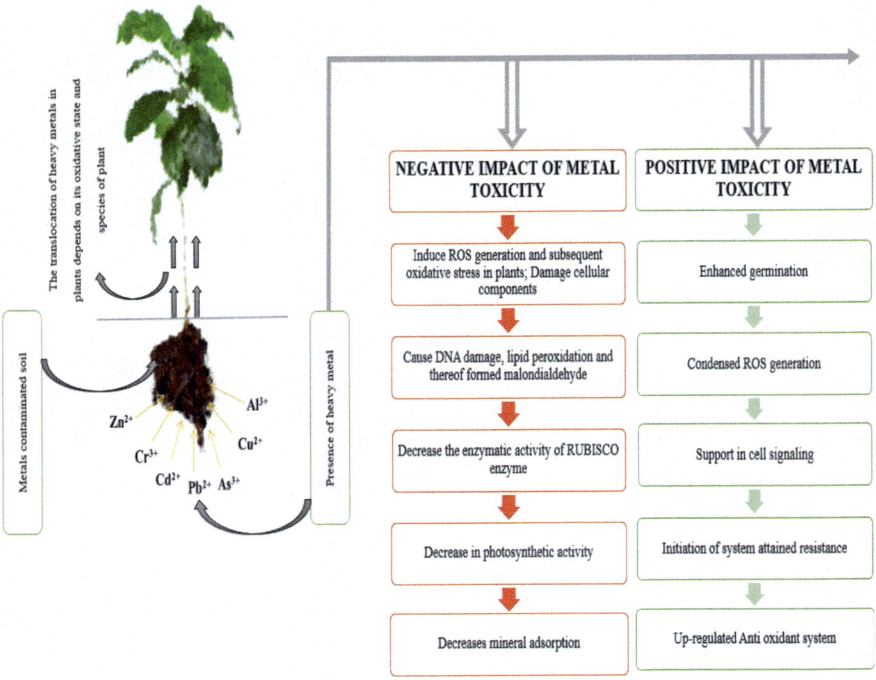

Fig. 21.3 Impact of heavy metals on plants

References

Adriano DC, Weber JT (2001) Influence of fly ash on soil physical properties and turf grass establishment. J Environ Qual 30:596–602

Adriano DC, Page AL, Elseewi AA, Chang AC, Straughan IR (1980) Utilization and disposal of fly ash & other coal residues in terrestrial ecosystems: a review. J Environ Qual 9:333–343

Adriano DC, Page AL, Elseewi AA, Chang AC (1982) Cadmium availability to sudangrass grown on soil amended with sewage sludge and fly ash. J Environ Qual 11(2):197–203

Basu M, Pande M, Bhadoria PBS, Mahapatra SC (2009) Potential fly-ash utilization in agriculture: a global review. Prog Nat Sci 19:1173–1186

Bharti B, Matte DB, Badole WP, Deshmukh A, Pillewan S (1999) Effect of fly on physico-chemical properties of versitol and yield of green gram. J Soil Crop 9(2):255–257

Bhattacharya M, Choudhari MA (1994) Effect of lead and cadmium on the biochemical changes in the leaves of terrestrial (Vigna) and aquatic (Hydrilla) plants under solution culture. Indian J Plant Physiol 37:99–103

Carlson CL, Adriano DC (1993) Environmental impacts of coal combustion residues. J Environ Qual 22:227–247

Chou SJ, Chou MM, Stucki JW, Warnock D, Chemler JA, Pepple MA (2005) Plant growth in sandy soil/compost mixture and commercial peat moss both amended with Illinois coal fly ash. In: World of coal ash, Lexington

Cline JA, Bijl M, Torrenueva A (2000) Coal fly ash as a soil conditioner for field crops in Southern Ontorio. J Environ Qual 29:1982–1989

Dash AK, Pradhan A, Das S, Mohanty SS (2015) Fly ash as a potential source of soil amendment in agriculture and a component of integrated plant nutrient supply system. J Ind Pollut Control 29:1982–1989

Dhindsa HS, Sharma RD, Kumar R (2016) Role of fly ash in improving soil physical properties and yield of wheat (*Triticum aestivum*). Agric Sci Dig 36(2):97–101

Dwivedi S, Tripathi RD, Shrivastava S, Mishra S, Shukla MK, Tiwari KK, Singh RR, Rai UN (2007) Growth performance and biochemical responses of three rice (*Oryza sativa* L.) cultivars grown in fly ash amended soil. Chemosphere 67:140–151

Eisenberg SH, Tittlebaun ME, Eaton HC, Soroczal MM (1986) Chemical characteristics of selected fly ash leachates. J Environ Sci Health 21:383–402

Gonzalez A, Navia R (2009) Fly ashes from coal and petroleum coke combustion: current and innovative potential applications. Waste Manag Res 27:976–987

Gupta DK, Tripathi RD, Rai UN, Dwivedi S, Mishra S, Srivastava S, Inouhe M (2006) Changes in amino acid profile and metal content in seed of Cicer arientinum L. (chick pea) grown under various fly ash amendments. Chemosphere 65:939–945

Gupta DK, Tripathi RD, Rai UN, Mishra S, Srivastava S, Dwivedi S, Maathuis FJM (2007) Growth and biochemical parameters of Cicer arietinum L. grown on amended fly ash. Environ Monit Assess 134:479–487

Gupta AK, Rohit K, Mishra Sinha S, Lee B-K (2010) Growth, metal accumulation and yield performance of *Brassica campestris* L. (cv. Pusa Jaikisan) grown on soil amended with tannery sludge/fly ash mixture. Ecol Eng 36(8):981–991

Hamp R, Beulich K, Ziegler H (1976) Effects of zinc and cadmium on the photosynthetic CO_2-fixation and hill activity of isolated spinach chloroplasts. Z Pfanzenphysiol 77:334–336

Hodgson DR, Holliday R (1965) The agronomic properties of pulverized fuel ash. Chem Ind 20:785–790

Hortensteiner S (1999) Chlorophyll breakdown in higher plants and algae. Cell Mol Life Sci 56:330–347

Inam A (2007) Use of fly ash in turnip (*Brassica rapa* L.) cultivation. Pollut Res 26:39–42

Jala S, Goyal D (2006) Fly ash as a soil ameliorant for improving crop production-a review. Bioresour Technol 97:1136–1147

Jaramillo-Lopez PF, Powell MA, Hayden DB (2011) The influence of soil amendments (fly ash and stabilized biosolids) on Meloidogyne halpa in microplots with tomato (*Lycopersicon esculentum*). Nematropica 41:141–149

Juwarkar AA, Jambhulkar HP (2007) Restoration of fly ash dumps through biological interventions. Environ Monit Assess 139:355–365

Kalra N, Jain MC, Joshi HC, Choudhary R, Kumar S, Pathak H, Sharma SK, Kumar V, Kumar R, Harit RC, Khan SA, Hussain MZ (2003) Soil properties and crop productivity as influenced by fly ash incorporation in soil. Environ Monit Assess 87(1):93–109

Kamenova S, Georgieva V, Merkchiiska M, Georgieva N, Biochmova M, Paunova S (1995) Direct and residual effect of fly ash from a thermal-electrical station on some biological responses of bean plants grown on different soil types. Rasteniev'dni Nauki 32(3):51–54

Katiyar D, Singh A, Malaviya P, Pant D, Singh P, Abraham G, Singh SK (2012) Impact of fly ash amended soil on growth and yield of crop plants. Int J Environ Waste Manag 10:150–162

Kene DR, Lanjewar SA, Ingole BM (1991) Effect of application of fly ash on physico-chemical properties of soils. J Soil Crop 1:11–18

Khan MR, Khan MW (1996) The effect of fly ash on plant growth and yield of tomato. Environ Pollut 92:105–111

Kumar D, Singh B (2003) The use of coal fly ash in acidic soil reclamation. Land Degrad Dev 14:285–299

Kumar T, Prasad SVK, Kanungo VK (2017) Impact of coal fly ash on growth of Withania somnifera (L.) Dunal. Indian J Sci Res 13(2):25–28

Kumari D, Prasad B (2014) Analysis of heavy metals on ornamental plant by use of fly ash and amended soil-an experimental approach. Int J Eng Technol Res 2(5):82–87

Mishra M, Sahu RK, Pandey RN (2005) Growth, yield, metabolism and elemental status of Green gram (*Phaseolus aureus*) and Til (*Sesamum indicum*) grown in soils amended with fly ash. Fresenius Environ Bull 14:559–564

Mishra M, Sahu RK, Pandey RN (2007) Growth, yield, metabolism and elemental status of rice (Oryza sativa) grown in fly ash amendment soil. Ecotoxicology 16:272–278

Mishra S, Prasad SVK, Kanungo VK (2017) Impact of coal fly ash as soil amendment on Physico-chemical properties of soil. Indian J Sci Res 13(2):15–20

Mittra BN, Karmakar S, Swain DK, Ghosh BC (2003) Fly ash a potential source of soil amendment and component of integrated plant nutrient supply. In: International ash utilization symposium, vol 84. University of Kentucky, Lexington, pp 1447–1451

Page AL, Elseewi AA, Straughan IR (1979) Physical and chemical properties of fly ash from coal-fired power plants with reference to the environmental impacts. Residue Rev 71:83–120

Pandey V, Mishra J, Singh SN, Yunus M, Ahmad KJ, Singh N (1994) Growth response of Helianthus annuus L. grown on fly ash amended soil. J Environ Biol 15(2):117–125

Pani NK, Samal P, Das R, Sahoo S (2015) Effect of fly ash on growth and yield of sunflower (Helianthus annuus L.). Int J Agro Agric Res 7:64–74

Pillai P, Chaturvedi A (2012) Effect of fly ash on reproductive biology of oil yielding plant Brassica juncea var. pusa bold. Glob J Biosci Biotechnol 1(2):175–177

Pitchel JR, Hayes JM (1990) Influence of fly ash on soil microbial activity and populations. J Environ Qual 19:593–597

Prabakar J, Dendorkar N, Morchhale RK (2004) Influence of fly ash on strength behavior of typical soils. Constr Build Mater 18:263–267

Pradhan A, Sahu SK, Dash AK (2013) Changes in pigment content (chlorophyll and carotenoid), enzyme activities (catalase and peroxidase), biomass and yield of rice plant (*Oryza sativa* L.) following irrigation of rice mill wastewater under pot culture conditions. Int J Sci Eng Res 4 (6):2706–2717

Pradhan A, Dash AK, Das S, Mohanty SS (2015) Potential use of fly ash in floriculture: a case study on the photosynthetic pigments content and vegetative growth of *Tagetes erecta* (Marigold). Ecol Environ Conserv 21:369–376

Punjwani J, Krishna R, Kalpana S, Gupta KK (2011) Application impact of coal fly ash, and water hyacinth on cultivation of tomato. Int J Res Chem Environ 1:71–76

Rai UN, Tripathi RD, Singh N, Kumar A, Ali MB, Pal A, Singh SN (2000) Amelioration of fly-ash by selected nitrogen fixing blue green algae. Bull Environ Contam Toxicol 64:294–301

Rai UN, Pandey K, Sinha S, Singh A, Saxena R, Gupta DK (2004) Revegetating fly ash landfills with Prosopis juliflora L. - impact of different amendments and rhizobium inoculation. Environ Int 30:293–300

Raj S, Mohan S (2018) Influence of metal uptake from fly ash on the growth of *Jatropha curcas* plant: bulk utilization approach. Int J Pharm Biosci 9(2):154–159

Rauser WE, Dumbroff EB (1981) Effects of excess cobalt, nickel and zinc on the water relations of *Phaseolus vulgaris*. Environ Exp Bot 21:249–255

Rautaray SK, Ghosh BC, Mittra BN (2003) Effect of fly ash, organic wastes and chemical fertilizers on yields, nutrient uptake, heavy metal content and residual fertility in a rice-mustard cropping sequence under acid lateritic soils. Bioresour Technol 90:275–283

Rizvi R, Khan AA (2009) Response of eggplant (*Solanum melongena* L.) to fly ash and brick kiln dust amended soil. Biol Med 1(2):20–24

Rizwan M, Ali S, Adrees M, Rizv H, Rehman M, Hannan F, Qayyum M, Hafeez F, Ok YS (2016) Cadmium stress in rice: toxic effects, tolerance mechanisms and management: a critical review. Environ Sci Pollut Res 23:17859–17879

Robab MI, Hismmaudin TA (2010) Impact of fly ash on vegetative growth and photosynthetic pigment concentrations of *Solanum nigrum* L. Nanobiotech Univ 1(2):133–138

Saba Pande D, Iqbal M, Srivastava PS (2000) Effect of $ZnSO_4$ and $CuSO_4$ on regeneration and lepidine content in *Lepidium sativum* L. Biol Plant 43:253–256

Sajwan KS, Ornes WH, Youngblood TV (1996) Growth and elemental composition of Sorghum Sudangrass grown on fly ash/organic waste amended soils. J Environ Sci Health 31 (7):1729–1739

Sale LY, Naeth MA, Chanasyk DS (1996) Growth response of barley on unweathered fly ash amended soil. J Environ Qual 25:684–691

Satyanarayana G, Bhatnagar A, Acharya UH (1988) Effect of fly ash pollution on *Datura innoxia*. Environ Ecol 6(1):92–95

Setia RC, Bala R, Setia N, Malik CP (1993) Photosynthetic characteristics of heavy metal treated wheat (*Triticum aestivum* L.) plants. J Plant Sci Res 9:47–49

Sharma SK, Kalra N, Singh GR (2001) Fly ash incorporation effect on soil health and yield of maize and rice. J Sci Indus Res 60:580–585

Shrivastava SK, Prasad SVK, Kanungo VK (2008) Impact of coal fly ash on properties of soil and germination & productive behavior of *Andrographis paniculata* (Burm.f.) Wall. ex Nees. Ph.D. Thesis, Pt R S University Raipur C.G

Singh N, Yunus M, Ahmad KJ (1994) Growth response and element accumulation in *Beta vulgaris* L. raised in fly ash amended soils. Ecotoxicology 3(4):287–298

Singh N, Yunus M, Iqbal M, Srivastava PS, Siddiqui TO (2000) Environmental impacts of fly ash in environmental hazards: plant and people. CBS, New Delhi, pp 60–79

Sinha S, Gupta AK (2005) Assessment of metals in leguminous green manuring plant of *Sesbania cannabina* L. grown on fly ash amended soil: effect on antioxidants. Chemosphere 61:1204–1214

Sinha D, Sharma S, Dwivedi MK (2013) The impact of fly ash on photosynthetic activity and medicinal property of plants. Int J Curr Microbiol Appl Sci 2(8):382–388

Thomas A, Kumar K, Tandon L, Prakash O (2015) Effect of fly ash on engineering properties of soil. Int J Adv Mech Civil Eng 2(3):2394–2827

Tripathi NK, Upadhyay MK, Tripathi M (2014) Comparative accounts of chlorophyll content, antioxidant and phenolic contents in Coleus forskohlii and Andrographis paniculata. Int J Curr Pharmaceut Clin Res 4(2):84–87

Wong MH, Wong JWC (1986) Effects of fly ash on soil microbial activity. Environ Pollut Ser A 40:127–144

Wong MH, Wong JWC (1990) Effects of fly ash on yields and elemental composition of two vegetables-*Brassica parachinensis* and *B. chinensis*. Agric Ecosyst Environ 30:251

Yeledhalli NA, Prakash SS, Gurumurthy SB, Ravi MV (2007) Coal fly ash as modifier of physico-chemical and biological properties of soil. Karnataka J Agric Sci 20(3):531–534

Yeledhalli NA, Prakash SS, Ravi MV, Narayana RK (2008) Long-term effect of fly ash on crop yield and soil properties. Karnataka J Agric Sci 21(4):507–512

Yu L, Deng Q, Jian S, Li J, Dzantor K, Hu D (2019) Effect of fly ash application on plant biomass and elemental application. Environ Pollut 250:137–142

Yue Y, Liu Z, Zhang J, Lu M, Zhou J, Qian G (2019) Rapid potential of leaching of heavy metals from municipal solid waste incineration fly ash. J Environ Manag 238:144–152

Digital Knowledge Ecosystem: A New Weapon to Achieve Sustainable Food Waste Management

22

Saumya Chaturvedi, Vandana, Stuti Sharma, and Anshika

Abstract

Food waste is something that occurs at every stage of the process ranging from the moment it is grown to the time it reaches the consumer. One-third of food produced globally is wasted and that equates to 1.3 billion tonnes. The food that is present in the supermarkets goes through an extensive production process to get on the shelves. During this process, a wide number of resources are used, all of which can have their own impact on the environment. Food waste is also a major global problem because of the number of people who are suffering from starvation in the world. Around 821 million people do not have access to food in the same way as people in developed countries. This means that 1 in 9 people in the world are starving or are malnourished. Considering the magnitude of social, environmental, and economic problems caused by food wastage there arises an urgent need to find ways to tackle this colossal issue. Digital technology comes as an efficient tool as they are advancing at a rapid rate they also re-establish and recast the relationships between the customer, workers, and employers as the silicon chips permeates almost everything that is done on the industrial scale. This digital global world will contribute for new systems regarding this move. Many digital applications are proving to be a saviour for reducing significant amount of food waste and helping to provide that food to the needy. They collect food waste and save tonnes of food to reduce the quantity of food waste and lower the malnourishment, globally. This chapter aims to highlight the various recent technological advancements which are providing sustainable food waste management processes.

S. Chaturvedi (✉) · Vandana · S. Sharma · Anshika
Department of Food Technology, Shaheed Rajguru College of Applied Sciences for Women, University of Delhi, New Delhi, India

M. Thakur et al. (eds.), *Sustainable Food Waste Management*,
https://doi.org/10.1007/978-981-15-8967-6_22

Keywords

Digital technology · Food waste · Sustainable management process

22.1 Introduction

Food waste refers to food appropriate for human consumption being discarded, whether or not after it is being kept beyond its expiry date or left to spoil. Frequently this is because the food has spoiled but it can be for other reasons such as extra supply due to markets or individual consumer shopping/eating habits. Owing to globalization and wider availability of food, the food supply chain get longer and more networked which eventually leads to increased food waste. On that often-long journey from farm to table a majority of the part that was meant for consumption gets wasted. Even though the world produces sufficient food to feed twice the present population, there are billions of people who are malnourished. According to the FAO (2011a, b), food production for human consumption, per capita was approximately 460 kg per year in under developed regions, compared to 900 kg in developed countries, due to this abundance of food production in industrialized countries contributes to significant food wastage. The industrial and agricultural sectors turn out to be major producers of waste which is produced due to a variety of reasons like bad weather, processing problems, overproduction, and unstable market and finally poor planning and confusion contribute to food waste in stores and in homes all over the world. With emerging technologies and an increased awareness among people about the economic, environmental, and social impacts of food waste, people all over the world have developed new management techniques that limit this waste or have presented novel technologies to channelize this waste to produce biofuel and compost.

22.2 Food Waste Generated by Different Industries

Food loss and wastes take place simultaneously throughout the entire production process. Different industries have different kinds of food waste, contributing to many ecological problems. Thus, there is a need to understand the kinds of waste generated in different food industries and measures to reduce it.

22.2.1 Brewery Industry

Brewery industry is the one where alcoholic beverage is produced, like beer, malt liquor, etc. About 200 billion litres of beer are produced, globally, in a year; which brings us to think for a huge amount of waste generated through this industry. Every step or operation in the industry contributes towards the waste production. 137–173 tons of solid waste may be produced on 1000 tons of beer produced (Jurado and

Sorensen 2020). The main forms of waste are a lot of wastewater, spent grain, trub (slurry containing unstable colloidal protein, worts, etc.), brewery mash, waste yeast, etc. The process of malting (generally barley) is done and then malted grain is mashed to extract protein, sugar, etc. The remaining solid waste is the spent grain. Spent grain or waste yeast can be given as livestock feed or reused to certain extent. Wastewater, too, shall be treated well, i.e. solid waste should be separated. Reduction of these wastes is necessary. It can be either done by reusing the waste for some purposes or by-products must be recovered by dewatering process.

22.2.2 Fruits and Vegetable Industry

The most used commodity in terms of food are fruits and vegetables and thus, this industry is the second largest waste generator (Joshi et al. 2012). Different processes involved generate huge amount of food waste. Earlier, these were used raw or minimally processed but with development, a lot of change is demanded and thus, processing is growing significantly. Canning of fruits and vegetables produces most amount of waste, both solid waste and wastewater. They undergo peeling, coring, trimming, etc. to get a good finished product, leading to a lot of solid waste. On the other hand, water used for cleaning the raw material adds up to wastewater generation. Waste in these industries differs for varying fruits and vegetables. They can be seeds, leaves, pulp, skin, roots, stones, cores, etc. They can be separated in different bins for easy usage. These wastes can be minimized by using it as a livestock feed or treating it properly before disposal. Reusing or getting by-products by different processes can also help.

22.2.3 Animal Slaughter Industry

India is bestowed with vast livestock wealth and it is growing at the rate of approximately 6% per annum (Jayathilakan et al. 2012). The contribution of livestock industry including poultry and fish is contributing substantially to GDP of the country. Thus, the meat industry turns out to be one of the largest producers of organic waste in the food processing sector and also forms an interface between livestock production and a hygienically safe product for use in both human and animal food preparation (Fig. 22.1).

Fig. 22.1 Animal Slaughter house waste

Water is used in the slaughter house for carcass washing after hide removal from cattle calves, and sheep and after hair removal from cattle hogs. It is also used to clean the inside of the carcass after evisceration, and for cleaning and sanitizing equipment and facilities. The waste water has a high strength in terms of biological oxygen demand and chemical oxygen demand, suspended solids, nitrogen, and phosphorus, compared to domestic waste water. Where blood and meat proteins are the most significant sources of nitrogen in the waste water and rapidly give rise to ammoniacal nitrogen as breakdown occurs (Galanakis 2015). The waste water contains a high density of total coliform, faecal coliform, and faecal streptococcus groups of bacteria due to the presence of manure material and gut content. The solid part of the slaughter house wastes constitutes the inedible parts of animals derived from the production of meat and other animal by-products. Inedible animal tissues (organs, integument, ligaments, tendons, blood vessels, feathers, bone) can comprise up to 45% or more of the solid waste.

Treatment methods for the slaughter houses waste water are comparable to those used in the municipal wastewater treatment and include primary, secondary, and tertiary treatment other methods of treatment include coagulation-flocculation and sedimentation, dissolved air floatation, electrocoagulation, and the anaerobic treatment. The solid waste treatment and uses include medicinal and pharmaceutical uses of blood, utilization of hides and skins, gelatin from hides and skin uses of bones and the bone marrow (Franke-Whittle and Insam 2013).

22.2.4 Cereal Industry

Cereal production and processing are one of the most important sectors of Agri-food industry and that the cereal products cover over 20% of the daily diet. They are the edible seeds derived from plants, which are good sources of carbohydrates, they contribute to 60% of the total world food production with the main seeds being maize and wheat waste in the cereal processing industry is produced in the harvesting period, postharvesting period, and the production period. Presently these products are used as animal feed. However, they need to be utilized more effectively as they comprise of proteins, dietary fibres, and small amount of unsaturated fatty acids. For example, Rice Bran is an important cereal industry by-product which is generated during the milling process and it is a rich source of nutrients, proteins, and peptides with a wide range of nutritional and functional applications. The cereal waste is also recycled through the bio refinery concept which means use the Agri-food waste products, to produce commercial value-added products and energy. This concept is seen as a sustainable option by the scientific community. The products like biofuels, fertilizers, biomass or chemical compounds are obtained by bio refinery process where this Agri-food waste via anaerobic digestion, fermentation, and composting technologies (Galanakis 2015).

22.3 Impacts of Food Waste

Be it the kitchen (domestic) waste or the large amount of waste generated by food industries, it leads to huge impacts, which can be classified into three main domains, i.e. social, environmental, and economic impacts.

22.3.1 Social Impact

Food security is one of the grappling issues globally, causing malnutrition. Even after producing nearly twice the amount of food required to feed everyone, the food systems are not able to satisfy the hunger needs of people. This is because food wastage has taken over and left behind millions to suffer.

In developed nations, people buy food in bulk and this raises the food prices and makes it more problematic for poor people to survive. And for an under developed nation, the lack of advance technology for producing food becomes the major issue. A study estimated that about 18% of the total food produced get wasted every year. Only if, this food wastage can be prevented, statistics would not be showing such high numbers, of people, dying due to hunger. Almost 200 million children (below 5 years of age) are malnourished, globally. The food wasted at parties or restaurants or throwing off because of their expired dates clearly shows the failure of food systems and humanity.

22.3.2 Environmental Impact

The most alarming effect of wastage of food is the burden its causes on the environment. The major consequences are faced by climate, land use, water, etc. According to FAO's study which is on Food Wastage Footprint, the carbon footprint of wasted food is estimated to generate 3.3 billion tons of greenhouse gases in a year. And the food produced, but not consumed, takes up almost 1.4 billion hectares of land, i.e. nearly 30% of the agricultural land in the world. Thus, it impacts on the land use pattern too. The food waste is either buried in landfills or incinerated and both these methods have large negative effects on the environment. These convert organic waste into methane, i.e. a major greenhouse gas. And also, because of accelerating quantities of food waste, landfills are filling up fast and consequently land use is starting to become affected. On the other hand, incineration leads to evolution of gases and causes air pollution and contributing to acid rains. Apart from the impact land and climate, water levels have also plummeted. The food, which ultimately gets wasted, is produced by using approximately 25% of fresh water. While, still millions of people have no access to drinking water.

22.3.3 Economic Impact

Food wastage, starting from farmer/producer levels to grocery shops, throwing away rotten/unsold products and coming down to the consumer level, in throwing unused foods leads to a huge exploitation of economy. UN estimates of about 1.3 billion tons of food getting wasted per year lead to high economic losses (FAO 2011a, b). Food wastage results in rise of the food prices, worsening the situation of people, below poverty line. Not only the costs of wasted food counts, but the maintenance of landfills used, transportation, treatment plants, and other management area costs too, add up the burden on economic crisis.

According to the waste and resources action programme (wrap), an average of USD 665 per household per year can be reduced by reducing the food wasted that can be avoided otherwise (Food Wastage: Economic, social and environmental impact 2018). With sustained efforts, these impacts can be largely reduced. Technology undoubtedly would help to achieve this but a humanitarian and intelligent approach would be a big step to mitigate the effects of food wastage.

22.4 Digital Knowledge Ecosystem: A Framework for Food Waste Management

According to the Food and Agriculture Organization of the United Nations (FAO 2010), the current population of the planet is over seven billion people of which approximately 925 million are starving. Yet there is an annual loss and food wastage enough to feed three billion people. In order to change the current trajectory what is required is transparency, inclusion and accountability in food management systems. A revolution in the food production, supply, distribution, management, consumer usage and sincere efforts of public and private sectors is needed to jointly generate high quality data and insights for the global public good while avoiding technology and data monopolies. Social media networks are shaping consumer preferences and political outcomes across the globe. There is still an opportunity to change the current scenario by using data and information wisely, to build awareness about number of people not getting adequate food and suffering from diseases like malnutrition, to influence consumer behaviour, to inform markets, and to reform governance systems. A global digital ecosystem for the food supply chain has the potential to put us on a path towards a sustainable future. This will necessitate action from citizens, governments as well as the private sector to collect, share, and process data, create analytical insights and information. Citizens must be engaged in using and collecting data. Digital technologies have advanced more rapidly than any innovation in our history, by enhancing connectivity, financial inclusion, access to trade and public services. In the present day and age, technologies such as data pooling and artificial Intelligence are used to track and diagnose issues related to agriculture, health and the environment.

22.4.1 Using Internet of Things or the IoT

The internet of things is an arrangement of unified computing devices and digital machines that are given some unique identifiers (UIDs) and have the capability to transmit data over a network without the requirement of human-to-human or human-to-computer interaction. With multiple sensors and cameras spread everywhere, IoT helps in collecting evocative data in real time (Internet of things, IoT n.d.).

Food producers may not realize it, but a massive amount of food is wasted during the food manufacturing process itself. Some factors such as spilling, handling, and storing contribute to the problem. To solve this problem, the first step is to identify the actual causes of food waste. And for this, tracking food waste in real time is obligatory. By implanting multiple sensors and cameras on the field, simultaneous details of the farm can be gathered. Using these sensors, farmers can track where the food is getting wasted and take suitable measures to reduce the same. Besides, there is a great apprehension of food getting wasted in the distribution process. Due to improper monitoring, the chances of perishable food getting decayed are very high. Besides, shelves must be properly handled and maintained all the time. IoT in retail can be used in these setups. Smart shelves, prepared with multiple sensors, can be used to collect details on how the food is being stored and maintained. The collected data can be managed in real time to check whether the food is in the proper condition or not (Fig. 22.2).

This technology tries to offer solutions that progress the supply chain as it uses a wide variety of sensors, some of them with high precision measurements and machine learning algorithms for Big Data analysis. An example of a company working in this direction is Centaur Analytics, which started at Volos, Greece in 2014 (Mouratidis 2018). A group of sensors records temperature and humidity fluctuations, which are majorly responsible for posing serious threats to the stored crops. The accuracy of forecast increases with measurements of oxygen and carbon dioxide concentrations. Furthermore, a phosphine sensor enables safe use of insect pesticides in the silos and insect eradication carried out in a proper way. The same sensor systems are able to administer the crops during transportation, casing all the links starting with harvesting till distributing the product to the manufacturer or the retailer's locations. After sensors have finished their job, machine learning algorithms, customized by Centaur, take over analysis on data being sent to Centaur's cloud. Training on simulation models as well as drawing on prior

Fig. 22.2 Smart shelves

experience from data processing in existing businesses cases help algorithms foresee dangers and consult the auditor to prevent product deterioration or to hasten the process of sending it to the market shelves in faultless condition.

22.4.2 Using Artificial Intelligence

Artificial intelligence (AI) refers to the replication of human intelligence in machines that are programmed to think like humans and imitate their actions. A retail shop having plethora of food items has to be restocked whenever the need arises. If the stock is filled before the food items are still unsold, the susceptibility of pest attacks become higher. One of the many reasons for this is inappropriate food stock monitoring and forecast. Instead of using manual methods, using AI can make this process quick and efficient. By entering pre-recorded details on food stock outs, food that previously got wasted, and real-time shelf information, AI tools can come up with precise predictions. This way, retailers can get a clear understanding of when a particular food item will be out of stock and with this accurate information, retailers can then order only the essential quantities of food items (Närvänen et al. 2020).

Another reason for food waste is inefficient food supply chain management. Most of the time, without checking the quality of the food, it is packed and distributed to the retailers. Retailers, without acknowledging this issue, fill their shelves with low-quality products and this increases the probability of such food getting wasted. To prevent such food wastage, the producers are using an AI-driven food analysis system that checks the food quality before it reaches the consumers.

There are systems that can freely monitor and assess the quality of food. The Artificial Intelligence, Machine Learning, and Deep Learning component understand the hidden patterns that may not be visible to the human eye. In case a particular batch of food is about to spoil, the manufacturers can send this stock first to the distribution centres. An example where artificial intelligence is being used to combat food waste is at Winnow which is an electronic system that makes it swift and easy to record precisely how food is being wasted through smart meter technology attached to the food waste bin in use. The programme helps the food service and hospitality industry, cut down on food waste by making the kitchen smarter. With their AI-enabled tool, kitchens can automatically track food waste. Winnow uses computer vision to help chefs easily pinpoint waste, cut costs, and save time. The system takes photos of wasted food as it is thrown away and, using the images, the machine trains itself to recognize what has been thrown in the bin. This means that, over time, food will be thrown in the bin and the data will be captured automatically. This increases data precision and ease of use. Simple and in-built to use, the Winnow Waste Monitor includes a digital scale and a connected tablet which can be attached to any basic food waste bin (Mouratidis 2018).

22.4.3 Different Apps Strengthening Digital Knowledge Ecosystem

As technologies such as smart phone applications exponentially increase the accessibility and new opportunities to share food and prevent waste through technology mediated interaction. Online spaces have facilitated new relationships around food and events that propose solutions to minimize food wastes. It is now becoming a global problem with many serious consequences. Around 1.3 billion tonnes of food get wasted every year, across the globe. People need to be aware of what can be done to alleviate it and this information too can be shared digitally. All across the world various mobile applications are providing help to reduce food waste. Some of them are listed below:

22.4.3.1 Love Food Hate Waste

It was launched in (2010) Scotland and is for both Android and Apple users. It helps in organizing the kitchen and planning meals. It has all details of the products in the kitchen and send reminders of '*Use by*' for reduced wastage (Fig. 22.3).

22.4.3.2 Love Your Leftovers

An efficient application that helps to use leftovers to make amazing recipes and planning meals. It helps both in reducing waste to the landfills and saving money.

22.4.3.3 Olio

A very useful application to reduce food waste launched in 2015. It helps to share kitchen leftovers with those who are in need. The food must be in an edible condition. This application is serving around 49 countries worldwide. Many supermarkets in UK have this app as an 'online food bank' for people to donate food and reduce wastage (Fig. 22.4).

Fig. 22.3 Love food hate waste app

LOVE FOOD hatewaste

Fig. 22.4 Olio app

22.4.3.4 Share Waste

This application helps to share leftovers or unwanted kitchen food to the people who can use it for composting or as animal feed. It helps in connecting people to recycle kitchen scraps, organic waste for composting and making fertile soil for produce.

22.4.3.5 Too Good to Go/Meal Saver/Food Print

These three apps have the same aim, i.e. to search restaurants or cafes giving excess food at a lesser price and to reduce sending food at landfills. Too good to go is an app that helps to get edible food from stores which were not bought and are waste to feed other people. This app helps around 15 countries to reduce wastage of food worldwide. Meal Saver too helps in this kind. It is a German organization for a waste less food system. Also, they give food in paper bags or bio-degradable boxes and is helping for the betterment of environment.

22.4.3.6 Seasons

Another way to lessen the food waste is to shorten the supply chain. And this app does the same. It searches farmer's markets and help us to get to know about seasonal produce in an area. It serves in the USA, Canada, UK, and many such countries.

22.4.3.7 Karma

This app was found in 2016 and awarded as the Start-up of the Year, it serves in 225 cities across Sweden, France, and UK. It provides unused food at half its price here. They have saved about three million meals and 1200 tonnes of food till now.

22.4.3.8 Food Cloud

This is a UK and Ireland based start-up, helping businesses to provide surplus food to 7500 charitable groups and feed the needy people (Tips, tricks and digital tools for reducing food waste in daily life n.d.). Their mission is to get no food wasted.

22.4.3.9 Copia

A US based platform, helping to get surplus food from businesses to the non-profit organization and feed the needy.

22.4.3.10 Wasteless

Israeli start-up helping for a more data driven approach and reducing prices for items on the shelf, with its use by date coming closer. They use AI powered dynamic pricing to increase profit and reduce waste by ensuring sales of products before they get expired.

22.5 Conclusion

With the ever-rising statistics around food waste, now is the time for the industry to implement a more digitalized method across the supply chain. Employing integrated and innovative systems will permit all key stakeholders, from field to fork, to resourcefully manage food and reduce the risk of waste and loss. Through implementing technology and digitalizing across the food supply chain, a step can be taken towards reducing the amount of food waste and loss within the industry. Not only that, but it can also help to ensure that there are enough resources to sustainably support the future global population. Digitalization will help in sustainable food waste management and also provide opportunities for businesses to initiate their profitability by redefining processes and improving efficiencies. It becomes the solution to the push–pull paradox for commercial benefit and fighting global food waste.

References

Galanakis CM (2015) Food waste recovery processing technologies and industrial techniques. Academic Press, Cambridge, pp 22–26

FAO (2010) 925 Million people in chronic hunger worldwide. Retrieved from http://www.fao.org/news/story/en/item/45210/icode/

FAO (2011a) Cutting food waste to feed the world. Retrieved from http://www.fao.org/news/story/en/item/74192/icode/

FAO (2011b) Global food losses and food waste-extent, causes and prevention. Retrieved from http://www.fao.org/3/a-i2697e.pdf

Food Wastage: Economic, social and environmental impact (2018). Retrieved from https://citytoday.news/food-wastage-economic-social-and-environmental-impact/

Franke-Whittle IH, Insam H (2013) Treatment alternatives of slaughterhouse wastes, and their effect on the inactivation of different pathogens: a review. Crit Rev Microbiol 39(2):139–151. https://doi.org/10.3109/1040841X.2012.694410

Internet of Things (IOT) (n.d.). Retrieved from https://internetofthingsagenda.techtarget.com/definition/Internet-of-Things-

Jayathilakan K, Sultana K, Radhakrishna K, Bawa AS (2012) Utilization of by-products and waste materials from meat, poultry and fish processing industries: a review. J Food Sci Tech Mys 49 (3):278–293. https://doi.org/10.1007/s13197-011-0290-7

Joshi VK, Kumar A, Kumar V (2012) Antimicrobial, antioxidant and phyto-chemicals from fruit and vegetable waste: a review. Intl J Food Ferment Technol 2(2):123–136

Jurado J, Sorensen H (2020) Towards zero waste in beer production-New Trends for brewery solutions. Retrieved from http://www.ibdlearningzone.org.uk/article/show/pdf/55/. Accessed 23 Feb 2020

Mouratidis Y (2018) How IoT and AI reduce food waste written by Yiannis. Retrieved from www.forbes.com

Närvänen E, Mesiranta N, Mattila M, Heikkinen A (2020) Food waste management solving the wicked problem. Palgrave Macmillan, Cham, pp 367–389

Tips, tricks and digital tools for reducing food waste in daily life (n.d.). Retrieved from https://en.reset.org/act/tips-tricks-and-digital-tools-reducing-food-waste-daily-life-09122019

The manufacturer's authorised representative in the EU is Springer Nature Customer Service Centre GmbH, Europaplatz 3, 69115 Heidelberg, Germany. If you have any concerns regarding our products, please contact ProductSafety@springernature.com

Printed and bound by CPI Group (UK) Ltd, Croydon, CR0 4YY
30/09/2026
02225247-0014